New Frontiers in Regional Science: Asian Perspectives

Volume 11

New Frontiers in Regional Science: Asian Perspectives

This series is a constellation of works by scholars in the field of regional science and in related disciplines specifically focusing on dynamism in Asia.

Asia is the most dynamic part of the world. Japan, Korea, Taiwan, and Singapore experienced rapid and miracle economic growth in the 1970s. Malaysia, Indonesia, and Thailand followed in the 1980s. China, India, and Vietnam are now rising countries in Asia and are even leading the world economy. Due to their rapid economic development and growth, Asian countries continue to face a variety of urgent issues including regional and institutional unbalanced growth, environmental problems, poverty amidst prosperity, an ageing society, the collapse of the bubble economy, and deflation, among others.

Asian countries are diversified as they have their own cultural, historical, and geographical as well as political conditions. Due to this fact, scholars specializing in regional science as an inter- and multi-discipline have taken leading roles in providing mitigating policy proposals based on robust interdisciplinary analysis of multifaceted regional issues and subjects in Asia. This series not only will present unique research results from Asia that are unfamiliar in other parts of the world because of language barriers, but also will publish advanced research results from those regions that have focused on regional and urban issues in Asia from different perspectives.

The series aims to expand the frontiers of regional science through diffusion of intrinsically developed and advanced modern regional science methodologies in Asia and other areas of the world. Readers will be inspired to realize that regional and urban issues in the world are so vast that their established methodologies still have space for development and refinement, and to understand the importance of the interdisciplinary and multidisciplinary approach that is inherent in regional science for analyzing and resolving urgent regional and urban issues in Asia.

Topics under consideration in this series include the theory of social cost and benefit analysis and criteria of public investments, socio-economic vulnerability against disasters, food security and policy, agro-food systems in China, industrial clustering in Asia, comprehensive management of water environment and resources in a river basin, the international trade bloc and food security, migration and labor market in Asia, land policy and local property tax, Information and Communication Technology planning, consumer "shop-around" movements, and regeneration of downtowns, among others.

More information about this series at http://www.springer.com/series/13039

Suminori Tokunaga • Budy P. Resosudarmo
Editors

Spatial Economic Modelling of Megathrust Earthquake in Japan

Impacts, Reconstruction, and Regional Revitalization

Springer

Editors
Suminori Tokunaga
Faculty of Economics and Business
Administration, Reitaku University
Kashiwa, Chiba
University of Tsukuba
Tsukuba, Ibaraki
Japan

Budy P. Resosudarmo
Arndt-Corden Department of Economics
Australian National University
Canberra, Australia

ISSN 2199-5974 ISSN 2199-5982 (electronic)
New Frontiers in Regional Science: Asian Perspectives
ISBN 978-981-10-6492-0 ISBN 978-981-10-6493-7 (eBook)
https://doi.org/10.1007/978-981-10-6493-7

Library of Congress Control Number: 2017957819

Printed on acid-free paper

This Springer imprint is published by Springer Nature
The registered company is Springer Nature Singapore Pte Ltd.
The registered company address is: 152 Beach Road, #21-01/04 Gateway East, Singapore 189721, Singapore

Preface

The Great East Japan Earthquake of March 11, 2011, had a massive economic impact, primarily on the affected areas in Japan. The research projects that this book is based on started in July 2011 after the Great Earthquake. The Research Institute of Economy, Trade and Industry (RIETI) took initiative in the following projects: (1) "Studies on the Structure of Japanese Economic Space and Japanese Supply Chains Sustaining Growth Under Globalization and Disaster Risks [July 2011–June 2013, Project Leader: Nobuaki Hamaguchi (Faculty Fellow)]" and (2) "Sustainable Regional Development: New Industrial Clusters and Division of Functions [July 2013–June 2015, Project Leader: Ryohei Nakamura (Faculty Fellow)]." In addition, we organized a special session of the Japan Section of Regional Science Association International (JSRSAI) on "Impacts of Disaster on Regional Economy" with Dr. Okiyama and Dr. Kunimitsu (Tokunaga as organizer) in 2013 and 2014 and a special session of JSRSAI on "Impacts of Regional Recovery Policy on Regional Economy" with Dr. Okiyama and Dr. Kunimitsu (Tokunaga as organizer) in 2015. Moreover, we organized a special session of the 23rd Pacific Conference of the Regional Science Association International (RSAI, PRSCO) on "Macroeconomic Policies, Global Crisis, and Recovery" with Budy Resosudarmo in Bandung, Indonesia, from July 2 to 4, 2013.

Many of the papers contained in this book were presented at the workshops of RIETI and special sessions of JSRSAI and PRSCO. Subsequently, all papers have been thoroughly reviewed and revised. We thank the members of research projects in RIETI including Masahisa Fujita (president for 2007–2016), Masayuki Morikawa (vice president), Nobuaki Hamaguchi, Ryohei Nakamura, and Tatsuaki Kuroda and the members of RSAI such as Kazuhiko Kakamu (Kobe University), Yoshihiro Kameyama (Saga University), Mitsuru Ota (University of Tsukuba), and Kingsley E. Haynes (George Mason University) for their valuable advice and comments. We would like to thank Mr. Takashi Maeno of Bunshin-do for his generous support on the publication of the Japanese version in 2014. We also acknowledge the financial support given by JSPS KAKENHI (B) Grant Number 24330073 and Reitaku University Research Grant Number 1470.

Tokyo, Japan
March 2017

Suminori Tokunaga
Budy P. Resosudarmo

Contents

1 **A Spatial and Economic Analysis of Megathrust Earthquakes** 1
Suminori Tokunaga

Part I Impacts of the Great East Japan Earthquake

2 **Economic Analysis of Regional Renewal and Recovery from the Great East Japan Earthquake** 13
Suminori Tokunaga, Maria Ikegawa, and Mitsuru Okiyama

3 **Impact of the Great East Japan Earthquake on Production Loss Using an Inter-Regional Social Accounting Matrix** 65
Mitsuru Okiyama

4 **Analysis of Supply Chain Disruptions from the Great East Japan Earthquake in the Automotive Industry and Electronic Parts/Devices** 95
Suminori Tokunaga and Mitsuru Okiyama

Part II Reconstruction and Regional Renewal After the Great East Japan Earthquake

5 **Economic Analysis of Fiscal Measures for Reconstructing the Tohoku Region After the Great East Japan Earthquake: Using a Dynamic Two-Regional CGE Model** 123
Mitsuru Okiyama and Suminori Tokunaga

6 **Measuring Economic Gains from New Food and Automobile Industry Clusters with Coagglomeration in the Tohoku Region** . . . 163
Suminori Tokunaga and Mitsuru Okiyama

7 Production Recovery of Fishery and Seafood Manufacturing After the Disaster in Japan: Economic Evaluation Using Dynamic CGE Model 193
Yuko Akune

8 Economic Ripple Effects of a Biogas Electricity Power Plant as Part of Earthquake Disaster Restoration in the Coastal Area of Iwate Prefecture 221
Yoji Kunimitsu

Part III Evaluating Regional Impacts of Megathrust Earthquakes and Tsunamis

9 Economic Impacts of Population Decline Due to the Great East Japan Earthquake: An Inter-regional Input–Output Approach 237
Yoshifumi Ishikawa

10 An NEG Analysis of Megathrust Earthquakes in Japan 259
Ryusuke Ihara

11 Evaluating Dynamic, Regional, and Economic Impacts of the Tokai Earthquake 289
Hiroyuki Shibusawa

12 Reconstruction and Rehabilitation After Large-Scale Natural Disasters: Lessons from the Indian Ocean Tsunami in Aceh and Nias, Indonesia 313
Budy P. Resosudarmo

Appendix: The Two-Regional Computable General Equilibrium (2SCGE) Model 335

Index 361

Contributors

Yuko Akune Faculty of Economics and Business Administration, Reitaku University, Kashiwa, Chiba, Japan

Ryusuke Ihara Faculty of Economics, Asia University, Tokyo, Japan

Maria Ikegawa Policy Research Institute, Ministry of Agriculture, Forestry and Fisheries, Tokyo, Japan

Yoshifumi Ishikawa Faculty of Policy Studies, Nanzan University, Nagoya, Aichi, Japan

Yoji Kunimitsu Regional Resource Engineering Division, National Agriculture and food Research Organization, Tsukuba, Ibaraki, Japan

Mitsuru Okiyama Reitaku Institute of Political Economics and Social Studies, Reitaku University, Kashiwa, Chiba, Japan

Budy P. Resosudarmo President of Regional Science Association International and Arndt-Corden Department of Economics, Australian National University, Canberra, Australia

Hiroyuki Shibusawa Toyohashi University of Technology, Toyohashi, Japan

Suminori Tokunaga Faculty of Economics and Business Administration, Reitaku University, Kashiwa, Chiba, Japan

Professor emeritus of University Tsukuba, Tsukuba, Japan

Chapter 1
A Spatial and Economic Analysis of Megathrust Earthquakes

Suminori Tokunaga

Abstract The Great East Japan Earthquake, which struck on March 11, 2011, had a massive economic impact, primarily on the affected areas in Japan. The Indian Ocean Tsunami hit Aceh and Nias, Indonesia, on December 26, 2004 and destroyed life and infrastructure within 10 km of the coastline in north and western Aceh. In this book, we examine the economic and human damage inflicted by megathrust earthquakes and tsunamis and analyze the economic effects of policies for reconstruction and regional renewal after such catastrophes, using a spatial and economic model. The contents of this book are as follows: Part I, Impacts of the Great East Japan Earthquake, is composed of three chapters that form a foundation on which further earthquake research can be based. Part II, Reconstruction and Regional Renewal after the Great East Japan Earthquake, focuses on reconstruction and regional renewal after the Great East Japan Earthquake amid a declining Japanese population, using reconstruction budgets and industrial clusters. Part III analyzes the impact and reconstruction after the Great East Japan Earthquake, in a depopulating society, as summarized in Parts I and II. Part III evaluates regional impacts of megathrust earthquakes and tsunamis, such as the Great East Japan Earthquake, the Nankai Megathrust Earthquake, the Tokai earthquake in Japan, and the Indian Ocean Tsunami in Asia.

Keywords Megathrust earthquakes • Tsunami • Depopulating society • Reconstruction • New industry cluster • Regional renewal

1.1 Introduction

In the last 200 years, Japan had approximately 20 major tsunamis, such as the Meiji Sanriku Tsunami in 1896, the Showa Sanriku Tsunami in 1933, the Southwestern Hakkaido Coast Tsunami in 1993, and the Great East Japan Earthquake in 2011.

S. Tokunaga (✉)
Faculty of Economics and Business Administration, Reitaku University, 2-1-1, Hikarigaoka, Kashiwa, Chiba 277-8686, Japan

University of Tsukuba, Tsukuba 305-8572, Japan
e-mail: tokunaga@reitaku-u.ac.jp

S. Tokunaga, B.P. Resosudarmo (eds.), *Spatial Economic Modelling of Megathrust Earthquake in Japan*, New Frontiers in Regional Science: Asian Perspectives 11,
https://doi.org/10.1007/978-981-10-6493-7_1

Japan also had megathrust earthquakes that reoccurred approximately every 100–150 years, such as the Ansei-Tokai Earthquake and the Ansei-Nankai Earthquake in 1854, the 1944 Tonankai Earthquake, the 1946 Nankaido Earthquake, the 1993 Hokkaido Earthquake, and the 1994 offshore Sanriku Earthquake (Karan and Suganuma, eds., 2016). In particular, the Great East Japan Earthquake, on March 11, 2011, had a massive economic impact in the region where the disaster occurred, with damage resulting from the earthquake itself (the largest ever recorded), from the resulting massive tsunami, and from accidents at a nuclear reactor within the region, in addition damage to production sites outside the disaster-affected region. More than 5 years have passed, and it is likely that a disaster similar to the Tokai Earthquake or the Nankai Trough Megathrust Earthquake will recur.

Since about 2010, Japan has seen its overall population decrease, caused by a declining birth rate and aging. These changing demographics, in which the core labor force aged 20–64 is shrinking while the population aged 65 and above is growing, are not uniform across Japan because of inter-regional disparities. Population forecasts can be seen in *Population Projections for Japan* (March, 2013), published by the National Institute of Population and Social Security Research (IPSS) in Japan. According to these projections, Japan's population is expected to decrease by approximately 970,000 per year from 2035 to 2040, when 36% of its total population will be of 65 years and above. While the elderly population will be growing in all regions of Japan, with this trend becoming more pronounced every year, the young will be shrinking in number. This is particularly noticeable in Hokkaido and Tohoku, where the population aged 20–64 is projected to decline by 40% relative to 2010, and the population aged 65 and above is projected to rise, despite regional disparities between 2025 and 2030. For instance, the 20–64 years population in Tohoku is projected to peak in about 2025 (before declining) while the population in Hokkaido aged 65 and above is projected to remain virtually flat.[1] Even as Japan's population goes down, we must strive to rebuild from the Great East Japan Earthquake and to revitalize the region.

1.2 Conceptual and Empirical Developments

This book aims to empirically assess (i) the economic impact of megathrust earthquakes, such as the Great East Japan Earthquake and the Nankai Trough Earthquake in a depopulating society, and (ii) the subsequent economic impact of policies regarding regional revitalization and industrial reconstruction, using spatial economic models. (iii) the regional impacts of megathrust earthquakes and tsunamis, such as the Nankai Megathrust Earthquake, the Tokai earthquake in Japan, and the Indian Ocean Tsunami in Asia.

[1] The IPSS Japanese population forecast (2012, 2013) does not consider population disruptions caused by the Great East Japan Earthquake.

The book also asks five questions. First, what is the status of the massive economic damage resulting from the Great East Japan Earthquake and the industrial and fiscal reconstruction measures executed after this earthquake, in a society with a declining population? To answer this question, an inter-regional social account matrix (SAM), an input–output (I–O) table, a geographically weighted regression (GWR) model and a two-regional computable general equilibrium (2SCGE) model are used to empirically explain the economic impact of the Great East Japan Earthquake in the afflicted and other regions (Chaps. 2 and 3). Chapter 4 examines supply chain disruptions from the Great East Japan Earthquake in the automotive and electronic parts and devices industries.

Prior studies have primarily analyzed the economic impact of earthquakes or the subsequent rebuilding in a growing population as opposed to a declining one. Thus, the second question is whether fiscal reconstruction measures or regional industrial reconstruction policies are possible in a declining population. Using a dynamic two-regional computable general equilibrium model (D2SCGE) for this population, we measure the economic impact of a reconstruction budget (Chap. 5) and new industry cluster policies, based on the New Economic Geography(NEG) model[2] as the effects of regional and industrial reconstruction policies (Chap. 6) on the regional economies in the disaster-affected region amid a declining population. This idea of Chap. 6 is that in addition to agglomeration within a same industry, it is essential to form "new industry clusters" using innovation to promote agglomeration in different industries (co-agglomeration). By using a D2SCGE model that incorporates the Dixit–Stiglitz model, which is characteristic of an NEG, we examine whether reconstruction and regional revitalization after an earthquake are possible by forming new industry clusters, using (1) subsidies and tax reduction policies, (2) a commodity-grade scenario with products (particularly intermediate goods) that have varying levels of inter-regional elasticity of substitution, and (3) a scenario with innovative TFP improvement. A dynamic computable general equilibrium (DCGE) model that assesses production recoveries after the disaster in the fishing and seafood industries (where damage was severe) is analyzed in Chap. 7. Using an I–O table, the process of reconstruction after the disaster, in the construction of a biogas electric power plant, is analyzed in Chap. 8.

The third question examines the impacts of an earthquake and of population migration in a declining population on regional economies. Because the aforementioned D2SCGE model does not incorporate an explicit inter-regional population migration, and therefore the first part of the questions is answered by analyzing the impact of further population declines due to a major earthquake, using an I–O table

[2]For economics of agglomeration, see Marshall (1890). For the NEG mode and agglomeration, see Krugman (1991), Fujita et al. (1999), and Fujita and Thisse (2013). For the industrial cluster, see Porter (1998, 2000), Tokunaga and Okiyama (eds) (2014); Tokunaga et al. (2014), and Tokunaga and Okiyama (2017). According to Porter (1998), the formation of an industrial cluster depends on the relative strength of three forces of localization: economies, price competition, and transport costs. Fujita and Thisse (2013, p. 350) also point out that firms must be able to serve almost all markets equally to enjoy the local advantages associated with the formation of a cluster.

of 47 prefectures (Chap. 9). The second part of the questions is answered using an NEG model that studies migrations caused by the Great East Japan Earthquake and a Nankai Trough Earthquake (Chap. 10).

The fourth question asks whether, in the event of a major earthquake, advance measures impact the recovery process. This question is answered using a forward-looking, dynamic inter-regional CGE model, which analyzes different post-disaster recoveries for a Tokai Earthquake (Chap. 11).

The Indian Ocean Tsunami that occurred after the Sumatra Earthquake in December 2004 is used as a successful example of reconstruction after a major tsunami. The fifth question asks how central and local governments, NPOs, and the international community should manage a reconstruction after a major natural disaster. This section proposes a set of policies that respond to damages caused by major natural disasters that are likely to occur in Asia in the future (Chap. 12). It also analyzes the Sumatra Earthquake's impact on the Indonesian state of Aceh and its surrounding regions and conducts a time-series analysis of the reconstruction policies implemented after a large natural disaster and for regional revitalization. Figure 1.1 shows the chapters' details and the causal relations between them.

1.3 Contributors to the Book

Part I, *Impacts of the Great East Japan Earthquake*, comprises three chapters that form a foundation on which further earthquake research can be based. In Chap. 2, entitled "Economic Analysis of Regional Renewal and Recovery from the Great Earthquake," Suminori Tokunaga, Maria Ikegawa, and Mitsuru Okiyama examined the impacts of the Great East Japan Earthquake on affected areas in Japan, using a simple spatial econometrics (GWR) model. They also evaluated economic policies for regional renewal and recovery from the Great Earthquake using a static regional computable general equilibrium (SCGE) model.

Chapter 3, entitled "Impact of the Great East Japan Earthquake on Production Loss Using an Inter-Regional Social Accounting Matrix," written by Mitsuru Okiyama, estimates the magnitude of production losses in regions of Japan both affected and not affected by this earthquake, primarily damage to agriculture and fisheries' production and loss of capital stock in the manufacturing industry.

Chapter 4, "Analysis of Supply Chain Disruptions from the Great East Japan Earthquake in the Automotive Industry and Electronic Parts/Devices," was written by Suminori Tokunaga and Mitsuru Okiyama. This chapter simulates negative supply shocks for both upper-level sectors between regions and lower-level sectors within the regions, using a two-region computable general equilibrium (2SCGE) model.

Part II, *Reconstruction and Regional Renewal after the Great East Japan Earthquake*, focuses on reconstruction and regional renewal in a declining population after the Great East Japan Earthquake, using reconstruction budgets and industrial clusters. This part comprises four chapters. Chapter 5, entitled "Economic Analysis of Fiscal Measures for Reconstructing the Tohoku Region after the Great East Japan Earthquake," written by Mitsuru Okiyama and Suminori

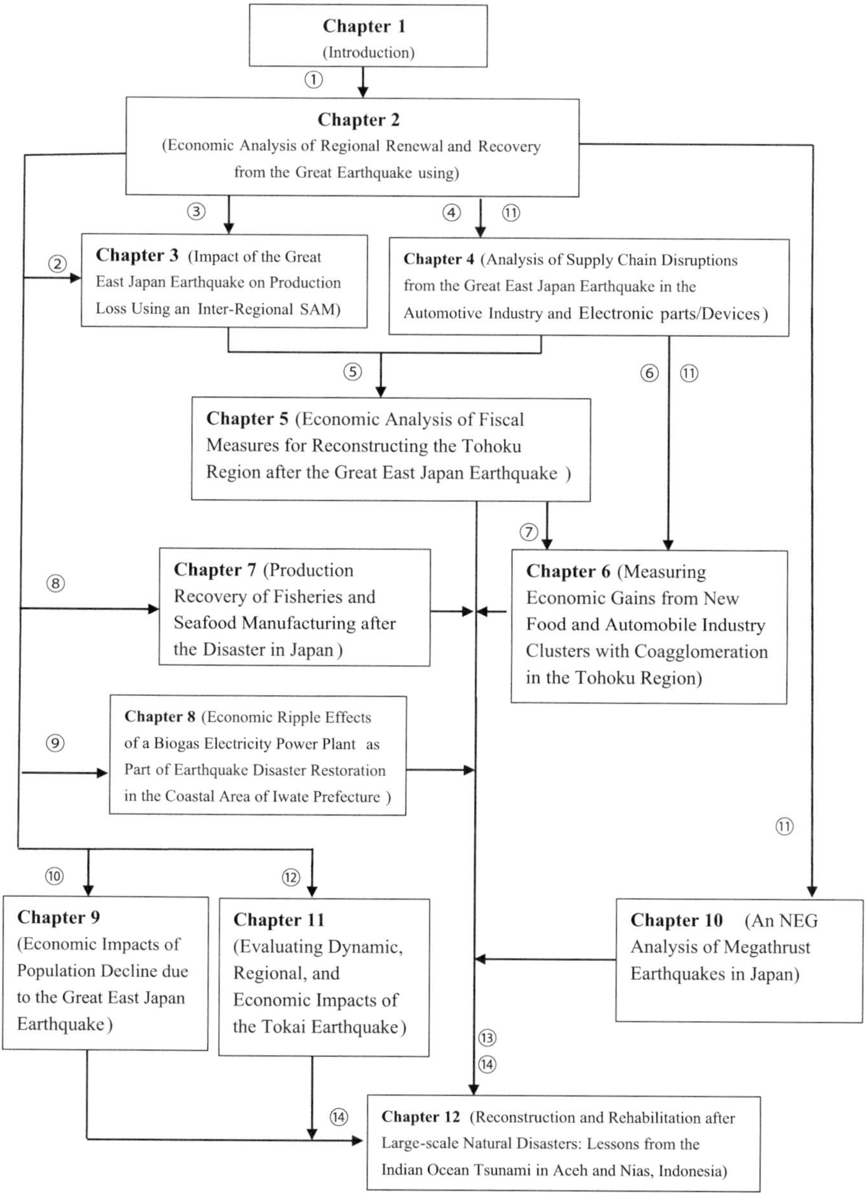

Fig. 1.1 Flowchart of the book (Note: ① Economic analysis using GWR and CGE model, ② Economic analysis using an inter-regional SAM, ③ SAM multiplier analysis of production loss, ④ Economic analysis of supply chain disruption using 2SCGE model,⑤ Economic analysis of fiscal measure using D2SCGE model, ⑥ Economic gain from industry clusters, ⑦ Economic evaluation of industry clusters using D2SCGE model, ⑧ Production recovery of fishers and seafood manufacturing by DCGE model, ⑨ Economic analysis of ripple effects of a biogas power plant using a regional Input-output tables, ⑩ An inter-regional Input-output analysis of population decline, ⑪ Economic analysis of the megathrust earthquakes using an NEG model, ⑫ Economic analysis of the Tokai earthquake using Dynamic spatial CGE model, ⑬ Lessons from the Indian Ocean Tsunami in Indonesia, ⑭ Policy recommendation)

Tokunaga, measures the economic impact of the reconstruction budget and illustrates the regional economies' transition in the disaster-affected region in a declining population, using a recursive dynamic regional CGE model. One of the most fascinating findings in this chapter is that even if the Great East Japan Earthquake had not occurred, the disaster-affected region would have experienced weaker growth in its sub-regional economies simply because of its declining population. It is advisable that the government provide fiscal measures to the disaster-affected region after the reconstruction period by adopting a different form of reconstruction budget; that is, by revising the distribution ratios of local allocation tax grants. However, this measure alone would not allow the regional economy in the disaster-affected region to recover and exceed its level prior to the earthquake.

Chapter 6, entitled "Measuring Economic Gains from New Food and Automobile Industry Clusters with Coagglomeration in the Tohoku Region," written by Suminori Tokunaga and Mitsuru Okiyama, provides, in the context of NEG, a new type of industrial cluster to assist recovery from the Great East Japan Earthquake amid a declining population. Using Porter's (1998, 2000) clusters, they analyze the feasibility of ongoing economic development in disaster regions, utilizing two new industry cluster models. The two clusters are the automobile industry, which targets the entire disaster region and where innovation comes from coagglomeration with different industries in megaregions, and the food industry, which focuses on leveraging local resources and targets individual disaster-struck prefectures. Using a dynamic two-region computable general equilibrium (D2SCGE) model, they conduct a simulation analysis of scenarios for these two new clusters, whose positive and high productivity reveals the two effects: (1) economies of agglomeration from vertical and horizontal coagglomerations boost real gross regional product (GRP) and productivity at the macro level when the two new industry clusters are formed jointly rather than separately and (2) the clusters contribute to long-term, sustained growth in disaster economies, thus reducing the gap between their growth and that of other regional economies.

Chapter 7, entitled "Production Recovery of Fisheries and Seafood Manufacturing after the Disaster in Japan: Economic Evaluation Using a Dynamic CGE Model," written by Yuko Akune, shows the economic effects of recoveries in the fisheries and seafood industries after the disaster in Japan with respect to depopulation, employing a DCGE model. She confirms the clear contrast between the production recovery at seafood manufacturers in Iwate and Miyagi Prefectures after the earthquake. From these five simulations, the author concludes the following: (1) The earthquake accelerated the problem caused by depopulation; (2) prompt capital restoration and production recovery contributed to shortening the period of production loss; (3) the degree of production recovery of the downstream industry was faster than that of the upstream industry; (4) no scenario based on the GEJE evidence was sufficient to reach a base scenario without a disaster; (5) stepwise production recovery contributed to increases not only in specific industries but also in economic welfare in the long term.

Chapter 8, "Economic Ripple Effects of a Biogas Electricity Power Plant as Part of Earthquake Disaster Restoration in the Coastal Area of Iwate Prefecture" was contributed by Yoji Kunimitsu. This chapter focuses on the coastal area of Iwate Prefecture, which was seriously damaged by the Great East Japan Earthquake. Using the input–output tables of this region from before and after the earthquake, the author shows the impacts of a particular revitalization measure, the construction of a biogas electricity power plant, and finds the following three points: (1) value-added production in 2011 was higher than the previous year because of recovery investments, but both intermediate inputs and total production decreased; (2) while investment demand increased in 2011 and 2012, intermediate inputs and private and public consumption decreased because of the earthquake; (3) if it had been constructed before the earthquake, the multiplier value of induced production for the construction of a biogas electricity power plant would have been 0.64. However, the value after the earthquake was 0.17 because the industrial linkage was damaged.

After analyzing the impacts of the reconstruction after the Great East Japan Earthquake in a depopulating society in the preceding two parts, Part III turns to *evaluating regional impacts of megathrust earthquakes and tsunamis*, such as the Great East Japan Earthquake, the Nankai Megathrust Earthquake, the Tokai earthquake in Japan, and the Indian Ocean Tsunami in Asia.

Yoshifumi Ishikawa contributed Chap. 9, which is titled "Economic Impacts of Population Decline Due to the Great East Japan Earthquake: An Inter-Regional Input–Output Approach." In the aftermath of the Great 2011 Earthquake, Fukushima Prefecture in particular is still suffering from its impacts, specifically, the nuclear power plant accident and the consequent evacuation of numerous people. Using a 47-region, inter-regional input–output table at the prefecture level, the author analyzes the economic impacts of the population decline for the 5 years after the Great Earthquake and its long-term economic impacts, using population projections that consider the impact of the earthquake.

Chapter 10, entitled "An NEG Analysis of Megathrust Earthquakes in Japan," was written by Ryusuke Ihara. This contribution focuses on the Great East Japan Earthquake, the Nankai megathrust earthquakes, and the associated tsunami. The author considers how these events changed the regional economies of Japan. Constructing an NEG model composed of the 47 prefectures of Japan, simulation results show that the predicted labor distribution approaches the actual distribution, and as transportation costs decrease, labor distribution changes from dispersion to agglomeration in metropolitan areas and then to re-dispersion in rural areas. In addition, adapting the damage data from megathrust earthquakes to this model, the author predicts the impact of the Great East Japan Earthquake and a Nankai Megathrust Earthquake on regional potential and labor distribution among the prefectures.

Chapter 11, "Evaluating Dynamic, Regional, and Economic Impacts of the Tokai Earthquake," was contributed by Hiroyuki Shibusawa. This chapter investigates the negative economic impacts of an earthquake, using a dynamic spatial computable general equilibrium model with a decentralized economy, utility-maximizing consumers, and value-maximizing firms in a dynamic context. The author estimates the impacts of a hypothetical earthquake, which is expected to occur in the near future, on Tokai's regional economy and shows both dynamic and spatial economic impacts before and after such an earthquake.

Finally, Chap. 12, written by Budy P. Resosudarmo, is entitled "Reconstruction and Rehabilitation after Large-scale Natural Disasters: Lessons from the Indian Ocean Tsunami in Aceh and Nias, Indonesia." It observes that the Indian Ocean Tsunami hit Aceh and Nias, Indonesia, on the morning of December 26, 2004 and destroyed life and infrastructure within 10 km of the coastline in north and western Aceh, which were a zone of armed conflict and all islands offshore. The author shows the impact of this huge natural disaster in a conflict zone and analyzes the responses to the Indian Ocean Tsunami of 2004 to estimate the recovery from this mega-disaster and to propose policies for regional renewal. This final contribution emphasizes the regional revitalization theme of this book.

1.4 Concluding Remarks

To conclude this introduction, we note the following. Looking back on the March 11, 2011 Great East Japan Earthquake, 5 years later, a number of research organizations calculated almost 18 trillion yen in damages to the social capital stock. Project expenditures in the 5 years (2011–2016), which were devoted to restoration and reconstruction in the areas hit by the disaster, already exceeded the amount of damages to social capital stock (Keynes 1936/1973). While it cannot be denied that businesses and factories have been restored to their previous condition after 5 years, areas outside those impacted by the Fukushima nuclear reactor disaster are entering a new reconstruction stage that will construct new full-scale towns. The purpose of this book is not to recommend a simple restoration to a pre-disaster state but rather to strive for a reconstruction and regional revitalization plan, amid a declining population, and to analyze the impact of policies necessary to achieve it. It is hoped that the various lessons learned from the Great East Japan Earthquake and the Indian Ocean Tsunami and the findings presented in this book, particularly, the new industrial agglomerations discussed in Chap. 6, specifically, policies for "forming new industry clusters" (Porter 1998, 2000) and "innovation" (Schumpeter 1934) should be linked to countermeasures and responses in preparation for megathrust earthquakes such as the Tokai and Nankai Trough Earthquakes, which are predicted to occur in the future.

References

Fujita M, Thisse JF (2013) Economics of agglomeration, 2nd edn. Cambridge University Press, New York

Fujita M, Krugman P, Venables AJ (1999) The spatial economy: cities, regions, and international trade. MIT Press, Cambridge

Karan PP, Suganuma U (eds) (2016) Japan after 3/11: global perspectives on the earthquake, tsunami, and fukushima meltdown. University Press of Kentucky, Kentucky
Keynes JM (1936/1973) The general theory of employment, interest and money, the collected writing of John Maynard Keynes, vol VII. Macmillan Press Ltd, London
Krugman P (1991) Increasing returns and economic geography. J Polit Econ 99:483–499
Marshall A (1890) Principles of economics. Macmillan, London. 8th edition published in 1920
National Institute of Population and Social Security Research in Japan (2012) Population projections for Japan (January 2012): 2011 to 2060. http://www.ipss.go.jp/site-ad/index_english/esuikei/gh2401e.asp. Accessed 17 Feb 2015
National Institute of Population and Social Security Research in Japan (2013) Regional population projections for Japan: 2010–2040. Population Research Series. http://www.ipss.go.jp/pp-shicyoson/j/shicyoson13/6houkoku/houkoku.pdf. Accessed 17 Feb 2015 (in Japanese)
Porter M (1998) On competition. Harvard Business School Press, Cambridge, MA
Porter M (2000) Location, competition, and economic development: local clusters in a global economy. Econ Dev Q 14(1):15–34
Schumpeter JA (1934) The theory of economic development. Harvard university Press, Cambridge, MA
Tokunaga S, Okiyama M (eds) (2014) Reconstruction the disaster-affected region of the great east Japan earthquake and recovery of the regional economy. Bunshindou, Tokyo (in Japanese)
Tokunaga S, Okiyama M (2017) Impacts of industry clusters with innovation on the regional economy in Japanese depopulating society after the great east Japan earthquake. Asia-Pacific J Reg Sci 1(1). https://doi.org/10.1007/s41685-017-0041-5
Tokunaga S, Kageyama M, Akune Y, Nakamura R (2014) Empirical analyses of agglomeration economies in japanese assembly-type manufacturing industry for 1985–2000: using agglomeration and coagglomeration indices. Rev Urban Reg Dev Stud 26(1):57–79

Part I
Impacts of the Great East Japan Earthquake

Chapter 2
Economic Analysis of Regional Renewal and Recovery from the Great East Japan Earthquake

Suminori Tokunaga, Maria Ikegawa, and Mitsuru Okiyama

Abstract The Great East Japan Earthquake, which struck on March 11, 2011, had a massive economic impact, primarily on the affected areas in Japan. In this chapter, we examine the economic and human damage inflicted on Iwate, Miyagi, Fukushima, and Ibaraki Prefectures by the Great East Japan Earthquake, as well as the current situation of industrial recovery, based on several statistical sources and a geographically weighted regression (GWR) model. In the latter part of this chapter, we will show the extent of fiscal transfers to date from the government for reconstruction and renewal of stricken areas and analyze the economic effect of the formation of new industrial clusters for reconstruction and renewal on these areas using a static two-regional computable general equilibrium (2SCGE) model. Our findings are as follows: (1) if production subsidies to support industries form new industry clusters, positive effects on regional economies could appear in the disaster regions; however, these impacts are weak and (2) formation of new industry clusters with productivity improvement has a positive effect on real gross regional product (GRP) and economic welfare in these regions, reducing the economic welfare gap between disaster and non-disaster regions.

Keywords Great East Japan Earthquake • Economic and human damage • Geographically weighted regression (GWR) model • Two-regional computable general equilibrium (2SCGE) model • New economic geography (NEG) model • New industry clusters • New industrial agglomeration

---

S. Tokunaga (✉)
Faculty of Economics and Business Administration, Reitaku University, 2-1-1, Hikarigaoka, Kashiwa, Chiba 277-8686, Japan

University of Tsukuba, Tsukuba 305-8572, Japan
e-mail: tokunaga@reitaku-u.ac.jp

M. Ikegawa
Policy Research Institute, Ministry of Agriculture, Forestry and Fisheries, Tokyo 100-0013, Japan

M. Okiyama
Reitaku Institute of Political Economics and Social Studies, Reitaku University, Kashiwa, Chiba 277-8686, Japan

S. Tokunaga, B.P. Resosudarmo (eds.), *Spatial Economic Modelling of Megathrust Earthquake in Japan*, New Frontiers in Regional Science: Asian Perspectives 11, https://doi.org/10.1007/978-981-10-6493-7_2

2.1 Introduction

Compared with past earthquake disasters in Chap. 1, the Great East Japan Earthquake, which struck on March 11, 2011, had a massive economic impact, primarily on the affected areas. The damage inflicted directly by the earthquake was compounded by damage from the tsunami arising from the earthquake and by the nuclear power plant incident. In the former part of this chapter, we examine the economic and human damage inflicted on Iwate, Miyagi, Fukushima, and Ibaraki Prefectures by the Great East Japan Earthquake, as well as the current situation of industrial recovery. In the latter part, we show the impacts of fiscal measures for reconstruction and industrial cluster in disaster-affected region for reconstruction on regiona economy using a simple computable general equilibrium model. Section 2.2 presents the economic and human damage wrought by the earthquake and the current situation of industrial recovery, based on several statistical sources and surveys such as industrial production indices. In particular, we discuss the impact of the nuclear power plant catastrophe in Fukushima Prefecture, distinguishing it from the damage in Iwate, Miyagi, and Ibaraki Prefectures. Section 2.3 then discusses which elements were prioritized in the allocation of government recovery funding measures to deal with the economic damage, the status of implementation, and additional challenges. In Sect. 2.4, we estimate impacts of this earthquake on firms' output, using a spatial econometric model. Section 2.5 moves on to present an economic model for natural disaster assessment and labor migration: a) the regional input-output and two-regional computable general equilibrium (2SCGE) models used in this tract to assess the policies of recent years, to analyze economic and human loss from the quake, and to assess recovery policies; b) a new economic geography (NEG) model that facilitates explicit analysis of population and labor migrations; and c) research about new industrial agglomerations and clusters. Section 2.6 uses the static 2SCGE model to analyze the earthquake's destruction and to describe a simulation analysis of regional renewal based on new industrial agglomerations and clusters. Finally, Section 2.7 presents conclusions for this chapter and discusses future topics.

2.2 Economic and Human Loss in the Disaster Areas and the Current State of Industrial Recovery

2.2.1 Economic Loss in the Disaster Areas and the Current State of Industrial Recovery

The Great East Japan Earthquake that struck on March 11, 2011 caused the greatest damage to the Japanese economy since World War II. Production activities in the automotive and electronics industries were hit not only in the affected areas but also throughout the country, as damage to factories in the stricken areas and power shortages disrupted the supply of products and raw materials. Industrial production plummeted due to this enormous supply shock. In this section, we shall consider supply shocks to major industries in the stricken areas immediately after the quake occurred and the extent to which production activities have subsequently recovered.[1] We begin with indices of industrial production. Fig. 2.1 shows these indices, and we can see that production in each prefecture took a 30%–50% fall immediately after the quake. Subsequently, the speed of the recovery differed among prefectures, with Miyagi Prefecture initially lagging behind but catching up 1 year after the quake. In Fukushima Prefecture, which suffered the largest impact from the nuclear power plant disaster, the impact of the quake was prolonged, with production levels stalling at approximately 90% in 2012 and subsequent years. At present, industrial production in the three affected prefectures, excluding Ibaraki, remains below national industrial production indices.

Table 2.1 shows the drop in manufacturing output by sector and the subsequent recovery. It is obvious that in the 6 months immediately following the disaster, production in all sectors declined in the affected areas, notably in Fukushima Prefecture. Therefore, as in Fig. 2.2, we see that while manufacturing sites damaged by the earthquake were concentrated in the coastal regions of Miyagi and Ibaraki Prefectures (where the automotive and electrical-related industries are concentrated) and in the coastal regions of Iwate and Miyagi Prefectures (where marine processing-related industries are concentrated), damaged factories were dispersed over a broad area, which caused enormous damage to the manufacturing industries of affected regions.

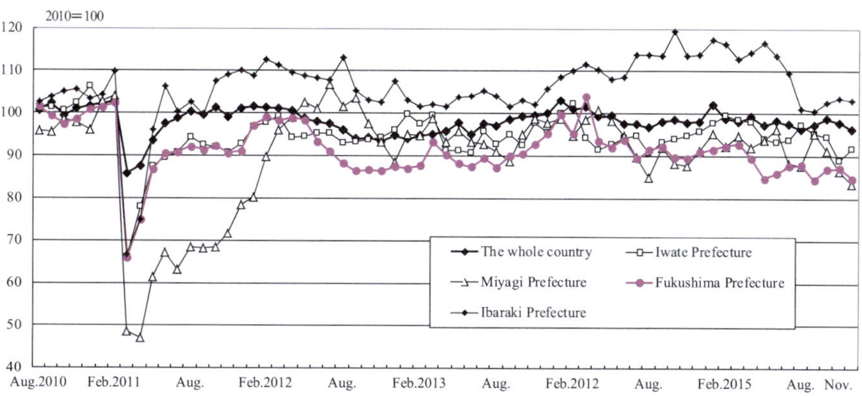

Source: Economic survey of Tohoku Bureau of Economy, Trade and Industry (2012 through 2016).

Fig. 2.1 Drop and recovery of production in manufacturing by sector in the disaster-affected prefectures (Source: Economic Survey of Tohoku Bureau of Economy, Trade and Industry (2012 through 2016))

[1]This section is an expanded version of Sect. 1.2 from Tokunaga and Okiyama (eds) (2014).

Table 2.1 Change amount of production value from September 2010 through December 2015

Change amount of Average production value from September 2010 through February 2011 (unit: Billion yen, %)		From March 2011 through June 2011 (The three months period after the earthquake)	Change % from pre-the earthquake level	From July 2011 through Septermber 2011 (From three months through six months period)	Change % from pre-the earthquake level	From Octber 2011 through January 2012 (From six months through one year period)	Change % from pre-the earthquake level	From February 2012 through January 2013 (From one year through two years period)	Change % from pre-the earthquake level	From January 2013 through December 2015 (after four years period)	Change % from pre-the earthquake level
Fukushima Prefecture		−1,240	−24.8	−509	−10.2	−378	−7.6	−459	−9.2	−696	−13.9
	Foods and tobacco	−223	−40.5	−186	−33.7	−119	−21.6	−26	−4.7	−139	−25.3
	Non-durable goods	−114	−42.7	−21	−8.0	−15	−5.6	−14	−5.3	−6	−2.1
	Petroleum products and chemicals	−188	−30.1	−102	−16.2	−111	−17.7	−99	−15.8	−82	−13.0
	Steel and metals	−218	−34.7	−74	−11.8	−33	−5.3	−31	−4.9	−58	−9.2
	General machinery	−19	−4.3	−2	−0.5	−19	−4.2	−15	−3.3	22	5.0
	Electronic parts and devices	−278	−17.6	−58	−3.7	−32	−2.0	−154	−9.7	−239	−15.1
	Motor vehicle and Motor vehicle parts	−93	−28.6	−16	−5.1	0	−0.1	−64	−19.6	−105	−32.4
	Other durable goods	−46	−14.1	−23	−7.1	−29	−8.8	−44	−13.2	−60	−18.2
	Miscellaneous manufacturing goods	−60	−24.6	−26	−10.6	−19	−7.9	−13	−5.1	−30	−12.3

The Stricken three prefectures		−4,505	−25.7	−1,746	−10.0	−969	−5.5	−44	−0.2	−137	−0.8
	Foods and tobacco	−521	−18.7	−156	−5.6	−18	−0.7	19	0.7	334	12.0
	Non-durable goods	−258	−29.1	−159	−18.0	−120	−13.5	−64	−7.2	−24	−2.7
	Petroleum products and chemicals	−1,870	−46.4	−897	−22.3	−771	−19.1	−199	−4.9	−155	−3.8
	Steel and metals	−938	−25.0	−289	−7.7	−161	−4.3	−141	−3.8	−313	−8.4
	General machinery	−243	−10.7	230	10.2	338	14.9	172	7.6	413	18.2
	Electronic parts and devices	−262	−21.2	−338	−27.3	−276	−22.4	−128	−10.3	−345	−27.9
	Motor vehicle and Motor vehicle parts	−249	−30.5	−26	−3.2	9	1.1	278	34.0	94	11.6
	Other durable goods	−60	−5.6	−65	−6.1	15	1.4	9	0.8	−111	−10.4
	Miscellaneous manufacturing goods	−103	−15.8	−45	−6.9	15	2.3	10	1.5	−31	−4.7

(continued)

Table 2.1 (continued)

Change amount of Average production value from September 2010 through February 2011 (unit: Billion yen, %)		From March 2011 through June 2011 (The three months period after the earthquake)	Change % from pre-the earthquake level	From July 2011 through Septermber 2011 (From three months through six months period)	Change % from pre-the earthquake level	From Octber 2011 through January 2012 (From six months through one year period)	Change % from pre-the earthquake level	From February 2012 through January 2013 (From one year through two years period)	Change % from pre-the earthquake level	From January 2013 through December 2015 (after four years period)	Change % from pre-the earthquake level
Other Region		−16,587	−6.1	−22	0.0	1,958	0.7	−7,258	−2.7	2,826	1.0
	Foods and tobacco	80	0.2	303	0.9	209	0.6	−172	−0.5	889	2.7
	Non-durable goods	−20	−0.2	−192	−1.5	39	0.3	−446	−3.6	−577	−4.6
	Petroleum products and chemicals	257	0.5	368	0.8	−196	−0.4	−836	−1.7	−1,057	−2.2
	Steel and metals	−1,060	−2.5	−1,157	−2.7	−964	−2.3	−1,059	−2.5	−2,073	−4.9
	General machinery	328	1.2	1,913	6.8	1911	6.8	−1,384	−4.9	848	3.0
	Electronic parts and devices	−3,554	−11.5	−3,482	−11.3	−5,550	−18.0	−6,429	−20.9	−5,130	−16.6
	Motor vehicle and Motor vehicle parts	−13,803	−32.7	−472	−1.1	4,177	9.9	1,536	3.6	555	1.3
	Other durable goods	1,095	4.6	2,794	11.7	2,368	9.9	1,678	7.0	10,052	42.2
	Miscellaneous manufacturing goods	90	0.8	−97	−0.9	−37	−0.3	−146	−1.3	−681	−6.2

Whole country		−22,332	−7.6	−2,277	−0.8	611	0.2	−7,760	−2.6	1,993	0.7
	Foods and tobacco	−664	−1.8	−39	−0.1	72	0.2	−179	−0.5	1,084	2.9
	Non-durable goods	−392	−2.9	−373	−2.7	−96	−0.7	−524	−3.9	−606	−4.5
	Petroleum products and chemicals	−1,801	−3.4	−630	−1.2	−1,078	−2.0	−1,134	−2.1	−1,293	−2.4
	Steel and metals	−2,216	−4.7	−1521	−3.2	−1,158	−2.5	−1,230	−2.6	−2,444	−5.2
	General machinery	65	0.2	2141	6.9	2,230	7.2	−1,227	−4.0	1,284	4.1
	Electronic parts and devices	−4,094	−12.2	−3878	−11.5	−5,859	−17.4	−6,710	−19.9	−5,714	−17.0
	Motor vehicle and Motor vehicle parts	−14,145	−32.6	−515	−1.2	4,186	9.6	1,751	4.0	544	1.3
	Other durable goods	989	3.9	2706	10.7	2,354	9.3	1,643	6.5	9,882	39.2
	Miscellaneous manufacturing goods	−74	−0.6	−168	−1.4	−41	−0.3	−148	−1.2	−742	−6.2

Source: Tokunaga and Okiyama (eds) (2014, p. 4) and economic survey of Tohoku Bureau of Economy, Trade and Industry (2012 through 2016)
Unit: billion yen

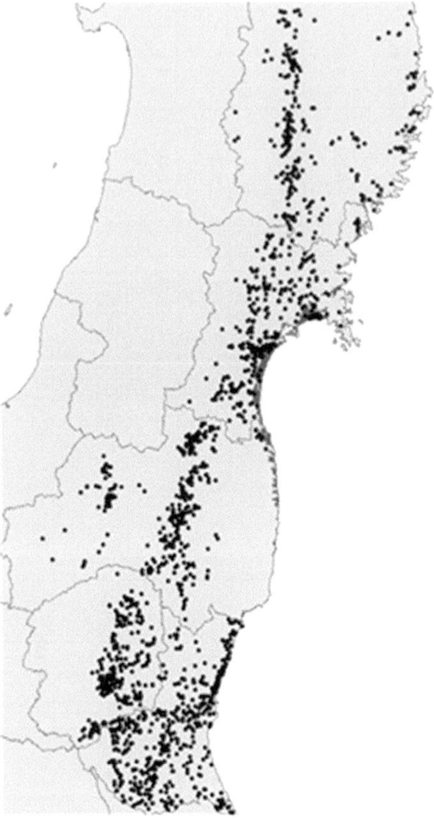

Fig. 2.2 Sites Damaged by the Great Earthquake (Manufacturing Industry) (Source: Tokunaga and Okiyama (eds) (2014, p. 5). Note: The authors composed this based on the results of the "2011 Investigation of Damage to Firms due to Great East Japan Earthquake" survey conducted by the Economic and Industrial Research Laboratory as well as Figure 2.4 by Hamaguchi (2013))

However, visible differences can be seen in the recoveries for Fukushima Prefecture and the other three affected prefectures. This holds true both for the same industry and among industries. It is difficult to say whether this similarity is due to the nuclear power plant incident, the relative impact of fiscal measures, or non-earthquake factors such as the rise in the yen's value, subsidies that stimulated demand for eco-friendly products, or some combination of these factors. From Table 2.1, we see that in the affected regions, for the 3 month period after the quake, manufacturing production reached 74.5% of pre-quake levels, rising to 90.0% during a 3–6 month period. Between 6 and 12 months after the quake, manufacturing production reached 94.0%, moving up to 97.8% between the first and second years. Four years after the quake, manufacturing production was 96.3%.

Some industries, such as the food and tobacco industry in Fukushima, showed noteworthy production recovery. Nevertheless, its recovery from the drop in production lagged behind other prefectures and other industries in Fukushima. It is conjectured that this was caused by negative rumors about radioactive pollution relating to the nuclear power plant incident. However, fiscal measures are effective against harmful rumors, and 1 year after the disaster, production levels in these industries had recovered to within 5% of pre-quake levels. Four years after the quake, however, production fell by 25%.

We also see contrasting patterns of recovery in the automotive and electronics equipment industries. Production after the quake dropped sharply in both industries by about the same amount. We can see from Fig. 2.3, which plots the automotive

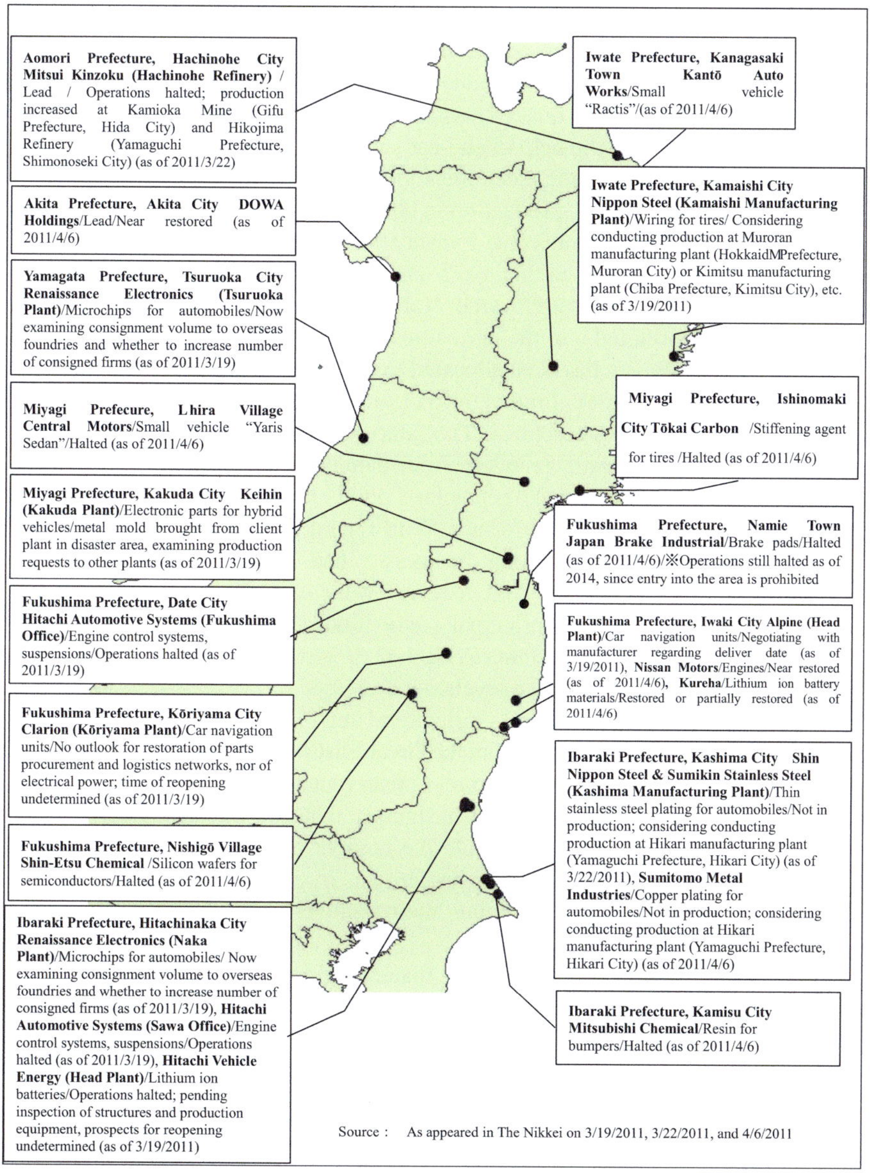

Fig. 2.3 Damaged plants for assembly and parts/components related to automobiles/electrical (Source: As appeared in The Nikkei on 2011/03/19, 2011/3/22, and 2011/4/6)

and electronics industries in the afflicted areas, how damaged factories were concentrated in the coastal area of the four affected prefectures where these industries were clustered. By contrast, damaged manufacturers of raw materials and parts were spread across a broad region, so supply shortages limited production at assembly firms. In particular, damage to the Naka plant of Renesas Electronics, which holds a 40% global share in microcontrollers, halted production of special-order microcontrollers, forcing automotive manufacturers, such as Toyota Motors and Honda, to reduce or stop production.[2] These supply chain interruptions in the affected areas caused automotive production in other geographic areas to fall even more. Three months after the quake, the automotive supply chain interruptions were being resolved, and with the help of eco-car subsidies, production recovered faster than in other industries. In the electronics industry, which suffered from the high value of the yen at that time, production in both affected and other areas was down by double digits, both 6 months and 4 years after the quake.

Let us now consider the earthquake's damage to non-manufacturing industries and how they subsequently recovered. Table 2.2 shows the recovery from the tsunami and the earthquake in the agriculture and fisheries industries. Although a budget of 509 billion yen has been allocated to date to support these two industries, with the exception of Ibaraki Prefecture, a clear disparity in the degree of recovery exists between affected prefectures. The table shows that even 2 years after the earthquake and the tsunami, recovery from damage to farmlands in Fukushima Prefecture was 15.2%, a full 20 percentage points behind the recovery in Iwate Prefecture, the second-worst performer. With regard to fisheries, 1 year after the quake, nearly half of the fishing businesses had reopened. By contrast, in Fukushima Prefecture, virtually no fisheries were open 2 years later. Because radioactive pollution was the principal factor hindering the recovery of farming and fishing production in Fukushima Prefecture, it will probably take time for these industries to recover to pre-quake levels.

The 2014 results of the "State of Agricultural and Fishery Businesses in Regions affected by the Tsunami Disaster from the Great East Japan Earthquake," published regularly by the Ministry of Agriculture, Forestry and Fisheries (MAFF) since the earthquake, show that the number of businesses that had intended to reopen but have not yet done so has shrunk, while the number of reopened businesses with turnover exceeding pre-quake levels has grown. In contrast, 40%–50% of businesses report that even after reopening, their turnover had not reached pre-quake levels.

Looking at Fig. 2.4, which uses public materials from Fukushima Prefecture, we see that production in the forestry and fisheries industries declined by 20.6% year-on-year to a production value of 185.1 billion yen in FY2011. Rice production dropped by 5% in 1 year but declined by 20% over 2 years. Fruits and vegetables also declined by 30% from the previous fiscal year. Agricultural production rose in fiscal 2012 and 2013, growing to the 200 billion-yen level, but with the national

[2]See Nikkei Shimbun from March 19, 2011 and March 26, 2011.

Table 2.2 Recovery from the tsunami and the earthquake in the agriculture and fisheries industries

	Amount of damage on farm products							Amount of damage on fishery						
	Immediately after quake	Six months period after quake	Restoration ratio	One year period after quake	Restoration ratio	Two years period after quake	Restoration ratio	Immediately after quake	Six months period after quake	Restoration ratio	One year period after quake	Restoration ratio	Two years period after quake	Restoration ratio
Fukushima Prefecture	44,770	41,757	6.7	38,744	13.5	37,944	15.2	20,256	20,256	0.0	19,982	1.4	19,873	1.9
Iwate Prefecture	10,459	9,165	12.4	8,417	19.5	6,707	35.9	39,313	32,869	16.4	18,346	53.3	11,008	72.0
Miyagi Prefecture	17,518	13,244	24.4	10,688	39.0	7,174	59.0	82,639	68,016	17.7	48,258	41.6	35,537	57.0
Ibaraki Prefecture	2,879	103	96.4	83	97.1	0	100.0	9,518	2,447	74.3	906	90.5	0	100.0
Stricken Region	75,626	64,268	15.0	57,931	23.4	51,825	31.5	151,725	123,588	18.5	87,492	42.3	66,417	56.2

Source: Tokunaga and Okiyama (eds) (2014, p. 10), MAFF (2015, 2016)

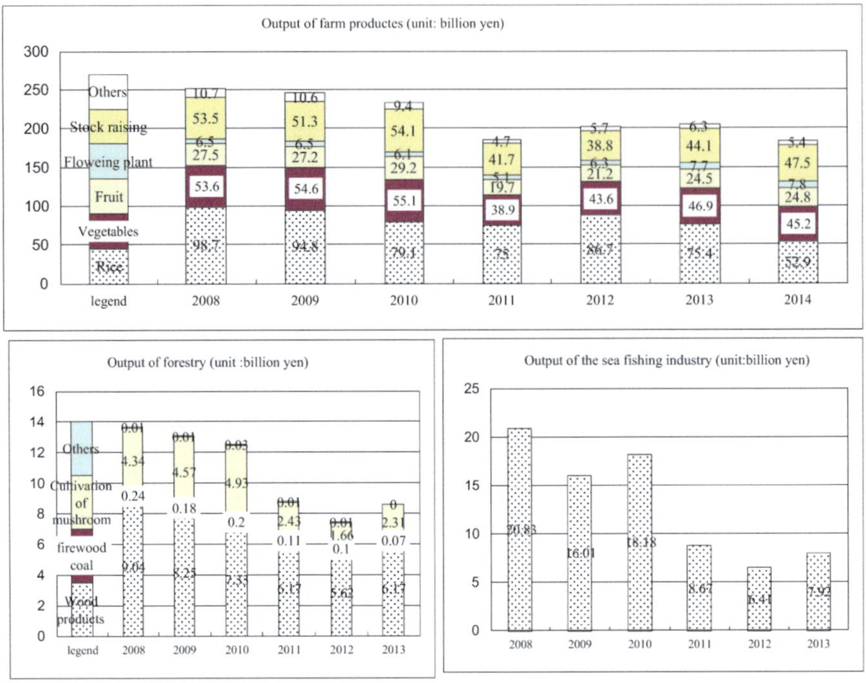

Fig. 2.4 Production in the forestry and fisheries industries of Fukushima Prefecture (2008–2014) (Source: Tokunaga and Okiyama (eds) (2014, p. 11), Reconstruction Agency in Fukushima (2008 through 2015))

decline in rice prices 2014, agricultural production dropped back to the 2011 levels. Forestry production in fiscal 2011 was 8.52 billion yen, 30% lower than in the previous fiscal year, and fisheries' production decreased to 52.3%. While forestry and fisheries' output dropped in 2012, both industries began to grow again in FY2013.

To understand the state of production recovery in the construction and tertiary industries, we refer to the "Survey of Recipients of Group Subsidies," carried out in September 2012 by the Tohoku Bureau of Economy, Trade and Industry (2013) (see Fig. 2.5). According to this publication, 70% of the businesses that managed to reopen have not reached pre-quake revenue levels and 30% show revenues of less than half their sales before the disaster. While half of the construction companies increased sales, nearly 30% of firms in the fisheries, foodstuffs, and hospitality industries reported sales less than 30% of pre-quake levels. From the degree of recovery in affected manufacturing industries and that in other manufacturing industries mentioned above, we analogized as a proxy for the degree of production recovery in the construction and tertiary industries. According to this measurement, through fiscal measures in the affected regions, the construction industry recovered more than 120% of its prior revenues, the transportation industry recovered all

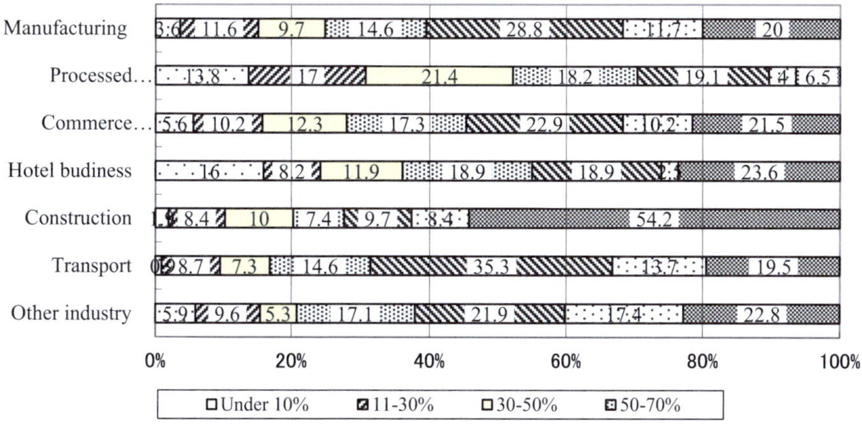

Fig. 2.5 State of production recovery from the great earthquake in 2012 (Source: Tokunaga and Okiyama (eds) (2014, p. 11), Tohoku Bureau of Economy, Trade and Industry (2013))

(100%), and the commerce and service industries recovered 95.3%. The Japanese-style inn and hotel, fisheries, and food processing industries, however, reported only approximately 60%–80% of pre-quake sales. The differences in recovery progress of affected regions and companies (the Tohoku Bureau of Economy, Trade and Industry reported that "there are notable regional disparities") seem to be caused by differences in the degree of destruction between and within the affected prefectures as well as non-quake factors such as the extent of fiscal measures and the high value of the yen; all these factors influenced the progress of recovery in different regions and industries.[3]

With regard to non-production recovery, according to the Reconstruction Agency's publication "Current State and Initiatives for Reconstruction" (2012), except for areas where houses were swept away and in nuclear power security districts, major lifelines such as electricity, natural gas, and water and nearly 100% of social infrastructure (other than ports and harbors) were restored on an emergency basis in the affected areas. The agency also noted that restoration of public services, such as communications, mail, hospitals, and schools, by and large was completed. The report also indicated that as of 2012, full restoration and recovery of the social infrastructure damaged in the disaster remained to be completed and that schedules for government and regional authorities' reconstruction projects were in the process of being drafted.

[3]For a rigorous spatial econometrics analysis, see Sect. 2.4.

2.2.2 *Current State of Production Activities in Regions Affected by the Nuclear Power Plant Disaster*

While production activities in the areas surrounding the Fukushima Dai-ichi Nuclear Power Plant incident resumed in some areas, where residents were permitted to return, production now has almost entirely halted. According to METI reports (2011), 619 manufacturing establishments and 1074 commercial establishments are located in the secured areas, the planned evacuation areas, and the emergency evaluation areas surrounding the Fukushima Dai-ichi Nuclear Power Plant. Total manufacturing shipments in 2008 were approximately 216.4 billion yen, and total sales by commercial establishments were approximately 89.2 billion yen. Even if only temporarily, these monetary gains from economic activities were lost due to the nuclear accident. We attempted to estimate production value, by industry, for one city, six towns, and three villages (excluding the bio-regional areas of Soma City and Shinchi Town) that were affected by the power plant disaster.[4] Our estimates were based on municipal reports for Fukushima Prefecture for FY2010 and FY2012 (on a value-added basis) and shipment values of manufactured items from the 2010 table of industrial statistics by the Ministry of Economy, Trade and Industry. Results are shown in Table 2.3. We estimated that production values for all industries in FY2010 and FY2012 in the regions affected by the nuclear power plant disaster were 1.4101 trillion yen and 707.0 billion yen, respectively. Of this, the total of electricity, natural gas, and water in FY2010 and FY2012 in the regions was 739.4 billion yen and 212.3 billion yen, respectively. Their share of the prefecture's total production dropped from 64.8% in FY2010 to 37.1% in FY2012, principally because of a 520 billion-yen reduction in electricity production caused by the Fukushima nuclear power plant incident. Agriculture, forestry, and fisheries industries shrank from 26.3 billion yen in FY2010 to 2.6 billion yen in FY2012, while manufacturing overall decreased by 30%, from 213.4 billion yen to 62.4 billion yen, and commerce lost 6.8 billion yen in revenues.

Looking at gross regional product (GRP) for the same regions in Table 2.4, we see that worker compensation was 247.6 billion yen and 200.2 billion yen in FY 2010 and FY2012, respectively, dropping as a ratio of the prefecture's total from 7.5% in FY2010 to 5.9% in FY2012. In addition, income from self-owned businesses dropped from 163.3 billion yen in FY2010 to 61.6 billion yen in FY2012, a decline of 101.7 billion yen, or 332.7 billion yen if corporate operating surpluses are included. The ratio to the prefectural total went down from 14.0% to 5.2%.

Human loss from the Great East Japan Earthquake was extensive; as of November 2013, 15,883 individuals died and 2651 were missing, mainly in the Iwate, Miyagi, and Fukushima Prefectures. In addition to this enormous loss,

---

[4]The one city, six towns, and three villages are Minami Soma City, Hirono Town, Naraha Town, Tomioka Town, Okuma Town, Futaba Town, Namie Town, Kawauchi Village, Katsurao Village, and Iitate Village. We define these municipalities as those affected by the nuclear power plant meltdown. The population of this area, as of March 2014, is approximately 13,600.

Table 2.3 Production values for all industries in FY2010 and FY2012

	FY 2010					FY 2012				
	Gross value-added			Estimated output		Gross value-added			Estimated output	
	The whole prefecture	The areas affected by the nuclear disaster	Ratio	The whole prefecture	The areas affected by the nuclear disaster	The whole prefecture	The areas affected by the nuclear disaster	Ratio	The whole prefecture	The areas affected by the nuclear disaster
Agriculture	122.4	13.2	10.8%	244.3	26.3	105.7	1.3	1.2%	210.9	2.6
Forestry	7.7	0.7	9.7%	13.1	1.3	5.5	0.2	4.1%	9.3	0.4
Fishery	8.6	0.5	5.4%	15.3	0.8	3.5	0.0	0.0%	6.3	0.0
Mining	3.6	0.2	5.3%	8.2	0.4	5.8	0.1	1.4%	13.4	0.2
Manufacturing	1,648.6	73.1	4.4%	4812.3	213.4	1,461.5	21.4	1.5%	4,266.2	62.4
Construction	348.9	33.7	9.7%	746.9	72.1	657.9	95.1	14.5%	1,408.5	203.6
Electricity power,gas and water supply	624.3	404.5	64.8%	1,140.9	739.4	313.2	116.2	37.1%	572.3	212.3
Commerce	610.6	16.3	2.7%	883.5	23.7	601.8	11.7	1.9%	870.7	16.9
Finance and insurance, Real estiate	1,007.3	66.0	6.5%	1,213.0	79.4	974.5	19.9	2.0%	1,173.5	24.0
Transport and Communication	295.2	14.7	5.0%	620.2	30.8	315.7	9.2	2.9%	663.3	19.4
Services	1,518.5	70.5	4.6%	2,704.0	125.5	1,556.7	38.9	2.5%	2,772.2	69.3
Pulic services	749.4	63.7	8.5%	1,141.7	97.0	759.9	62.9	8.3%	1,157.7	95.8
Total	6,945.0	757.0	10.9%	13,543.3	1,410.1	6,761.7	376.9	5.6%	13,124.2	707.0

Source: Economic survey of Tohoku Bureau of Economy, Trade and Industry (2012 through 2014)
Unit: billion yen

Table 2.4 GRP of the regions affected by the nuclear power plant disaster in FY2010 and FY2012

(unit billion yen)

	FY 2010			FY 2012		
	The whole prefecture	The areas affected by the nuclear disaster	Ratio	The whole prefecture	The areas affected by the nuclear disaster	Ratio
Prefectural gorss domestic product	6,945.0	757.0	10.9%	6,761.7	376.9	5.6%
Compensation of employee	3,314.3	247.6	7.5%	3,377.7	200.2	5.9%
Operating surplus	3,630.7	509.4	14.0%	3,384.0	176.7	5.2%
Private unincorporated enterprises	1,652.3	163.3	9.9%	1,522.0	61.6	4.0%
Household income	5,110.7	381.7	7.5%	5,246.8	309.1	5.9%

Source: Tokunaga and Okiyama (eds) (2014, p. 13), economic survey of Tohoku Bureau of Economy, Trade and Industry (2012 through 2014)

nationally 280,000 people were evacuated over a prolonged period. These evacuations included 60,000 people outside Fukushima Prefecture and an additional 10,000 people from other affected areas. Including those who died in the disaster, population in the three ill-fated prefectures dropped by approximately 90,000.[5]

2.3 Current State of Fiscal Reconstruction Measures in the Affected Areas

2.3.1 Current and Future Fiscal Measures for Reconstruction

Table 2.5 was prepared from the Reconstruction Agency's budget-related materials.[6] Emergency funds were used five times in FY2010 in response to the March 2011 disaster. In FY2011, budgeted expenditures in the first through third supplementary budgets for recovery and reconstruction were 14.9243 trillion yen, including emergency funds. Subsequently, in FY2012 and FY2013, another 9.7402 trillion yen and 7.5089 trillion yen, respectively, were budgeted as a Special Accounting Budget for Recovery. The total cumulative budgeted amount for FY2011–FY2014 reached 29.3946 trillion yen. Of this amount, 23.9132 trillion actually was disbursed, giving a disbursement ratio of 81.4%. The item with the

[5]We shall analyze the depopulation stemming from the disaster in detail in Chap. 9.

[6]See Reconstruction Agency (2012, 2013, 2014, 2015, 2016) and MAFF (2015, 2016).

Table 2.5 State of fiscal reconstruction measures in the affected regions for FY2011- FY2014

(unit :billion yen)

	The supplementary FY2011 budget (First-third)	percentage (%)	The FY2012 budget	percentage (%)	The FY2013 budget	percentage (%)	From FY2011 through FY2014 period accumlated	percentage (%)
Support for disaster-affected people	1,550	10.4	509	5.2	223	3.0	2,154	7.3
Restoration /Reconstruction of living	6,042	40.5	4,733	48.6	3,201	42.6	10,569	36.0
a)Removal and disposal of disaster waste	738	4.9	738	7.6	508	6.8	1,232	4.2
b)Reconstruction Grants	1,566	10.5	1,604	16.5	931	12.4	2,865	9.7
Reconstruction of industry, Securing employment	3,192	21.4	813	8.3	625	8.3	4,191	14.3
a) Support for Agriculture and fisheries industry	536	3.6	140	1.4	54	0.7	636	2.2
b) Support for small-and medium-sized companies and Creation of subsidies for the location of	524	3.5	272	2.8	344	4.6	933	3.2
Recovery and reconstruction from the nuclear disaster	1,241	8.3	866	8.9	1,236	16.5	3,695	12.6
a)Decontamination etc	579	3.9	656	6.7	996	13.3	2,409	8.2
Reconstruction Agency-related budget	-	-	6,922	71.1	5,294	70.5	20,635	70.2
Grants in response to the disaster	2,241	15.0	670	6.9	605	8.1	3,989	13.6
Other	658	4.4	2,148	22.1	1,609	21.4	4,771	16.2
Total	14,924	100.0	9,740	100.0	7,509	100.0	29,395	100.0

Source: Tokunaga and Okiyama (eds) (2014, p. 15), Reconstruction Agency in Fukushima (2012 through 2016)

lowest disbursement ratio was "Town Restoration and Recovery," including recovery grants, for which 10.5687 trillion yen was budgeted. Only 7.5809 trillion yen, or 71.7%, of this total was disbursed. 1.698 trillion yen was carried over to the next fiscal year, with 70% of public project funds allocated to disaster recovery and restoration remaining unused. Fukushima Prefecture's "Recovery and Renewal from the Nuclear Power Plant Disaster" project had a budget of 3.6952 trillion yen throughout FY2014, but only 2.7534 trillion yen was disbursed. This gave a disbursement ratio of 74.5%, with the majority of the extra 358.2 billion yen spent on pollution clean-up. A total of 3.9889 trillion yen was allocated throughout FY2014 by the Special Accounting Budget for Recovery as regional tax grants (the Extraordinary Disaster Recovery Grant Tax), of which 3.8 trillion yen was expended (a disbursement ratio of 95.3%). These grants added to local contributions and compensated for the reduction in local taxes in the Ministry of Internal Affairs and Communications' budget for restoration and recovery projects, which amounted to 589.8 billion yen in FY2015 and an estimated 347.8 billion yen in FY2016.

In terms of future fiscal measures for recovery, while the budget for FY2015 was 3.9087 trillion yen and the expected budget for fiscal 2016 is 3.2469 trillion yen, the numbers are going down. The "Requirements Policy for the Fiscal 2016 Recovery Agency Budget" presents the following four policies to move ahead steadily with initiatives required for recovery in the affected areas during the "Recovery and Creation Period": (1) Institute a budget for solving challenges faced by the affected areas, (2) accelerate the renewal of Fukushima Prefecture after the nuclear power plant disaster, (3) create a sustainable regional society, including a "new Tohoku;"

and (4) prioritize projects truly necessary for recovery. In the future, fiscal measures for recovery likely will be defined both qualitatively and quantitatively by these policies.

2.3.2 *Fiscal Measures for Recovery in Fukushima Prefecture*

Regarding fiscal measures for recovery in Fukushima Prefecture, the FY2014 budget detailed items listed explicitly in the "Recovery and Renewal after the Nuclear Power Plant Disaster" project, including, in addition to pollution removal costs and damage from negative rumors, items such as "Fukushima Renewal Accelerator Grants" and "Project to Support Restoring Hope in Local Areas." Between FY2014 and FY2016, the renewal budget in Fukushima increased from 660.0 billion yen (2014) to 780.7 billion yen (2015) to 1.167 trillion yen (2016). In particular, the FY2014 budget of 108.8 billion yen for "Fukushima Renewal Accelerator Grants" will remain almost constant at 100.0 billion yen for FY2015 and FY2016. Fiscal transfer from the government to Fukushima Prefecture reached nearly two trillion yen during the 2 years after the disaster, and fiscal measures in excess of one trillion yen annually are planned going forward, continuing the fiscal support for recovery in Fukushima Prefecture.

Concerning the effectiveness of these kinds of fiscal measures, Table 2.6 shows the rate of progress in restoring and developing the social infrastructure, based on publications from Fukushima Prefecture. We can see that more than 90% of public infrastructure projects have begun. Project completion is 90% for roads and bridges and 80% for rivers and harbors but only 50% for bays and fishing port facilities. These metrics clearly demonstrate how the impact of the nuclear power plant disaster has become a major barrier to recovery in Fukushima Prefecture.

Table 2.6 Fiscal measures for recovery in Fukushima Prefecture

Disaster recovery projects for public works (the figures for Dec.31,2015)	Planned cases	Started cases	Ratio of started cases (%)	Completed cases	Ratio of completed cases (%)
Total	2,133	1,996	93.6	1,678	78.7
Rivers/Erosion control	271	263	97.0	231	85.2
Seaside	156	148	94.9	44	28.2
Road/Bridge	798	753	94.4	726	91.0
Harbors	331	314	94.9	285	86.1
Fishing port	480	421	87.7	295	61.5
Drainage	3	3	100.0	3	100.0
Park/Public facility	5	5	100.0	5	100.0
Public housing	89	89	100.0	83	93.3

Source: Reconstruction Agency in Fukushima (2016)

2.4 Impact of the Great East Japan Earthquake on Firm Output

In the preceding section, we took a general view of the 5 years of recovery after the Great East Japan Earthquake, and we clarified that in the Iwate, Miyagi, Fukushima, and Ibaraki Prefectures, which suffered heavy damages, the effects of the earthquake varied greatly regionally and by industry type. In this section, we will focus on the effects of the tsunami and nuclear disaster on the manufacture of processed marine, food, general-purpose machinery (which suffered damage over a broad area), and motor vehicles and motor vehicle parts (which encompass a broad range of supporting industries). First, using data from 2011 when the Great Earthquake occurred, we will analyze the impact the earthquake had on production by these industries. Next, based on those results, using data from before and after the earthquake (from 2010 and 2012), we will analyze the process of recovery of those manufacturing firms' production according to a geographically weighted regression (GWR) model.[7] For these analyses, by measuring the spatial adjacency of individual similar firms in areas affected by the disaster, i.e., the degree to which similar firms are clustered, using a revised version of Moran's *I* (an index of spatial autocorrelation), for each industry type we will also analyze the relation between the degree of industrial concentration and the degree of impact by the earthquake.[8]

2.4.1 Impact on Production by Manufacturing Firms in Regions Affected by the Great Earthquake

From Fig. 2.6, which shows the tsunami-flooded areas of Iwate, Miyagi, Fukushima, and Ibaraki Prefectures (these areas also took heavy damage from the Great East Japan Earthquake), as well as the distribution of seismic intensity and the 20–50 km area around the Fukushima Daiichi nuclear plant, it is clear that the damage from the earthquake was on a large scale and extended over a broad area, from the coastal and central areas of the Tohoku region to the Kanto region. Next, if we compare that map with Fig. 2.2, which plots manufacturing business sites that

[7]The first author, Tokunaga, obtained fruitful advice at a spatial computation seminar by Professor L. Anselin held at Arizona State University in 2011 Dec.

[8]Moran's *I* is defined as $I = \frac{n}{S_0}\frac{\sum_{i=1}^{n}\sum_{j=1}^{n} w_{i,j} z_i z_j}{\sum_{i=1}^{n} z_i^2}$. Here, $S_0 = \sum_{i=1}^{n}\sum_{j=1}^{n} w_{i,j}$, z is deviation from the average value of feature i (which refers to the location of each region), $w_{i,j}$ is a spatial weight (which shows the relation of proximity between regions i and j), n is the total number of regions, and S_0 is the sum of all spatial weights. The z score of Moran's *I* is defined as $z_I = \frac{I - E[I]}{\sqrt{V[I]}}$. Here, $E[I] = -\frac{1}{n-1}$ and $V[I] = E[I^2] - E[I]^2$ (regarding Moran's *I* and the spatial econometrics, see LeSage and Pace 2009; Anselin L1988; Arbia 2014).

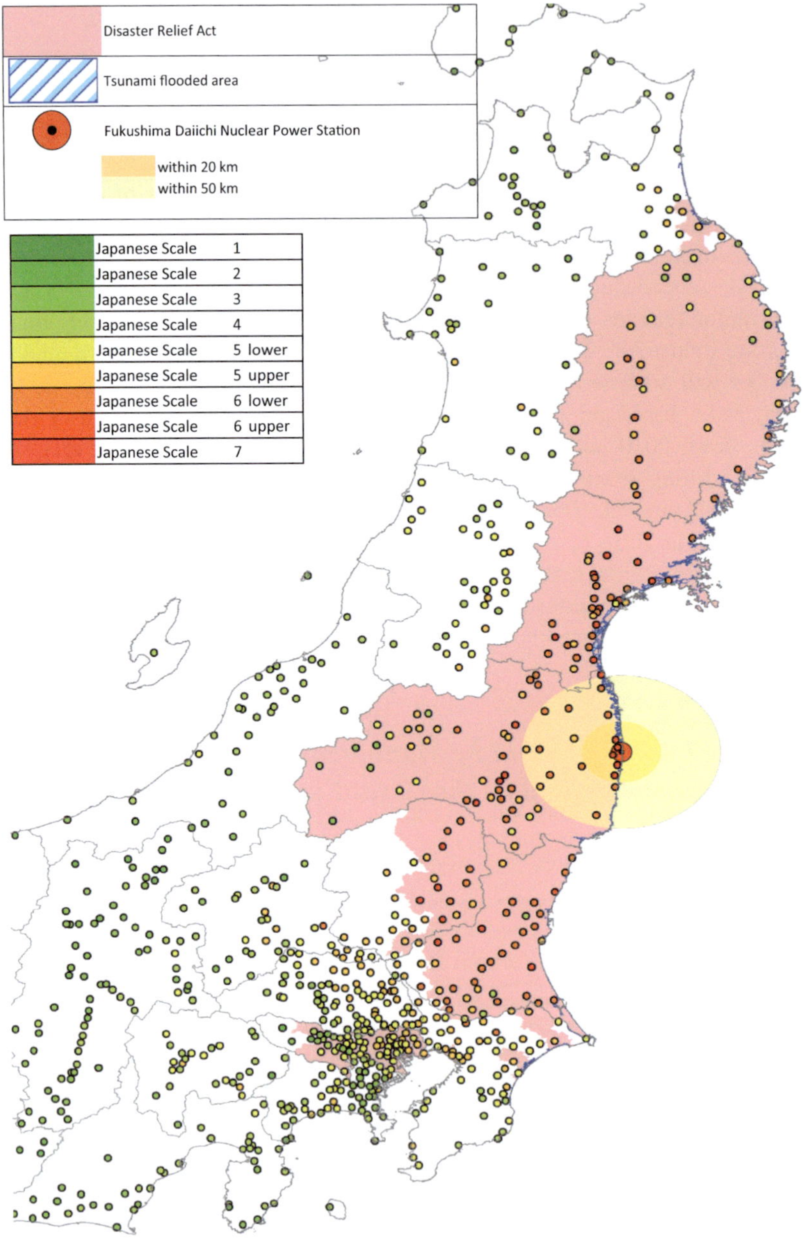

Sources: Japanese MAFF, the Japanese Nuclear Safety Commission, CSiS of the University of Tokyo.

Fig. 2.6 Geographical distribution of the damage by the great earthquake (Sources: MAFF Japan, the Japanese Nuclear Safety Commission, CSiS of the University of Tokyo)

were at least partially damaged by the earthquake, a correlation between the distribution of business sites that were damaged and the geographical data of the damage by the earthquake is apparent. Next, targeting the areas of "manufacture of food (9)," "manufacture of fisheries and seafood product manufacturers (processed marine)", "manufacturer of general-purpose machinery (25)," and "manufacture of automobile products," all of which suffered heavy financial damage, we will analyze the impact of decreases in employment due to the earthquake on firm production in the affected areas using individually contributed data from "Economic Census" that studied relevant business sites with thirty or more employees in 2011. The production function of an individual firm uses the volume of production as the explained variable and the number of employed workers as the explanatory variable. The estimation formula is the log linear of Eq. (2.1). Since the relation between the explanatory variable and the explained variable is nonstationary, it is estimated using a geographically weighted regression (GWR) model that gives regression coefficients that vary for each location (business site).[9]

$$\ln(output) = \alpha_i + \beta_i \ln(labor_i) + \varepsilon_i \tag{2.1}$$

Here, $output_i$ is the production volume of an individual firm in region i, $labor_i$ is the number of employees in an individual firm in region i, and ε_i is the error term.[10] The estimation result is illustrated in Fig. 2.7. This is mapped for each industry type, since the GWR estimates different coefficients for each location. It also shows the estimation result of the spatial autocorrelated Moran's I.[11] The labor coefficient (labor elasticity) is positive for all four industry types. However, the value of this elasticity varies greatly by region and industry type, so it can be seen that the impact on production from labor decreases due to the earthquake also varies greatly by region and industry type. The estimation result of Moran's I for "manufacture of food (9)" and "manufacture of processed marine" in 2011 shows that similar firms were spatially adjacent at a significance level of 1% and that its value is around 1 or higher. In contrast, the value of Moran's I for both "manufacturer of general-purpose machinery (25)" and "manufacture of automobile products" was positive, showing that similar firms are spatially adjacent at a significance level of 1%, i.e., industrial concentration exists. However, that value was below 1. This shows that damage from the earthquake was significant for areas that experienced seismic shock above 6 on the Richter scale, tsunami-flooded areas, areas within 20–50 km of the Fukushima Daiichi nuclear reactor, and areas of industrial concentration.

[9]We used "area-to-point kriging" in the geographical data system ArcDIS10.3.

[10]In this section, we analyze the impact of decreases in employment due to the earthquake on production, but in Ch. 3.4., the impact on production from private capital stock loss is elucidated according to the SAM multiplier analysis. Moreover, in Ch. 9, we analyze the impact of decreases in population due to the earthquake on regional economics, and in Ch. 10, we analyze population movements due to the earthquake.

[11]Here, we consider the index showing this clustering to represent the degree of recovery in the production aspect by individual firms.

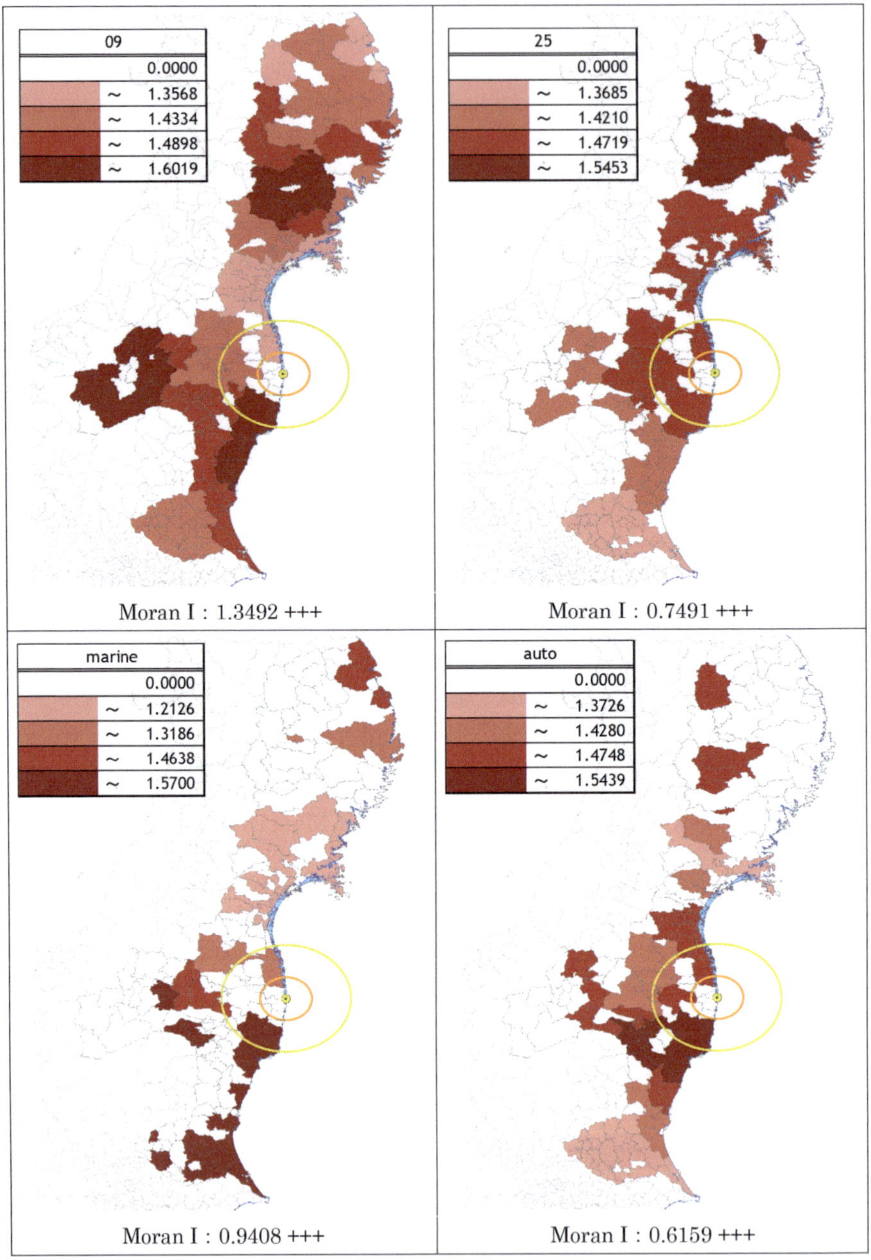

Fig. 2.7 Estimation results of coefficients for labor by GWR (2011)

2.4.2 Recovery Process of Production by Manufacturing Industry Firms in Areas Affected by the Great Earthquake

Next, based on these results and using data from before and after the earthquake, we will analyze the industrial concentration conditions and the recovery process of manufacturing firms in the affected areas using the GWR model. Here, we use individually contributed data from "industrial statistical surveys" (for business sites employing four or more workers) from before the 2010 earthquake and after the 2012 earthquake. The analysis targets the four industry types mentioned above. First, for business sites in four of the affected prefectures (Iwate, Miyagi, Fukushima, and Ibaraki), we extracted data from across the country and performed geocoding from the address data. For geocoding, we used an address matching service within Tokyo University's "Geographic Spatial Information System"; then, we converted the address data into longitude and latitude data. For the analysis, we used samples from this that matched at the street block level or higher (match level 6 or higher). The match rates for all industry types as a whole were 84.5% for 2010 and 84.7% for 2012. As for administrative districts, we sought to consolidate them using MLIT's "Land Numerical Data, Administrative Districts 2015 Version" as the standard. The geographical coordinate system was GCS-JGD-2000, and the geodetic survey standard system was D-JGD-2000 (the Japanese standard system).

Next, to analyze the conditions of business sites being restored/continuing to be used, continuing with data from 2010 to 2012, we created a "continuation/restoration" dummy in which sites that were restored or are still being used received a 1 and all others received a 0 for both 2010 and 2012. Also, using these data and with the production volumes of individual firms being the explained variable, we made labor and the continuation/restoration dummy the explanatory variables as in Eq. (2.2) and estimated these using GWR.[12]

$$\ln(output) = \alpha_i + \beta_i \ln(labor_i) + \delta_i \ln(Dummy_i) + \varepsilon_i \tag{2.2}$$

Here, i represents location$i$. Forthecoefficient $\ln(labor)$, estimated by GWR, the regression coefficient is averaged by municipality, as depicted in the maps Fig. 2.8a (2010) and Fig. 2.8b (2012). The labor coefficients were positive for both 2010 and 2012 for the four industry types, and production volume increased as labor increased. For "manufacture of food (9)" and "manufacture of processed marine," the comparison of the value of labor elasticity from before and after the earthquake showed an increase; moreover, we found that the intervals between maxima and minima varied greatly by region. However, for "manufacturer of general-purpose

[12]In this section, in measuring Moran's I of the GWR model, we used the reciprocal of the distance; however, further improvement is necessary. Note that we cannot compare with Fig. 2.7 for 2011 and Fig. 2.8b for 2012 as we use the different data of "Economic Census" for 2011 and "industrial statistical surveys" for 2012.

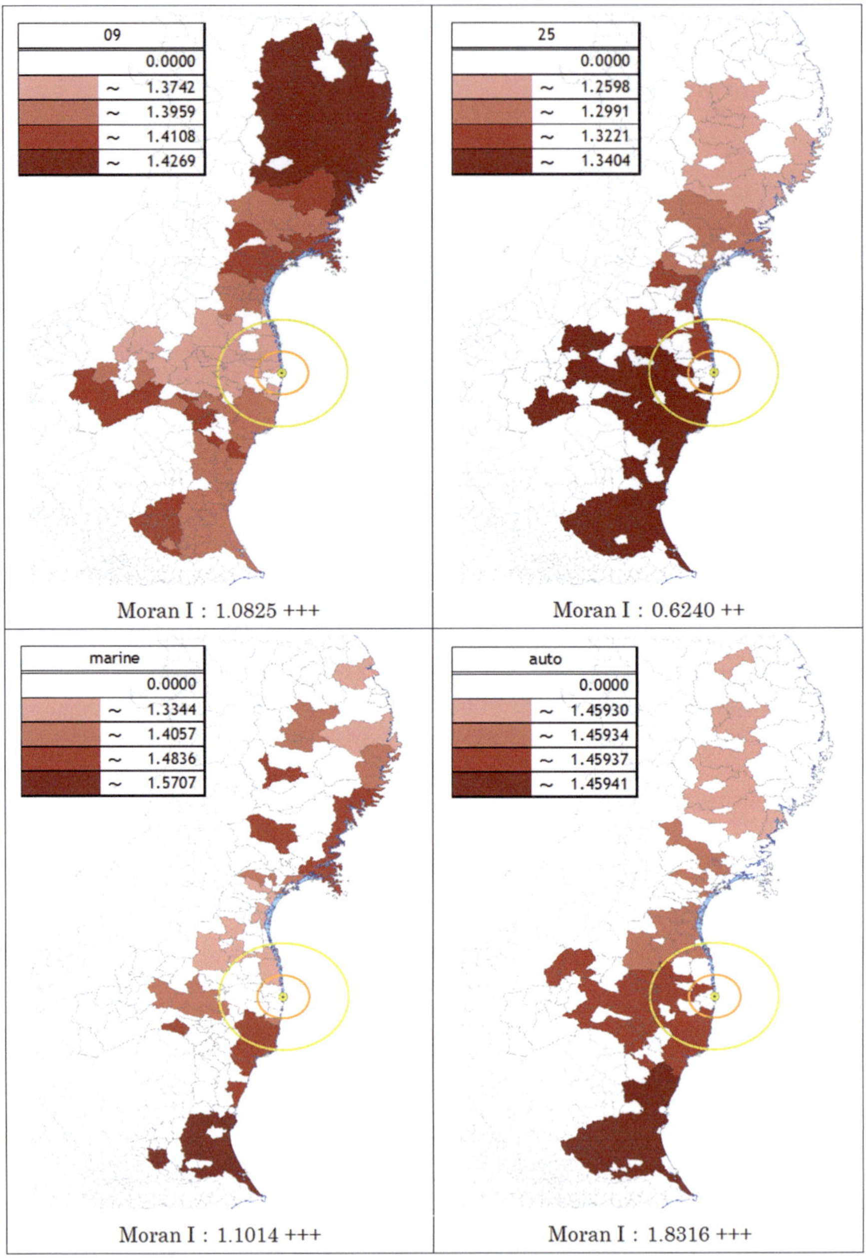

Fig. 2.8a Estimation results of coefficients for labor by GWR (2010)

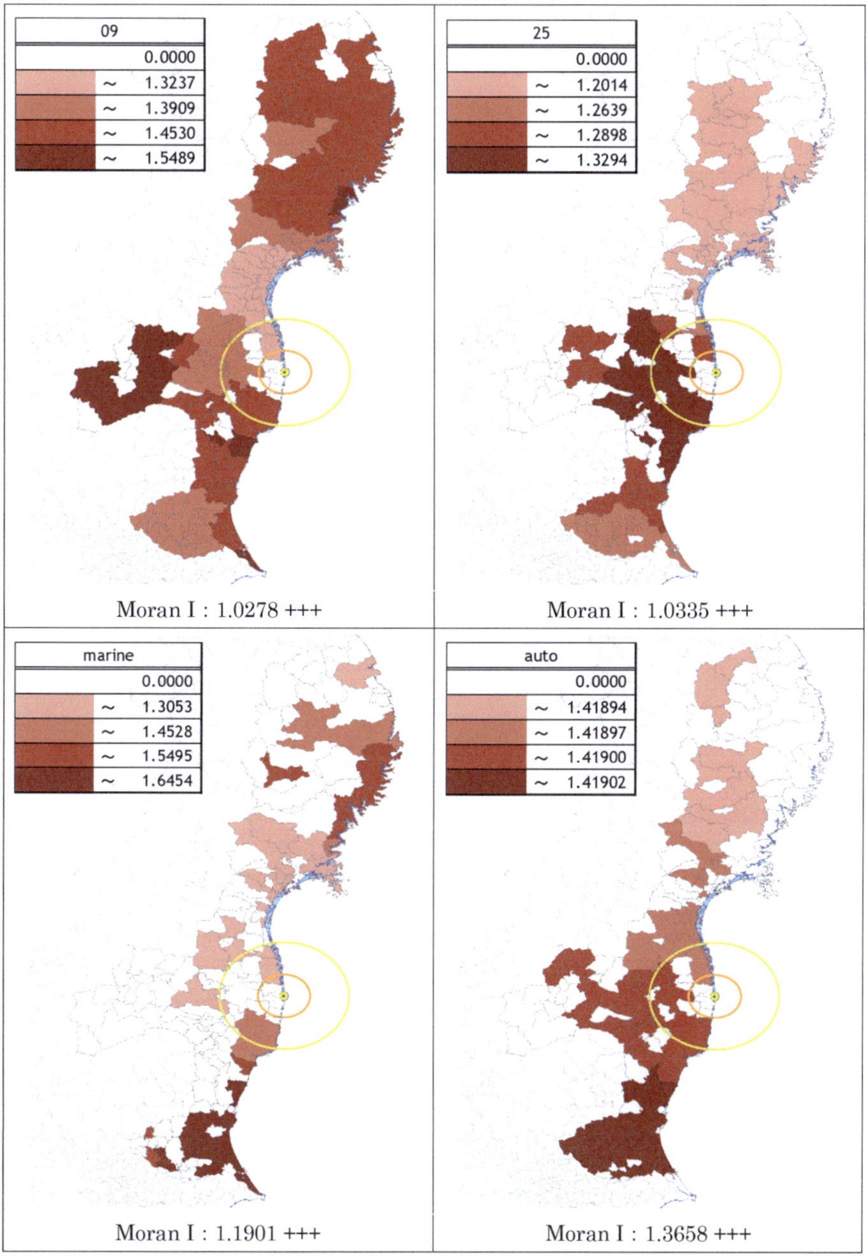

Fig. 2.8b Estimation results of coefficients for labor by GWR (2012)

machinery (25)" and "manufacture of automobile products," the value of labor elasticity fell, and though the intervals between maxima and minima were small, they still varied by region.

The estimation results of the spatial autocorrelated Moran's *I* for "manufacture of food (9)" in 2010 and 2012 for all target regions show that similar firms are spatially adjacent at a significance level of 1%, i.e., industrial concentration exists, but also that this value tends to remain constant or decrease somewhat. This was seen in 2010 in the industrial concentration present along the coastal regions of Miyagi and Fukushima Prefectures, but in 2012, only weak industrial concentration was observed in the same regions, which in turn coincided with a slight increase in new concentration in the inland region of the Fukushima Prefecture. For "manufacture of processed marine" as well, at both points in time for all target regions, we see that similar firms are spatially adjacent at a significance level of 1% and that this value tends to remain constant. In contrast, for "manufacturer of general-purpose machinery (25)," we see that when Moran's *I* tends to increase, the degree of concentration increases and tends toward recovery. We see that when Moran's *I* for "manufacturer of automobile" tends to decrease, the degree of concentration decreases and there is no clear tendency toward recovery.

Finally, the regression coefficient of the continuation/restoration dummy variable estimated by the GWR is averaged by municipality, as depicted in Fig. 2.8c (2012). The value of the coefficient of the continuation/restoration dummy ranges from negative to positive (including zero). The results varied greatly, by industry type as well as region. For "manufacture of food (9)," the estimation value of the regression coefficient of the continuation/restoration dummy was positive for Iwate, Miyagi, and Fukushima (excluding areas within 20 km of the Fukushima Daiichi nuclear reactor), and we see that they are headed toward recovery. However, for "manufacture of processed marine," although there were regions for which the estimation value of the regression coefficient of the continuation/restoration dummy partially took a positive value, such as the coastal regions of Iwate/Miyagi, it took a negative value in the northern part of the Iwate Prefecture and the southern part of the Fukushima Prefecture (including areas within 50 km of the Fukushima Daiichi nuclear reactor); this showed that recovery has not progressed very much. Conversely, for "manufacturer of general-purpose machinery (25)," it took a positive value in the inland area of the Fukushima Prefecture (excluding areas within 20 km of the Fukushima Daiichi nuclear reactor), and we see that recovery is progressing. Contrary to expectations, for "manufacture of automobile products" in 2012, the value of the regression coefficient of this continuation/restoration dummy was negative for parts of Iwate, Miyagi, Fukushima (excluding areas within 20 km of the Fukushima Daiichi nuclear reactor), and northern Ibaraki, and we see that there has still hardly been any progress toward recovery. From these data, for manufacturers of general-purpose machinery and the manufacture of automobile products, we can surmise that damage from the tsunami extended not only to regions that were directly hit by the tsunami but also that supply lines were

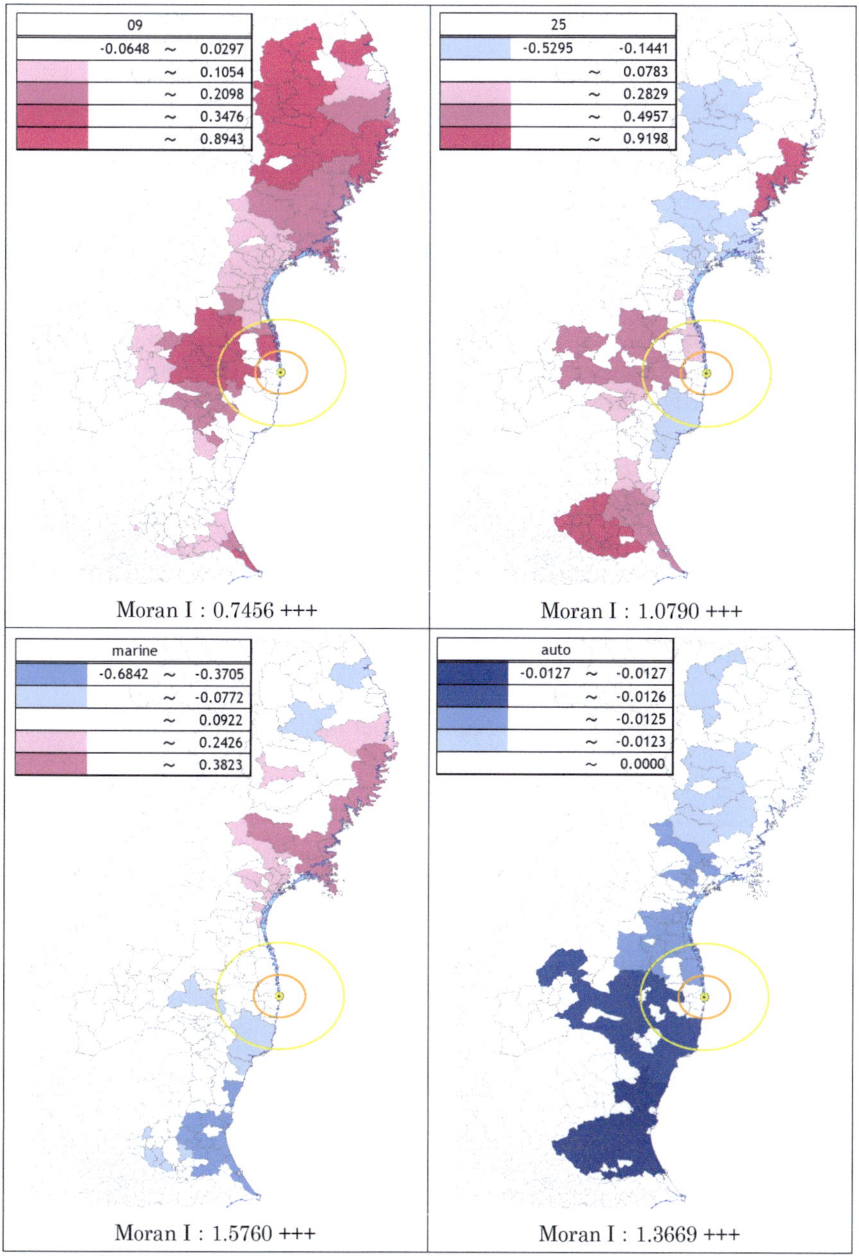

Fig. 2.8c Estimation results of coefficients for Dummy by GWR (2012)

disrupted in Miyagi, Fukushima, and Ibaraki Prefectures, which delayed the recovery of production.[13]

2.5 Economic Model for Natural Disaster Assessment and Labor Migration

In preceding sections, as we examined the economic and human damage inflicted on the affected areas by the Great East Japan Earthquake, as well as the current situation of industrial recovery, we move on to present an economic models for natural disaster assessment and labor migration: (a) the regional input–output and two-regional CGE models to analyze economic and human loss from the quake, and to assess recovery policies; (b) NEG model that facilitates explicit analysis of population and labor migrations; and (c) research about new industrial agglomerations and clusters as recovery policy in this section.

2.5.1 Regional Input–Output Model and Two-Regional SAM and CGE Models

First, we present the multiregional input–output model using a regional technical coefficients matrix $\mathbf{A}^r$. For simplicity, we consider the case of a small two-sector, two-region economy. The basic structure of inter-regional input–output models is as follows:

$$\begin{aligned}\left(\mathbf{I}-\hat{\mathbf{c}}^{rr}\mathbf{A}^r\right)\mathbf{x}^{\mathrm{r}}-\hat{\mathbf{c}}^{rs}\mathbf{A}^s\mathbf{x}^{\mathrm{s}}&=\hat{\mathbf{c}}^{rr}\mathbf{f}^{\mathrm{r}}+\hat{\mathbf{c}}^{rs}\mathbf{f}^{\mathrm{s}}\\-\hat{\mathbf{c}}^{sr}\mathbf{A}^r\mathbf{x}^{\mathrm{r}}+\left(\mathbf{I}-\hat{\mathbf{c}}^{ss}\mathbf{A}^r\right)\mathbf{x}^{\mathrm{s}}&=\hat{\mathbf{c}}^{sr}\mathbf{f}^{\mathrm{r}}+\hat{\mathbf{c}}^{ss}\mathbf{f}^{\mathrm{s}},\end{aligned} \tag{2.3}$$

where $c_i^{rs}\left(=z_i^{rs}/\sum_{r=1}^{2}z_i^{rs}\right)$ is the proportion of all of good i used in s that comes from each region r and $\hat{\mathbf{c}}^{rs}=\begin{pmatrix}c_1^{rs} & 0\\ 0 & c_2^{rs}\end{pmatrix}$, $\mathbf{x}$ is the regional total output vector, and $\mathbf{f}$ is the final demands vector. Thus, Eq. (2.3) can be represented as $(\mathbf{I}-\mathbf{CA})\mathbf{x}=\mathbf{Cf}$,

[13]In Chap. 4 of this book, along with the analysis of supply chain disruption due to the Great East Japan Earthquake in the electrical/electronics/automotive industries, we also analyze economic recovery by the formation of new industrial agglomeration. This is also discussed in Sect. 2.6 below and in Chap. 6.

where $\mathbf{A} = \begin{pmatrix} \mathbf{A}^r & \mathbf{0} \\ \mathbf{0} & \mathbf{A}^s \end{pmatrix}, \mathbf{C} = \begin{pmatrix} \hat{\mathbf{c}}^{rr} & \hat{\mathbf{c}}^{rs} \\ \hat{\mathbf{c}}^{sr} & \hat{\mathbf{c}}^{ss} \end{pmatrix}, \mathbf{x} = \begin{bmatrix} \mathbf{x}^r \\ \mathbf{x}^s \end{bmatrix}, \mathbf{f} = \begin{bmatrix} \mathbf{f}^r \\ \mathbf{f}^s \end{bmatrix}$, and the solution will be given by[14]

$$\mathbf{x} = (\mathbf{I} - \mathbf{CA})^{-1}\mathbf{Cf} \tag{2.4}$$

In Chap. 8 of this book, using the regional input–output model, Dr. Kunimitsu will analyze the economic ripple effects of a biogas electricity power plant as part of earthquake disaster restoration in the coastal area of Iwate Prefecture, Japan. Furthermore, in Chap. 9, based on an inter-regional input–output model, Prof. Ishikawa will analyze the economic impacts of population decline due to massive earthquakes using 47 inter-regional input–output tables at Japan's prefectural level.

Second, let us provide an overview of the two-regional social accounting matrix (SAM), which is the basis of our model. The base data used in the SAM is a 2005 inter-regional input–output table of all 47 prefectures, jointly created by Yoshifumi Ishikawa and the Mitsubishi Research Institute.[15] We calculated additional data that could not be obtained from this table, such as income and expenditures for households, enterprises, governments, and other segments within the two regions, using values from the 2005 National Economic Accounts and the 2005 Economic Accounts for each prefecture. In accordance with Ito's (2008) two-regional SAM framework, we created a 70 × 70 dimensional matrix, with production activities comprising 20 sectors for each of the two regions, production factors of labor and capital comprising two sectors for each region, and institution, savings, and investment comprising nine sectors for each region. We added seven additional sectors and one overseas sector. Our model partially modifies the above-mentioned SAM, using a simplified 58 × 58 dimensional SAM as the database.[16] In Chap. 3 of this book, using this two-regional social accounting matrix, we will analyze the impacts of the Great East Earthquake.

Third, the static two-regional computable general equilibrium (2SCGE) model was built using GAMS code provided by EcoMod Modeling Schools.[17] This code is a basic system that uses a static single-country open-economy model with the country divided into two regions, each region having economies made up of 14 agents: one household, 10 industries, one enterprise, one local government, and one investment bank. There are also 10 product markets, two production factors

---

[14]For input–output models at the regional level, see R. Miller and P. Blair (2009, in Chap. 3).

[15]An inter-industry relations table, with two regions of the four prefectures hit by the Stricken and other areas, was provided by Prof.Yoshifumi Ishikawa of Nanzan University.

[16]Please refer to the macro SAM based on this SAM, shown in Table 3.1 in Chap. 3.We partially modified the SAM to satisfy the homogenous zero prices in our model.

[17]The authors, Tokunaga and Okiyama, attended EcoMod Modeling School: Advanced techniques in general equilibrium modeling with GAMS (Singapore, 2011 and 2012) and obtained useful advice there. For the spatial CGE model, see Dixon and Rimmer (2002), Tokunaga et al. (2003), Hosoe et al. (2010), EcoMod Modeling School (2012), and Okiyama et al. (2014). For 2SCGE model in detail, see Appendix 1 of this book.

of labor and capital, and two additional agents for central government and overseas sectors. Labor and capital for each region can move inside the 10 industries of their respective region, and the endowment for each region is fixed.

While the overview for our model is shown in Fig. 2.9 for the model structure. We explain further the primary blocks of the model below. First, in the domestic production block in Fig. 2.10, we note that each production activity sector (i.e. industry) produces one product, and we assume multi-stage profit-maximizing behavior. In the first stage, industries operate within Leontief production technology constraints, producing aggregate intermediate goods with added value. In the second stage, aggregate intermediate goods are extracted from that region's Armington composite goods, along with Armington composite goods from regions unaffected by the disaster. The model also incorporates constant elasticity of substitution (CES) production technology constraints of a certain scale, and the value added is likewise created from labor and capital within each region, with CES production technology constraints of a certain scale. Earnings are equal to production expenditures as producer pricing meets the "zero profit condition," and return on capital and wage rates are equal for all industries because they can move between all industries in that region. Pricing for aggregate intermediate goods is derived from an expression that defines the supply and demand equilibrium of

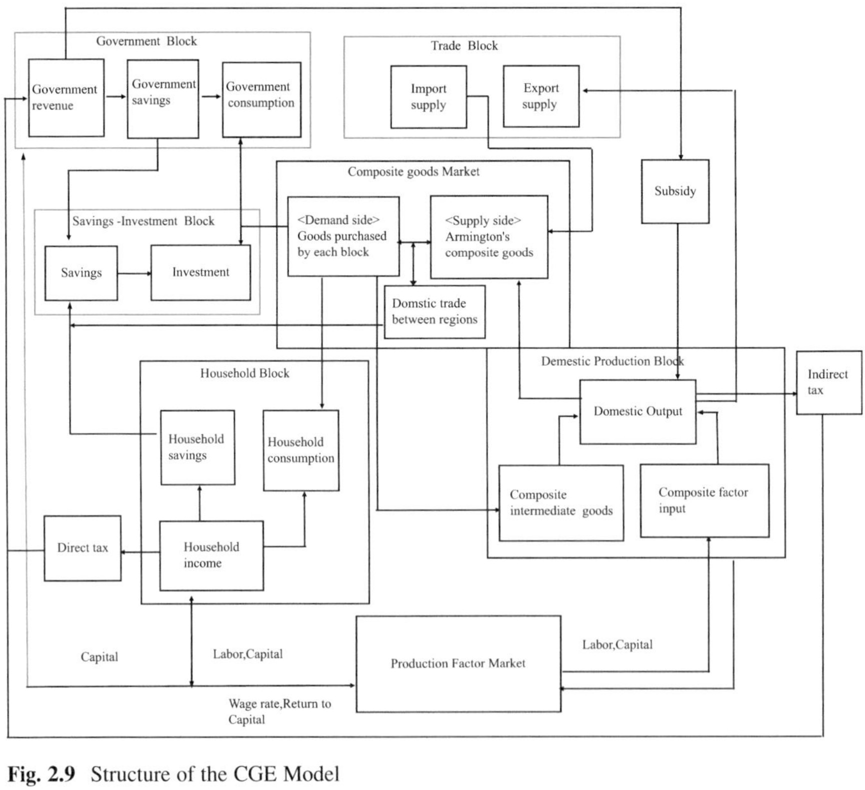

Fig. 2.9 Structure of the CGE Model

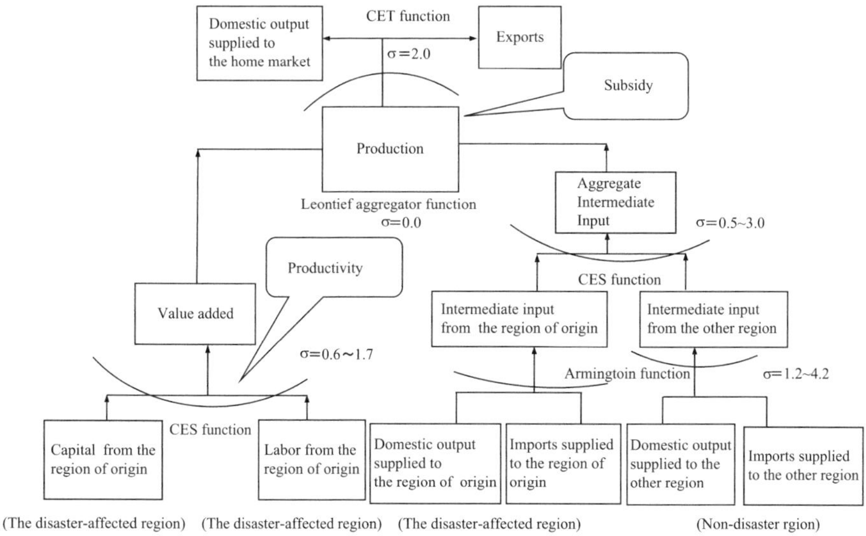

Fig. 2.10 Structure of the production block

intermediate goods. Moreover, within the trading block, the ratio of produced goods shipped domestically and those shipped overseas is derived from a constant elasticity of transformation (CET)-type function. Armington composite commodities, comprising both domestic and imported supplies of goods, were derived using a CES-type production function. Export pricing was calculated using international pricing, adjusted with an exchange rate, and import pricing included import taxes and tariffs. International pricing denominated in foreign currency is fixed in our model, and of the two regions, the disaster-affected region's foreign savings is modeled as an exogenous variable using a foreign trade balance formula.

Next is the household block with households exhibiting utility-maximizing behavior. As depicted in Fig. 2.11, in the first stage, households maximize aggregate goods within budgetary constraints using the Cobb–Douglas utility function. The propensity to save is a set value. In the second stage, aggregate goods are extracted from that region's Armington composite goods, along with Armington composite goods from the regions unaffected by the disaster, within CES production technology constraints of a certain scale. Aggregate pricing is derived from an expression that defines the supply and demand equilibrium of such goods.

Although the trade sector includes exports and imports between each region and the foreign sector, as shown in Fig. 2.12, trade also occurs through imports and exports between the disaster-affected region and non-disaster region. Specifically, the structure incorporated into the 2SCGE model allocates products produced in region for the domestic market and for export, with being derived by solving the problem of sales maximization under the constraint of the Constant Elasticity of Transformation function. In addition, for the composite commodity according to the Armington assumption, which is the composite commodity for domestic supply

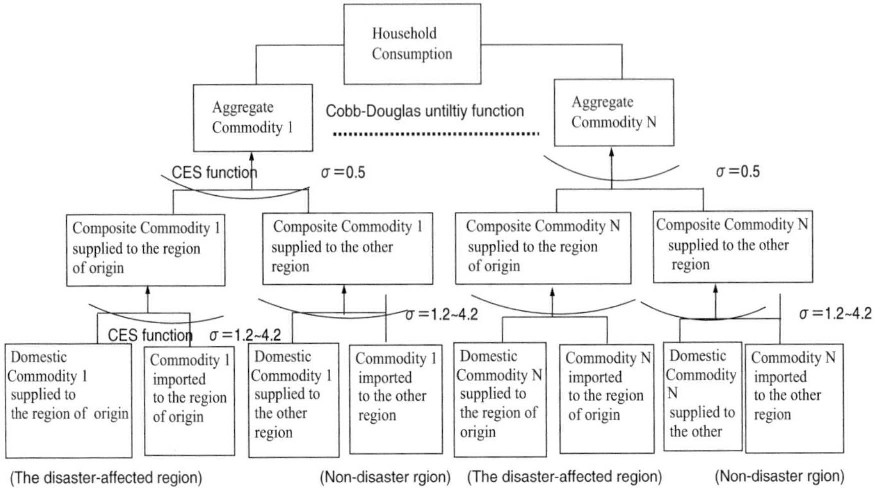

Fig. 2.11 Structure of the household block

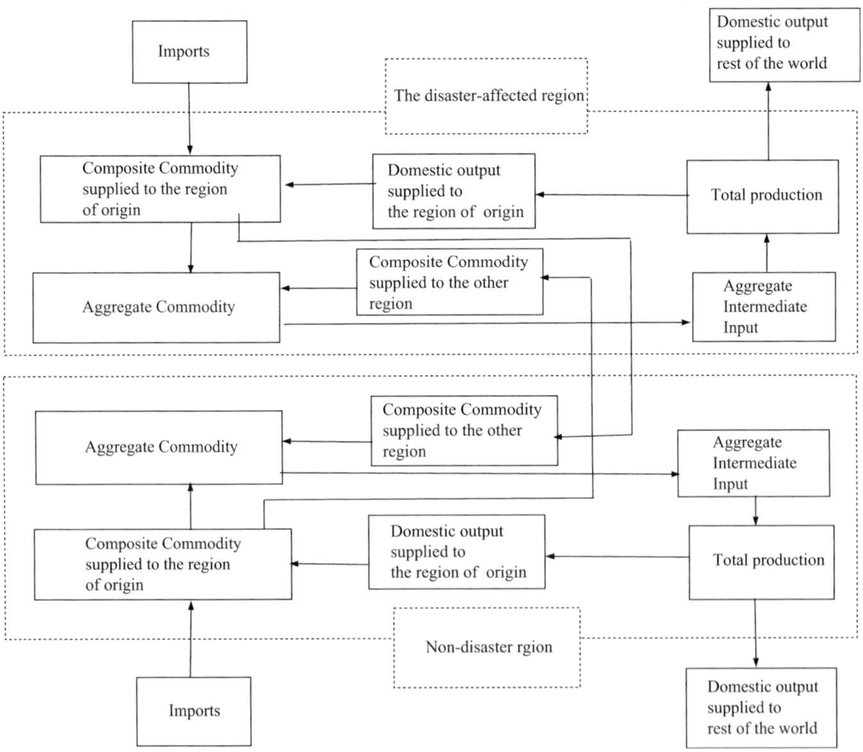

Fig. 2.12 Structure of the intra-regional and the inter-regional trade block

comprising producer goods for the domestic market and imported goods, this part is derived by solving the constrained optimization problem by minimizing its total costs subject to the CES function constraint. The 2SCGE model fixes the foreign-currency denominated international price and, in the trade balance formula, it sets net overseas transfers of labor and net overseas transfers of capital as exogenous variables for foreign savings in region, whereas the common exchange rate for the regions is set as endogenous variables.

To give a simplified explanation of the other blocks, savings within the government block for both central and local governments is obtained by adding a certain percentage to revenues. Local government expenditures for goods are allocated by adding a certain percentage to revenues after excluding transfers to savings and other system sectors. In our model, savings are allocated by agents called "banks" in response to investment demand from each of the 10 industries, according to a linearly homogeneous Cobb–Douglas utility function. Finally, within the market equilibrium conditions block, equilibrium conditions for the nine markets are given a set formula. In the system of equations described above, one equation becomes redundant due to Walras' law. We must therefore select one of the goods' prices as the numéraire. In this case, the wage rate in regions unaffected by the disaster is selected and fixed.

Previous sections provided an overview of the economic and human loss in the affected areas and the state of fiscal measures for recovery. In Sect. 2.6, we introduce a computable general equilibrium (CGE) model, which will be useful in assessing the damage from the earthquake as well as recovery policies using the formation of new industrial clusters. Futhermore, in Chap. 6, using a dynamic regional CGE model, we analyze the economic impact of the formation of new food and automotive industry clusters with productivity improvement. In Chap. 7, we also analyze the production recovery of fisheries and seafood manufacturing and the economic impacts of the Tokai Earthquake in Chap. 11 of this book using a dynamic regional CGE model.

2.5.2 New Economic Geography Model

Next, we introduce the NEG theory, which is useful in analyzing population and labor migration stemming and regional renewal from the earthquake and the nuclear power plant disaster. As opposed to traditional trade theories exemplified by Ricardo's theory of comparative advantage and the Heckscher–Ohlin model, the 1980s produced new trade theories within the framework of monopolistic competition according to Krugman (1980), Helpman and Krugman (1985), which introduced factors such as the production function of increasing returns, transport expenses, and product discrimination; intra-industry trade, namely, the mechanism by which trade is conducted between countries having the same industries, was elucidated. However, these new trade theories had models that assumed that international labor migration does not exist. Therefore, Krugman (1991) as well as Fujita et al. (1999), while employing the framework of these new trade theories, developed the new economic geography (NEG) model, which accounts for labor migration internationally (and regionally). In particular, Krugman (1991) presented a simple general equilibrium model for monopolistic competition assuming regional migration of labor, the so-called core–periphery (CP) model.

In this section, following the research by Fujita and Thisse (2013), we introduce the CP model, which is a spatial version of the Dixit–Stiglitz model (1977).[18] The economic space is made of two regions, and there are two sectors, agriculture (A) and manufacturing (M), with two production factors, the farmers (immobile, L) and workers (perfectly mobile, H). First, consumers have a Cobb–Douglas utility function, $U = Q^{\mu}A^{1-\mu}/\mu^{\mu}(1-\mu)^{1-\mu}0 < \mu < 1$, where Q is an index of the consumption of manufacturing goods and A is the consumption of agricultural goods given by $Q = \left[\int_0^M q_i^{\rho}\, di\right]^{1/\rho}$; $A = (1-\mu)Y/p^{\mathbb{A}}$. The parameter ρ is the inverse of the intensity of love for variety, and σ is the elasticity of substitution between any two varieties, defied by $\sigma \equiv 1/(1-\rho)$. The individual demand functions are $q_i = (\mu Y/p_i)\left(p_i^{-(\sigma-1)}/P^{-(\sigma-1)}\right) = \mu Y p_i^{-\sigma}P^{\sigma-1}$. The price index of the differentiated product is $P \equiv \left[\int_0^M p_i^{-(\sigma-1)} di\right]^{-1/(\sigma-1)}$. Next, for the behavior of producers, the agricultural technology ensures that one unit of output requires one farmer, and the manufacturing technology ensures that production of the quantity q_i requires workers l_i $(=f + cq_i)$, where f and c are the fixed and marginal labor requirements, respectively. Using the "iceberg" form for trade cost, the notation of the market access of exports from region s to region r as $\phi_{sr}\left(\equiv \tau_{sr}^{-(\sigma-1)}\right)$ is introduced, which shows the freeness of trade. Thus, if variety i is produced in region r and sold by p_r, then the delivered price in region $s(\neq r)$ p_{rs} is $p_{rs} = p_r\tau_{rs}$. The price index P_r in region r is $P_r = \left\{\sum_{s=1}^{R} \phi_{sr}M_s p_s^{-(\sigma-1)}\right\}^{-1/(\sigma-1)}$. Each firm sets its mill prices to maximize profits. Following Dixit and Stiglitz (1977), firms treat the elasticity of substitution, σ, as if it was the price elasticity of demand. The nominal wage rate of workers in region r sets w_r. As there is free entry and exit, zero profit occurs in equilibrium; thus, the income of region r is $Y_r = \lambda_r H w_r + v_r L$. Hence, the profit function of a firm producing variety i in region r is $\pi_r(i) = p_r(i)q_r(i) - w_r[f + q_r(i)] = (p_r - w_r)q_r - w_r f$ as the total demand of firm produced variety i is $q_r = \mu p_r^{-\sigma} \sum_{s=1}^{R} Y_s \phi_{rs} P_s^{\sigma-1}$.

Next, we explain how firms and workers are distributed between regions using these equations for the behavior of consumers and producers. In the short-run equilibrium for immobile workers among regions, the equilibrium wage for

[18]For an explanation of the CP model, see Fujita and Thisse (2013; Sect. 8.2, pp. 291–315).

workers in region r is $w_r^* = k_2\left[\sum_{s=1}^{R}\phi_{rs}Y_sP_s^{\sigma-1}\right]^{1/\sigma}$ $r = 1, \ldots, R,$ where $k_2 \equiv (\sigma - 1/\sigma)[\mu/(\sigma - 1)f]^{1/\sigma}$, and the real wage in region r is $V_r = \omega_r = \frac{w_r^*}{P_r^{\mu}}$ $r = 1, \ldots, R$. In contrast, in the long-run equilibrium, a spatial equilibrium arises when no worker may get a higher utility in another region.[19] Following migration modeling, the myopic evolutionary process in which workers are attracted by regions providing high utility levels is as follows: $\dot{\lambda}_r = \lambda_r(\omega_r - \overline{\omega})$ $r = 1, \ldots, R$, where $\dot{\lambda}_r$ is the time derivative of λ_r, ω_r is the equilibrium real wage corresponding to the distribution $(\lambda_1, \ldots, \lambda_R)$, and $\overline{\omega} \equiv \sum \lambda_s\omega_s$ is the average real wage across all regions. In this context, the equilibrium equations for the two regional CP model, assuming that farmers are equally split between two-region $(r = 1, 2)$ are as follows:

$$Y_r = \lambda_r H w_r^* + L/2 \quad r = 1, 2 \tag{2.5}$$

$$P_r = k_1\left[\lambda_r\left(w_r^*\right)^{-(\sigma-1)} + \tau^{-(\sigma-1)}\lambda_s\left(w_s^*\right)^{-(\sigma-1)}\right]^{-1/(\sigma-1)} \quad s \neq r \tag{2.6}$$

$$w_r^* = k_2\left(Y_rP_r^{\sigma-1} + \tau^{-(\sigma-1)}Y_sP_s^{\sigma-1}\right)^{1/\sigma} \quad s \neq r \tag{2.7}$$

$$\omega_r = w_r^* P_r^{-\mu} \quad r = 1, 2 \tag{2.8}$$

If $\lambda \equiv \lambda_1$, then $\lambda_2 = 1 - \lambda$. Thus, there exists a unique short-run equilibrium. Using the system (2.5, 2.6, 2.7 and 2.8) of nonlinear equations, Krugman (1991) got the following results of the agglomeration: for a large value of transport costs, there is one stable equilibrium corresponding to the full dispersion of the manufacturing sector ($\lambda^* = 1/2$), and for a low value of transport costs, the symmetric equilibrium becomes unstable; thus, the core–periphery structure is the only stable outcome.[20] The above is a summary of the basic CP model. Afterwards, the NEG model based on this CP model was developed by Venables (1996), Ottaviano et al. (2002), Baldwin and Krugman (2004) for agglomeration, Okubo et al. (2010) for heterogeneous firms, Tabuchi and Thisse (2002, 2011) and Mori and Smith (2011) for central place, Head and Mayer (2004) and Takatsuka and Zeng (2012) for the home market effect, and Fujita and Hamaguchi (2011, 2014) for coagglomeration of intermediate and final sectors. For the two-regional CGE model in Sect. 2.6, Chaps. 4, 5 and 6 of this book, we try to take some basic ideas of this NEG model in the two-regional CGE model to analyze the industrial

[19]For the long-run equilibrium, see Fujita and Thisse (2013, pp. 297–298).

[20]For the two regional CP model, see Fujita et al. (1999, Chap. 5) and Fujita and Thisse (2013, pp. 298–306).

agglomeration effects.[21] Furthermore, using NEG model based on this core–periphery model, we analyze the impact of labor migrations brought on by massive earthquakes in Chap. 10.

2.5.3 Trends in Research on New Industrial Agglomeration

Next, we focus on the formation of new industrial clusters as a strategy for recovery and local renewal from the earthquake in Sect. 2.6 and Chap. 6. Since the 1980s there has been much research into industrial clusters and regional planning, but starting in the 1990s, research into industrial clusters began to bloom. We present these trends here.

Classical research on industrial clusters began with Alfred Marshall (1890), who elucidated the origins of the economies of agglomeration. He argued that the three factors driving the emergence of economies of agglomeration are (1) spillover of new ideas based on the exchange of information and knowledge and face-to-face communications, (2) sharing of non-traded elements of production in the region, and (3) access to a large pool of similar and specialized workers in the region.[22] These Marshall-type external economies are categorized into economies of localization and economies of urbanization. Concerning these two types of external economies, Jacobs (1969) claimed that economies of urbanization are dominant. However, Porter (1998, 2000), in a series of studies, advocated for industrial clusters, arguing that in a global economy, the primary reason for the success of industrial clusters lies in the existence of strong economies of localization. The conclusion of this debate depends on the characteristics of the industry and scale of the region in question. The micro-foundation for the notion of external economies was provided by the NEG model of Krugman (1991) and others (Fujita et al. 1999; Belleflamme et al. 2000; Fujita and Thisse 2013). The seminal studies elucidating the economies of agglomeration in an empirical fashion were by Nakamura (1985) and by Henderson (1988). The latter half of the 1990s was marked by a great deal of empirical analysis about the economics of agglomeration, inspired by Krugman's (1991) study. Ellison and Glaeser's (1997) study also provided pioneering research. Additional papers validating the economies of agglomeration in Japanese manufacturing include those by Mori et al. (2005), Kageyama and Tokunaga (2006), Tokunaga and Kageyama (2008), Nakajima (2008), Duranton et al. (2010), Tabuchi (2014), and Tokunaga et al. (2014). To distinguish these studies from traditional industrial agglomeration research by Marshall (1890) and others,

---

[21]Especially, we calibrate the parameter of the CES production function for the intermediate goods and adopt the value of the inter-regional elasticity of substitution for intermediate goods in the CES function as commodity-grade products policy simulation of industry cluster in Sect. 2.6 and Chap. 6.

[22]See Marshall (1890, Chap. X), McCann (2013, Chap. 2), and Fujita and Thisse (2013, Chap. 1).

they are called studies in "new industrial agglomerations" or "new industrial cluster."

Porter examined competitive superiority under conditions of growing competition due to the rapid progress of globalization. This presented a concept whereby the clustering of diverse, heterogeneous companies increases exchanges, including face-to-face communication, between different fields, using ideas arising from the development of new products and technologies. This strategy finds not just external economies in clusters of firms but promotes regional revitalization in a systematic fashion, in conjunction with organizations near the firm, such as universities and research institutions. The IT industry cluster in America's Silicon Valley, and the wine and food industry clusters in France, are particularly famous examples of clustering. Studies analyzing industrial clusters in Japan and Asia include Kuchiki and Tsuji (2005), Tawada and Iemori (2005), and Kiminami and Nakamura (2016). The formation of industrial clusters is moving ahead not only in developed countries but also in developing countries, where it is revitalizing local economies. In Sect. 2.6, we use a static computable equilibrium model for a simulation analysis of local revitalization, stemming from the formation of industrial clusters. In Chap. 6, based on a dynamic CGE model based on the NEG model, we analyze the economic impact of the formation of new food and automotive industry clusters on reconstruction and local revitalization.

2.6 Impacts of the Disaster and Recovery Using the Static 2SCGE Model

Before explanation about simulation for impacts of the disaster and recovery from the Great East Japan Earthquake using a static 2SCGE model, we now review of previous literature for the models on impacts of the great earthquake. The economic impacts of the earthquake have been analyzed using the input-output, CGE, and econometric models (Okuyama 2004; Xie et al. 2014; Shibusawa and Miyata 2011). The CGE model is widely recognized as policy evaluation tool (Dixon and Rimmer 2002; Kehoe et al. 2005; Hosoe et al. 2010). For the Great East Japan Earthquake, Tokui et al. (2012, 2015) examined the economic impact of supply chain disruptions caused by the earthquake using regional input–output tables. Fujita and Hamaguchi (2011) and Hamaguchi (2013) studied the characteristics of the supply chain and the impact of the disaster based on a survey of manufacturing facilities located in areas affected by the earthquake. In addition, Saito, Y. (2012) and Todo et al. (2013) analyzed the nature of corporate networks in the supply chain, while Fujimoto (2011) and Otsuka and Ichikawa (2011) particularly assessed the supply chain in the automotive industry. In addition to the automotive industry, Nemoto (2012) discussed the reconstruction of local economies, including supply chains in the logistics and fishery industries. On the other hand, Okiyama et al. (2014) and Tokunaga and Okiyama (eds) (2014) analyzed the impacts of the Great East Japan

Earthquake on production loss using an inter-regional SAM and Ishikawa (2014) also showed the economic impacts of population decline due to the Great East Japan Earthquake using an inter-regional input-output data. Furthermore, Saito, M. (2015) and Karan and Suganuma (2016) provided a comprehensive account of the devastation caused by this earthquake, tsunami, and nuclear radiation. From these previous research surveys, we find that there are no studies to evaluate the impacts of industry clusters with innovation on the regional economy for regional reconstruction after the Great East Japan Earthquake. Thus, we construct a static 2SCGE model based on the idea of NEG model and evaluate its impacts of industry clusters on regional economy after the Great East Japan Earthquake. Based on the considerations outlined in Sects. 2.2, 2.3 and 2.4, we will now simulate the impacts of the damage from the disaster in the affected prefectures as the base scenario (Base Simulation for Great Earthquake) and supply chain disruptions under Base Simulation (Simulation I). We will examine how the affected agriculture, forestry and fisheries industries, and the manufacturing industry can rebuild themselves using the fiscal measures policy under Base Simulation (Simulation II); the fiscal measures policy and commodity-grade products policy under Base Simulation (Simulation III); and the fiscal measures policy, commodity-grade products policy, and the industrial clusters policy under Base Simulation (Simulation IV) using the simulations of the static 2SCGG model

2.6.1 Establishment of Four Scenarios

(a) Base Scenario for the Great Earthquake

The base scenario, that the Great Earthquake occurred, assumes that no fiscal measures or supply chain changes took place.[23] In the two regions, by comparing simulation results that either did or did not incorporate an earthquake, we found that labor and capital endowments decreased (see Sects. 5.2.1, 5.2.2 in Chap. 5). To reflect this in our model, we adjusted labor and capital endowments in the disaster-affected region with multipliers of 0.9858 and 0.9430, respectively, using 2005 labor and capital endowment data for the region. In non-disaster regions, labor and capital endowments were adjusted using multipliers of 0.9872 and 0.9866, respectively. These coefficients were estimated based on real GRP results of two simulations, one with and one without the Great Earthquake.

(b) Scenario for Supply Chain Disruptions under the Base Scenario

To verify the effects of supply chain disruptions with and without disasters, we focused on the automotive and electronics machinery industries.[24] Hence, the inter-

[23]See Japan Center for Economic Research (2015), Japan Institute for Labor Policy and Training (2014), and National Institute of Population and Social Security Research in Japan (2013).

[24]See Fujimoto (2011), Nemoto (2012), and Otsuka and Ichikawa (2011).

regional elasticity of substitution in the CES functions for the electronic devices/parts and parts of automotive industries (intermediate goods) was set at 0.5, as shown in Table 2.7. In other words, we assumed that these two industries' supply chains were more closely connected in the two regions in light of post-earthquake circumstances.

(c) Scenario for Fiscal Measures for Reconstruction under the Base Scenario

This scenario assumes that the Great Earthquake occurred and that fiscal measures were implemented for the disaster-affected region as follows: local governments in disaster-affected areas obtain 1.5 trillion yen in additional annual revenue from the central government by distributing a portion of local allocation tax grants to the non-disaster region. These grants make use of (1) the subsidy for inviting new firms and forming two new industry clusters, as shown in Chap. 6, (2) the partial restoration of public infrastructure, and (3) social security expenditures, such as a partial pension benefit, as shown in Table 2.8.[25]

(d) Scenario for New Industrial Agglomeration under the Base Scenario

In this scenario, we incorporate three factors for the formation of a new industry cluster and new industrial agglomeration.[26] The first is a subsidy policy after the intensive reconstruction period supporting agricultural, forestry, and fishing industries, as well as food and beverage industries that make up the new food industry cluster. The policy also supports the automotive and auto parts industry, the electronic parts and devices and electronic circuits industry, and other manufacturing and mining industries that make up the new automobile industry cluster. Revenues are sourced from local allocation tax grants paid to disaster regions. Second, we incorporate the key terms of commodity-grade products for intermediate goods and differentiated products for final goods in the formation of the new food and automobile industry clusters. More specifically, we alter the inter-regional elasticity of substitution for intermediate goods in the CES function from 2.0 to 3.0 for inter-regional products for the above six categories and set the inter-regional elasticity of substitution for final goods in the CES function at 0.5 as emphasized by the NEG model, as shown in Table 2.7. Third, we continue the round of policies and countermeasures mentioned above in the new food and automobile industry clusters, incorporating productivity increases by agglomeration effects in the aforementioned six categories, as shown in Table 2.8.[27]

---

[25]Subsidy and savings rates are set based on Simulations III-a and III-b in Chap. 6, after an intensive reconstruction period (Cabinet Office 2014a, b).

[26]See Mokudai and Ishiro (2013), Murayama (2013), Orihashi (2013)for automobile industrial cluster.

[27]For the emergence of industrial clusters in equilibrium, see Fujita and Thisse (2013, pp. 375–379).

Table 2.7 List of elasticity of substitution setup in this model

Production activries sectors	Elasticity of substitution between capital-labor in the CES function	Elasticity of transformation in CET function	Elasticity of substitution of ARMINGTON function	Inter-regional elasticity of substitution with final goods in the CES function	Inter-regional elasticity of substitution with intermediate goods in the CES function		
					Base simulation and Simulation II	Simulation I	Simulation III and Simulation IV
Agricultture and forestry	0.6	2.0	2.7	0.5	2.0	2.0	3.0
Fishery	0.6	2.0	1.2	0.5	2.0	2.0	3.0
Foods and Beverages	1.2	2.0	2.0	0.5	2.0	2.0	3.0
Electronic devices and parts	1.3	2.0	4.2	0.5	2.0	0.5	3.0
Motor vehicles/Motor vehicle parts	1.3	2.0	2.8	0.5	2.0	0.5	3.0
The others of Manufacturing and Mining	1.3	2.0	3.2	0.5	2.0	2.0	3.0
Construction	1.4	2.0	1.9	0.5	2.0	2.0	2.0
Electricity,gas and hiat-water supply	1.3	2.0	2.8	0.5	2.0	2.0	2.0
Commerce	1.3	2.0	1.9	0.5	2.0	2.0	2.0
Transport	1.7	2.0	1.9	0.5	2.0	2.0	2.0
Other tertiary industry	1.3	2.0	1.9	0.5	2.0	2.0	2.0

Table 2.8 Setting the change in subsidy rate, Savings rate and productivity increase

	Base Simuation and Simulation I	Simulation II and Simulation III and Simulation IV	Simulation IV
The disasater-affected regon	Subsidy rate per industry		The change in productivity per industry
Agriculture and forestry	1.4%	4.4%	2.0%
Fisheries	0.5%	7.0%	2.0%
Foods and Beverage	0.5%	4.5%	3.0%
Electrical devices and parts	0.0%	4.0%	1.3%
Motor vehicles and parts	0.0%	6.0%	1.7%
Other manufacturing products and Mining	0.0%	2.0%	1.0%
Commerce	0.0%	0.0%	0.0%
Transport	0.3%	0.3%	0.0%
Savings rate of local government	2.6%	4.9%	4.9%

2.6.2 *Simulation Designs and Results*

2.6.2.1 Simulation Designs

In this section, we implemented five simulations based on the above scenarios. First, we implemented a simulation based on the Base Scenario as the "Base Simulation for the Great Earthquake." Second, we implemented a simulation based on the scenario for supply chain disruptions under Base Simulation as "Simulation I." Third, we implemented a simulation based on the scenario for fiscal measures in a reconstruction under Base Simulation as "Simulation II." Fourth, we implemented a simulation based on assumptions about fiscal measures for reconstruction and commodity-grade product under Base Simulation as "Simulation III." Finally, we implemented a simulation based on assumptions about fiscal measures for reconstruction, commodity-grade products, and productivity increases under Base Simulation as "Simulation IV." Simulations II, III, and IV are simulations for the formation of an industry cluster and industrial agglomeration under Base Simulation. The contents of these simulations are summarized in Table 2.9. In this Table, each factor marked with a "○" is incorporated in the above simulations.

2.6.2.2 Simulation Results

Simulation results of changes from base value in 2005 are shown in Tables 2.10 and 2.11. First, we examine real GRP, the equivalent variation (EV), and the volume of unemployment for disaster-affected and non-disaster regions, under the Base Scenario for the Great Earthquake. Because of the earthquake, real GRP for the disaster-affected region declines to 3.86%, and in the non-disaster region it decreases to 1.31%. EVs for disaster-affected and non-disaster regions go down by roughly one trillion yen and 4.7 trillion yen, respectively. Changes in unemployment in the disaster-affected region rise to 8.21% compared with before the earthquake, while in the non-disaster region, it declines to 1.29%. In Table 2.11, production volume in the disaster-affected region shows a decline, from 2% to 6.7%, while in the non-disaster region, it also decreases by approximately 1%.

Looking at the industry's rate of change in production volume in Simulation I with its disrupted supply chains (semi-core parts) under Base Simulation, production volume in the automotive and electronic devices and parts industries in the non-disaster region drop by 1.05% and 0.75%, respectively, compared with 1.04% and 0.73%, respectively in the Base Scenarios shown in Table 2.11. Therefore, we regard the difference between Simulation I and the Base Simulation as supply chain disruptions.

Next, we evaluate fiscal measures for reconstruction and economic recovery under Base Simulation (after the earthquake), as shown in Simulation II, using indexes from Table 2.10. In Simulation II, real GRP in the disaster-affected region improves by 1.72% points and EV also gains 989 billion yen compared with the Base Simulation. Real GRP for the non-disaster region drops by 0.08% point and EV loses 500 billion yen compared with the Base Simulation. This explains why fiscal support is funded by a portion of locally allocated tax grants distributed to the non-disaster region. The employment level in the disaster-affected region improves, with unemployment decreasing by 48% points compared with the Base Simulation. In the non-disaster region, unemployment worsens, rising to 0.72%. Regarding production volume in the disaster-affected region in Simulation II, as shown in Table 2.11, each industry's production volume rate of change is uneven. The food and beverage, construction, motor vehicles and parts, and other tertiary industries recover. In particular, the construction industry's production records a robust increase of 33.9%, and the automotive industry's production rises to 7.37%. However, excluding the four mentioned above, most industries show worse results. In particular, production in the electronic devices and parts and fishery industries decreases by 6.98% and 6.77%, respectively. Because of these fiscal measures, imports from the non-disaster region increase and exports decrease due to commodity price increases in the disaster-affected region.

Finally, we evaluate the impact of the formation of an industry cluster from the results of Simulations II, III, and IV. In Simulation IV, based on assumptions about fiscal measures for reconstruction, commodity-grade products, and productivity increases for the full formation of a new industry cluster under Base Simulation,

Table 2.9 Contents of each simulation

	The factors are incorporated in each scenario	Base Simulation	Simulation I	Simulation II	Simulation III	Simulation IV
Base Scenario (Great Earthquake occurred)	Decrease in the labor and capital endowment	○	○	○	○	○
Scenario for the supply chain disruptions	Semi-core parts		○			
Scenario for fiscal measures for reconstruction	Local allocation tax grants Subsidy policy Public investment			○	○	○
Formation scenario for new food and automoblie cluster in diasater-affected region	Commodity-grade products				○	○
	productivity increase					○

Table 2.10 Spillover effects to the disaster-affected region and the non-disaster region

Changes in base value (2005) (%)		Wage rate	Return to capital	Total output	Real GRP	Normal GRP	Consumer price index commodies	Utiltty level for the household	Equivalent variation (Billion Yen)	Volume of unemployment	Exchage rate
Base Simulation	Disaster-affected region	−0.89	2.22	−3.82	−3.86	−3.33	0.38	−5.70	−1,011.8	8.21	0.011
	Non-disaster region	-	−0.04	−1.25	−1.31	−1.32	0.00	−1.75	−4,665.1	−1.29	
Simulation I	Disaster-affected region	−0.89	2.22	−3.82	−3.86	−3.33	0.39	−5.69	−1,016.4	8.19	0.012
	Non-disaster region	-	−0.04	−1.25	−1.31	−1.32	0.00	−1.75	−4,664.8	−1.29	
Simulation II	Disaster-affected region	15.14	20.22	−2.19	−2.14	12.66	9.53	2.12	−22.6	−40.28	0.616
	Non-disaster region	-	−0.07	−1.22	−1.39	−1.39	0.30	−2.09	−5,165.2	0.72	
Simulation III	Disaster-affected region	15.80	20.46	−2.35	−2.13	13.21	9.78	2.47	25.2	−42.98	0.647
	Non-disaster region	-	−0.05	−1.21	−1.40	−1.38	0.32	−2.10	−5,196.7	0.84	
Simulation IV	Disaster-affected region	16.35	21.12	−1.82	−1.71	13.91	9.96	2.88	95.7	−45.47	0.642
	Non-disaster region	-	−0.04	−1.21	−1.40	−1.38	0.32	−2.10	−5,195.8	0.84	

Table 2.11 Results of the change in industry's production volume by simulation

Changes in base value (2005) (%)			Agricultture and forestry	Fishery	Foods and Beverages	Electronic devices and parts	Motor vehicles/ Motor vehicle parts	The others of Manufacturing and Mining	Construction	Electricity, gas and hiat-water supply	Commerce	Transport	Other tertiary industry
Domestic production	Base Simulation	Disaster-affected region	−4.512	−3.725	−3.626	−2.197	−2.051	−3.548	−6.731	−3.830	−3.570	−3.605	−3.830
		Non-disaster region	−1.488	−1.667	−1.734	−0.734	−1.042	−1.038	−0.021	−1.444	−1.316	−1.376	−1.494
	Simulation I	Disaster-affected region	−4.517	−3.730	−3.628	−2.062	−1.907	−3.558	−6.763	−3.831	−3.574	−3.609	−3.829
		Non-disaster region	−1.487	−1.666	−1.734	−0.747	−1.047	−1.037	−0.019	−1.444	−1.315	−1.375	−1.494
	Simulation II	Disaster-affected region	−6.143	−6.777	−2.308	−6.981	7.371	−8.757	33.988	−9.174	−7.681	−8.324	−1.251
		Non-disaster region	−0.974	−1.077	−1.512	−0.389	0.457	−0.199	−2.943	−0.826	−0.983	−0.778	−1.770
	Simulation III	Disaster-affected region	−8.353	−9.568	−3.379	−8.817	8.305	−11.388	42.399	−9.447	−7.258	−8.247	−1.083
		Non-disaster region	−0.662	−0.786	−1.372	−0.355	0.378	0.099	−3.589	−0.788	−1.027	−0.774	−1.779
	Simulation IV	Disaster-affected region	−6.533	−7.505	−1.907	−7.889	9.330	−10.697	43.839	−9.415	−7.202	−8.136	−0.920
		Non-disaster region	−0.719	−0.841	−1.371	−0.379	0.370	0.089	−3.691	−0.765	−1.017	−0.759	−1.768

real GRP in the disaster-affected region improves by 2.15% points and EV also gains 1107 billion yen compared with the Base Simulation. Real GRP for the non-disaster region drops by 0.09% point and EV loses 531 billion yen compared with the Base Simulation. From the difference between Simulations III and IV for real GRP and EV, we see that Simulation IV outperforms Simulation III in the disaster-affected region because of productivity increases in the new food and automobile industries clusters. Thus, we regard the difference between Simulation III and Simulation IV as a productivity increase by agglomeration effects. In contrast, from the difference between Simulations II and III for real GRP and EV in the disaster-affected region, we find that real GRP is higher by 0.01% point, EV improves to 47.8 billion yen, and unemployment decreases by 2.7% points compared with Simulation II. This is because of the introduction of commodity-grade products. Therefore, we regard the difference between Simulation II and Simulation III as commodity-grade products for intermediate goods and differentiated products for final goods in the formation of industry clusters.

Looking at industries' production volume rate of change in Simulations II, III, and IV in detail, for the disaster-affected region in Simulation III, most industries' production volume rate of change, excluding the construction and automobile industries, declines compared with Simulation II, despite commodity-grade products in the food and beverage and electronic devices and parts industries. Often, commodity-grade products increase imports and lower exports. However, the reason for the construction industry's production volume increasing was higher investment demand in the disaster-affected region. Savings rise with higher income transfers from the non-disaster to the disaster-affected region due to the interregional current account balance. Therefore, even if products in the disaster-affected region had become commodity-grade under fiscal measures, the production volume of most industries would have declined. However, in Simulation IV, for the full formation of industry clusters, industrial productivity rose compared with Simulation III, and due to the formation of the two industry clusters, all industries' production volumes recovered. In particular, the food and beverage industry's production increased by 1.47% points compared with a 3.38% drop in Simulation III, and the automotive industry's production increased by 1.02% points compared with an 8.31% drop in Simulation III. In addition, real GRP for the disaster-affected region in Simulation IV recovered by 0.43% point compared with Simulation II. As a result, the difference between real GRP in disaster-affected and non-disaster regions declines by 0.31%, from 2.55% in the Base Simulation.

These simulations showed that the formation of these two industry clusters with productivity improvement in a disaster region has a positive effect on real GRP and economic welfare in these regions, reducing the economic welfare gap between disaster and non-disaster regions.

2.7 Conclusions

In this chapter, we have examined the economic and human damage inflicted on Iwate, Miyagi, Fukushima, and Ibaraki Prefectures by the Great East Japan Earthquake, as well as the devastation caused in Fukushima Prefecture by the nuclear power plant disaster. We used various materials, including industrial production indices, Fukushima prefectural statistics, and surveys done by the Tohoku Bureau of Economy, Trade and Industry. In the latter part of this chapter, we have shown the extent of fiscal transfers to date from the government for reconstruction and renewal in the stricken areas. In addition, we analyzed the economic effect of the formation of industrial clusters using a static two-regional computable general equilibrium (2SCGE) model and found that (1) if production subsidies to support industries form industry clusters, positive effects on regional economies could appear in disaster regions; however, these impacts will be weak and (2) formation of industry clusters with productivity improvement has a positive effect on real GRP and economic welfare in these regions, reducing the economic welfare gap between disaster and non-disaster regions.

Leaving aside the areas affected by the nuclear power plant catastrophe, fiscal recovery measures exceeded 20 trillion yen. Stricken prefectures and industries achieved steady recovery in the affected regions (with some differences in degree) compared to the pre-earthquake situation. However, almost 5 years after the earthquake, except for areas affected by the nuclear power plant incident, the livelihood of residents in the tsunami-hit areas in Iwate and Miyagi Prefectures has yet to stabilize despite some recovery of regional industries. While there are limits in extrapolating from prefecture-level, macroeconomic statistics, and reconstruction budgets, unless these two gaps are filled, one cannot claim to have done an economic analysis of the recovery from the earthquake.

Starting from Chap. 3 in this book, we will construct an inter-regional SAM, integrating various statistics such as prefecture-level input–output tables and reports on prefectural accounts, and conduct multiplier analysis. Based on this inter-regional SAM, from Chaps. 4, 5, 6 and 7, we will construct a dynamic regional computable general equilibrium model and empirically analyze the long-run impact of fiscal measures and industrial clusters on regional economies in Tohoku. Furthermore, in Chaps. 8, and 9, we will analyze the impact of a biogas electricity power plant and population decline on prefectures due to the Great East Japan Earthquake using an inter-regional input–output model. In addition, using a new economic geography (NEG) model and a dynamic regional CGE model, we will empirically analyze the massive economic impact, primarily on the affected areas, of the Nankai Megathrust Earthquake and the Tokay Earthquake in Chaps. 10, 11, respectively. Finally, we will analyze the impact of the Indian Ocean tsunami on Indonesia and the process of reconstruction and rehabilitation in Asia in Chap. 12.

References

Anselin L (1988) Spatial econometrics: methods and models. Kluwer Academic Publishers, Dordrecht

Arbia G (2014) A primer for spatial econometrics: with application in R. Palgrave Macmillan, Hampshire

Baldwin R, Krugman P (2004) Agglomeration, integration and tax harmonization. Eur Econ Rev 48:1–23

Belleflamme P, Picard P, Thisse J (2000) An economic theory of regional clusters. J Urban Econ 48:158–184

Cabinet Office (2014a) Basic policies for the economic and fiscal management and reform 2014-from deflation to an expanded economic virtuous cycle (in Japanese). http://www5.cao.go.jp/keizai-shimon/kaigi/minutes/2014/0624/shiryo_01.pdf

Cabinet Office (2014b) Japan revitalization strategy as revised in 2014 (in Japanese). http://www5.cao.go.jp/keizai-shimon/kaigi/minutes/2014/0624/shiryo_02_1.pdf

Dixit A, Stiglitz J (1977) Monopolistic competition and optimal product diversity. Am Econ Rev 67(3):297–308

Dixon PB, Rimmer MT (2002) Dynamic general equilibrium modeling for forecasting and policy. North-Holland, Amsterdam

Duranton G, Martin P, Mayer T, Mayneris F (2010) The economics of clusters. Oxford University Press, Oxford

EcoMod Modeling School (2012) Advanced techniques in cge modeling with gams global economic modeling network, Singapore, January 9–13

Ellison G, Glaeser EL (1997) Geographic concentration in U.S. manufacturing industries: a dartboard approach. J Polit Econ 105(5):898–927

Fujimoto T (2011) Supply chain competitiveness and robustness: a lesson from the 2011 Tohoku Earthquake and 'Virtual dualization.' MRC (Tokyo University Manufacturing Management Research Center) Discussion Paper Series, 354 (in Japanese)

Fujita M, Hamaguchi N (2011) Japan and economic integration in East Asia: post-disaster scenario. RIETI Discussion Paper Series, 11-E-079

Fujita M, Hamaguchi N (2014) Supply chain internationalization in East Asia: inclusiveness and risk. RIETI Discussion Paper Series, 14-E-066

Fujita M, Thisse J (2002, 2013) Economics of agglomeration: cities, industrial location, and globalization, 2nd edn. Cambridge University Press, New York

Fujita M, Krugman P, Venables AJ (1999) The spatial economy: cities, regions, and international trade. MIT Press, Cambridge

Hamaguchi N (2013) Impact of the great East Japan earthquake on local industries: synthesis and analysis from a survey. RIETI Discussion Policy Discussion Paper Series, 13-P-001 (in Japanese)

Head K, Mayer T (2004) Market potential and the location of Japanese investment in the European union. Rev Econ Stat 86(4):959–972

Helpman E, Krugman P (1985) Market structure and foreign trade. MIT Press, Cambridge

Henderson JV (1988) Urban development: theory, fact and illusion. Oxford University Press, Oxford

Hosoe N, Gasawa K, Hashimoto H (2010) Textbook of computable general equilibrium modeling. Palgrave Macmillan, New York

Ishikawa Y (2014) Economic impacts of population decline due to the great East Japan earthquake using an inter-regional input-output approach. In: Tokunaga S, Okiyama M (eds) Reconstruction the disaster-affected region of the great East Japan earthquake and recovery of the regional economy. Bunshindou, Tokyo, pp 219–238. (in Japanese)

Ito H (2008) Interregional SAM model and structure path analysis focusing on institution sector. J Commer Sci Kwansei Gakuin Univ 56(1):33–70. (in Japanese)

Jacobs J (1969) The economy of cities. Random House, New York

Japan Center for Economic Research (2015) Potential economic collapse in the late 2020s-adverse effects of the declining population after the Tokyo olympics. 42nd medium-term forecast for the Japanese economy: FY2035-FY2030 (Baseline Scenario)
Japan Institute for Labor Policy and Training (2014) Labor supply and demand estimates: policy simulations based on the labor supply and demand model (2013). Research Material Series129 (in Japanese)
Kageyama M, Tokunaga S (2006) An empirical analysis of co-agglomeration effect in the japanese food industry. J Rural Econ Spec Issue 2006:132–138. (in Japanese)
Karan PP, Suganuma U (eds) (2016) Japan after 3/11: global perspectives on the earthquake, tsunami, and fukushima meltdown. University Press of Kentucky, Kentucky
Kehoe TJ, Srinivasan TN, Whalley J (2005) Frontiers in applied general equilibrium modeling. Cambridge University Press, Cambridge
Kiminami L, Nakamura T (eds) (2016) Food security and industrial clustering in Northeast Asia. Springer Japan, Tokyo
Krugman P (1980) Scale economies, product differentiation, and the pattern of trade. Am Econ Rev 70:950–959
Krugman P (1991) Increasing returns and economic geography. J Polit Econ 99:483–499
Kuchiki A, Tsuji M (eds) (2005) Industrial clusters in Asia: analyses of their competition and cooperation. Palgrave Macmillan, New York
LeSage J, Pace PK (2009) Introduction to spatial econometrics. CRC Press, Boca Raton
Marshall A (1890) Principles of economics, 8th edn. Macmillan, London. Published in 1920
McCann P (2013) Modern urban and regional economics. Oxford University Press, Oxford
Miller R, Blair P (2009) Input-output analysis: foundations and extensions, second edn. Cambridge University Press, Cambridge
Ministry of Agriculture, Forestry and Fisheries (2015) Survey result 2014 about management condition of farmworker and fisherman in tsunami-disaster area by the great East Japan Earthquake (in Japanese). http://www.maff.go.jp/j/tokei/keikou/pdf/fukkou-20150911.pdf
Ministry of Agriculture, Forestry and Fisheries (2016) Commitment for reconstruction support of agriculture, forestry and fisheries after the great East Japan earthquake (in Japanese). http://www.maff.go.jp/j/kanbo/kihyo02/fukkou/pdf/1603-zen.pdf
Ministry of Economy, Trade and Industry (2011) Survey of the Fukushima Dai-ichi Nuclear Power Plant incident (in Japanese)
Mokudai T, Ishiro K (2013) The current situation and challenges facing automobile industry in Kyushu. In: Orihashi S, Mokudai T, Murayama T (eds) Tohoku region and automobile industry. Soseisha, Tokyo, pp 168–186. (in Japanese)
Mori T, Smith T (2011) An industrial agglomeration approach to central place and city size regularities. J Reg Sci 51(4):694–731
Mori T, Nishikimi K, Smith T (2005) A divergence statistic for industrial localization. Rev Econ Stat 87(4):635–651
Murayama T (2013) Automobile industry promotion by industry-academia-government collaboration in Iwate. In: Orihashi S, Mokudai T, Murayama T (eds) Tohoku region and automobile industry. Soseisha, Tokyo, pp 62–107. (in Japanese)
Nakajima K (2008) Economic division and spatial relocation: the case of postwar Japan. Journal of the Japanese and International Economies 22:383–400
Nakamura R (1985) Agglomeration economies in urban manufacturing industry. J Urban Econ 17:108–124
National Institute of Population and Social Security Research in Japan (2013) Regional population projections for Japan: 2010–2040. Population Research Series (in Japanese). http://www.ipss.go.jp/pp-shicyoson/j/shicyoson13/6houkoku/houkoku.pdf
Nemoto T (2012) Vitalizing local economies through supply chain restoration. Plann Adm 35 (2):13–16. (in Japanese)

Okiyama M, Tokunaga S, Akune Y (2014) Analysis of the effective source of revenue to reconstruct the disaster- affected region of the great East Japan earthquake: utilizing the two–regional CGE model. J Appl Reg Sci 18:1–16. (in Japanese)
Okubo T, Picard P, Thisse J (2010) The spatial selection of heterogeneous firms. J Int Econ 82:230–237
Okuyama Y (2004) Modeling spatial and economic impacts of an earthquake: input-output approaches. Disaster Prev Manag 13(4):297–306
Orihashi S (2013) The problem for the development of automobile industry in Tohoku. In: Orihashi S, Mokudai T, Murayama T (eds) Tohoku region and automobile industry. Soseisha, Tokyo, pp 248–258. (in Japanese)
Otsuka T, Ichikawa Y (2011) How do you evaluate Japan-type supply Chain? A collection of theses of Mizuho Research Institute 3 (in Japanese)
Ottaviano G, Tabuchi T, Thisse J (2002) Agglomeration and trade revisited. Int Econ Rev 43:409–436
Porter M (1998) Clusters and competition: new agendas for companies, governments, institutions. In: On competition. Harvard Business School Press, Boston, pp 155–196
Porter M (2000) Location, competition, and economic development: local clusters in a global economy. Econ Dev Q 14(1):15–34
Reconstruction Agency (2012, 2013, 2014, 2015, 2016) Current state and initiatives for reconstruction. 2012–2016 (in Japanese)
Saito M (2015) Political economy for reconstruction of the earthquake disaster. Japan: Nippon Hyoron Sha (in Japanese)
Saito (Umeno) Y (2012) Impact of East Japan earthquake disaster on companies in non-affected area: the structure of Inter-company network of supply chain and its implication. J JSS 42 (1):135–144. (in Japanese)
Shibusawa H, Miyata Y (2011) Evaluating the dynamic and spatial economic impacts of an earthquake: a CGE application to Japan. Reg Sci Inq 3(2):13–25
Tabuchi T (2014) Historical trend of agglomeration to the capital region and new economic geography. Reg Sci Urban Econ 44:50–59
Tabuchi T, Thisse T (2002) Taste heterogeneity, labor mobility and economic geography. J Dev Econ 69:155–177
Tabuchi T, Thisse T (2011) A new economic geography model of central places. J Urban Econ 69:240–252
Takatsuka H, Zeng D (2012) Trade liberalization and welfare: differentiated-good versus homogeneous-good markets. J Jpn Int Econ 26:308–325
Tawada M, Iemori N (2005) Industrial clusters of Tokai region and financial structure. Cyuou Keizai Sha (in Japanese)
Todo Y, Nakajima K, Matous P (2013) How do supply chain networks affect the resilience of firms to natural disasters? Evidence from the great East Japan earthquake. RIETI Policy Discussion Paper Series, 13-E-028
Tohoku Bureau of Economy, Trade and Industry (2012, 2013, 2014, 2015, 2016) Economic survey. (in Japanese)
Tohoku Bureau of Economy, Trade and Industry (2013) Survey of recipients of group subsidies. (in Japanese)
Tokui J, Arai N, Kawasaki K, Miyagawa T, Fukao K, Arai S, Edamura K, Kodama N, Noguchi N (2012) Economic impacts of the great East-Japan earthquake: comparison with the past great disasters, supply chain disruption, and effect of supply constraints in electric utility. RIETI policy discussion paper series, 12-P-004 (in Japanese)
Tokui J, Kawasaki K, Miyagawa T (2015) The economic impact of supply chain disruptions from the great East Japan earthquake. RIETI discussion paper series, 15-E-094
Tokunaga S, Kageyama M (2008) Impacts of agglomeration and co-agglomeration effects on production in the Japanese manufacturing industry: using flexible translog production function. Studies in Regional Science 38(2):331–337

Tokunaga S, Okiyama M (eds) (2014) Reconstruction the disaster-affected region of the great East Japan earthquake and recovery of the regional economy. Bunshindou, Tokyo. (in Japanese)

Tokunaga S, Resosudarmo BP, Wuryanto LE, Dung NT (2003) An inter-regional cge model to assess the impacts of tariff reduction and fiscal decentralization on regional economy. Stud Reg Sci 33(2):1–25

Tokunaga S, Kageyama M, Akune Y, Nakamura R (2014) Empirical analyses of agglomeration economies in Japanese assembly-type manufacturing industry for 1985–2000: using agglomeration and coagglomeration indices. Rev Urban Reg Dev Stud 26(1):57–79

Venables A (1996) Equilibrium locations of vertically linked industries. Int Econ Rev 37:341–359

Xie W, Li N, Wu J-D, Hao X-L (2014) Modeling the economic costs of disasters and recovery. Nat Hazards Earth Syst Sci 14:757–772

Chapter 3
Impact of the Great East Japan Earthquake on Production Loss Using an Inter-Regional Social Accounting Matrix

Mitsuru Okiyama

Abstract Using an inter-regional social accounting matrix, this chapter estimates the magnitude of production losses in both the disaster-affected and disaster-unaffected regions in Japan caused by production damage to agriculture and fisheries and by the loss of capital stock to the manufacturing industry. In addition, we also examine the cost-effectiveness of recovery support by comparing the initial production loss with that 1 or 2 years after the earthquake. Our estimates show that (1) the production loss caused by damages to agriculture and fisheries were 655.7 billion yen in the disaster-affected region and 2.69 trillion yen in the disaster-unaffected region; (2) the production losses caused by the loss of capital stock to the manufacturing industry were 937.2 billion yen in Fukushima Prefecture and 3.132 trillion yen in the other three disaster-affected prefectures (amounting to 5.9% and 6.0%, respectively, of the gross production value in the economic activity sector in 2005); and (3) 2 years after the disaster, the production loss in the disaster-affected region reduced by 311.4 billion yen, owing to the 2 years' worth of recovery support for agriculture and fisheries, thereby confirming the cost-effectiveness of this support. However, these estimations are based on the multiplier effect, on the assumption that the halt in production activities caused by the damage persists for 1 year.

Keywords Great East Japan earthquake • Production loss • Loss of capital stock • Recovery support • Multiplier effects • Inter-regional SAM

M. Okiyama (✉)
Reitaku Institute of Political Economics and Social Studies, Reitaku University, Kashiwa, Chiba 277-8686, Japan
e-mail: okiyama.m@gmail.com

S. Tokunaga, B.P. Resosudarmo (eds.), *Spatial Economic Modelling of Megathrust Earthquake in Japan*, New Frontiers in Regional Science: Asian Perspectives 11,
https://doi.org/10.1007/978-981-10-6493-7_3

3.1 Introduction

The Great East Japan Earthquake had a serious negative economic impact on the economy of Tohoku and other regions. The earthquake damaged production activities not only in the areas directly hit but also in other regions through disruptions to industrial linkages, including supply chains. We focus on the damage to two industries as follows: agriculture and fisheries as the key industries of the Tohoku region and the manufacturing industry because of the measureable impact of capital stock loss. In addition, other analyses focused on the relation between the recovery process from the earthquake and the recovery support for the disaster-affected firms.

This chapter utilizes an inter-regional social accounting matrix (SAM) to estimate the magnitude of production losses in both the disaster-affected and disaster-unaffected regions caused by production damage to agriculture and fisheries and by the loss of capital stock to the manufacturing industry. We also examine the cost-effectiveness of recovery support by comparing the immediate production loss with that 1 or 2 years after the earthquake.

We consider the multiplier effect caused by the decrease in economic activity (which was, in turn, induced by the direct external shocks resulting from the earthquake and tsunami) in the disaster-affected and disaster-unaffected regions, production activity, the household sector, and other sectors. Defining the disaster-affected region as the four prefectures most severely affected in terms of economic disruptions, namely, Iwate, Miyagi, Fukushima, and Ibaraki, and the disaster-unaffected region as all other prefectures, we first create a SAM for the two regions and employ it to estimate the magnitude of the negative linkage effect on production values in all industries (in both the disaster-affected and disaster-unaffected regions) caused by production damage to agriculture and fisheries in the four disaster-affected prefectures immediately following the earthquake, which we hereafter refer to as *production loss*. We also estimate the production loss from production damage that remains below pre-earthquake levels even after reconstruction and recovery works restore production to some extent. Second, we create a three-regional SAM exclusively for Fukushima Prefecture, the disaster-affected prefectures excluding Fukushima Prefecture, and other prefectures to estimate the production loss due to the reduction in manufacturing capital stock, which were destroyed in the earthquake. We also estimate the changes in gross production value in the disaster-affected region resulting from changes in tangible fixed asset values before and after the earthquake and calculate multiplier effects. Section 3.2 provides an overview of the two-regional SAM along with an explanation of the SAM multiplier analysis. Section 3.3 uses the two-regional SAM model to estimate the production loss inflicted by production damage to agriculture and fisheries in the disaster-affected region. Section 3.4 uses the three-regional SAM to estimate the production loss arising from the loss of manufacturing capital stock. Finally, Sect. 3.5 presents the results of our analysis.

3.2 Overview of the Two-Regional SAM and SAM Multiplier Analysis

3.2.1 Overview of the Two-Regional SAM

A SAM, systemized by Stone, R., is a square matrix that presents the accounts of all economic *actors* in the economy using the database of an input–output table.[1] An inter-regional SAM can be seen as an extension of a SAM using inter-regional IO tables and regional SNA. We review the literature in and outside of Japan for the construction of SAM. For a nationwide SAM applied to a pilot study, Pyatt and Roe (1977) constructed a SAM for Sri Lanka, wherein households were divided by income class. Thorbecke et al. (1992) applied a SAM for Indonesia to examine the distribution effects of Indonesia's structural adjustment, and Makino (1995) constructed a Japanese SAM and analyzed the Japanese economic structure. On the other hand, as examples of inter-regional SAMs, Nidaira (2000) constructed a three-regional SAM for Indonesia on the basis of the nationwide SAM to clarify the regional structure and the circulating structure among endogenous sectors, and Miyagi (1998) and Hisatake and Yamasaki (2006) constructed an inter-regional SAM for Japan to apply the regional Computable General Equilibrium (CGE) model. Further, Ito (2008) concretely explained the framework for constructing a two-regional SAM comprising Hokkaido and the rest of Japan. As explained below, our SAM is mostly based on Ito's (2008) framework.

First, we provide an overview of the two-regional SAM proposed by Ito (2008) on which our analysis is based. The base data used in this SAM is a 2005 inter-regional input–output table of all 47 prefectures created jointly by Yoshifumi Ishikawa and the Mitsubishi Research Institute. We calculate data that could not be obtained from this table, such as income and expenditures among households, enterprises, governments, and other segments within the two regions, using values from the 2005 National Economic Accounts and the 2005 Economic Accounts of each prefecture. In accordance with Ito's (2008) two-regional SAM framework, we create a 70×70 dimensional matrix with production activities comprising 20 sectors for both regions; production factors of labor and capital comprising two sectors for both regions; and institution (household, corporation, financial institution and prefecture etc.), savings, and investment comprising nine sectors for both regions, an additional seven other sectors, and one overseas sector. Table 3.1 is the macro-SAM based on this SAM.

[1]See Miller and Blair (2009) for details concerning input–output analysis.

Table 3.1 Two-regional macro SAM for the disaster-affected and disaster-unaffected regions

			The disaster-affecred Region						
				Production factor		Institution			
Billion Yen			Production activites	Labor	Capital	Household	Enterprises	Local Government	Saving-Investment
The disaster-affecred Region	Production activities		20,402			13,206	1,116	6,449	5,403
	Production facter	Labor	15,246						
		Capital	14,465						
	Institution	Household		16,045	4,532		575	3,612	
		Enterprises			8,900	604			
		Local government	−233		818	3,335		633	
	Saving-Investment					2,955	4,985	313	
The disaster-unaffected Region	Production activities		11,100			4,126	282	86	1,743
	Production facter	Labor	786						
		Capital	121						
	Institution	Household							
		Enterprises							
		Local government							
	Savings-Investment								
Other Sector	Direct taxs					1,080	834		
	Indirect taxes		3,074						
	Property income					646	5,128	638	
	Current transfer					656	534	611	
	Central government				283			17	
Rest of the world			3,895	3	56				
Total			68,856	16,048	14,588	26,609	13,454	12,358	7,146

The disaster-unaffected Region													
	Production factor		Institution				Other Sector						
Production activites	Labor	Capital	House-hold	Enter-prises	Local Govern-ment	Savings-Invest-ment	Direct taxs	Indirect taxes	Property income	Current transfer	Central Govern-ment	Rest of the world	Total
11,212			4,708	347	43	2,257						3,713	68,856
789												13	16,048
115												8	14,588
									1,338	508			26,609
									3,073	876			13,454
							1,029	1,602	272	56	4,844		12,358
											174	−1,281	7,146
432,002			263,809	24,246	84,656	112,559						59,684	994,291
246,881												144	247,810
210,111												86	210,318
	247,780	58,388		12,055	52,588				21,883	18,058			410,751
		137,070	11,325						73,491	15,899			237,785
−3,104		11,127	49,950		5,432		19,381	25,859	8,585	1,213	37,221		155,663
			23,157	111,126	−4,853						−2,566	−12,049	114,816
			23,498	16,330									41,742
39,115													42,189
			13,237	65,199	11,826							17,463	114,137
			21,066	8,482	5,806							1,871	39,026
		3,146			167		21,333	14,728					39,673
57,171	30	587							5,495	2,416			69,652
994,291	247,810	210,318	410,751	237,785	155,663	114,816	41,742	42,189	114,137	39,026	39,673	69,652	

3.2.2 Framework of the SAM Multiplier Analysis

The SAM multiplier analysis is based on the typical approach proposed by Pyatt and Round (1979, 2006) and can be used to estimate the effects of exogenous changes and injections, such as an increase in the demand for a given production activity, government expenditures, or exports, on the whole system.

Several important conditions underlying the SAM multiplier analysis stress its limits:

(i) The existence of excess capacity, which would allow prices to remain constant
(ii) Constant expenditure propensities of endogenous accounts
(iii) Production technology and resource endowment are given

SAM multipliers can be used to evaluate the potential impacts of the output changes on the corresponding production activity, factorial income distribution changes on labor and capital, and the household income distribution changes triggered by with an exogenous shock. These multipliers are based on a nationwide SAM.

In this chapter, we explain the framework of SAM multipliers based on a regional SAM.[2]

According to the two regional SAM shown in Table 3.1, we analyze demand-side linkages for how an exogenous shock in the rows changes the corresponding endogenous production activity, production factors, institution with household and government, saving/investment, and other sectors.

We look at the rows of the SAM to capture the effect and obtain Eq. (3.1).

$$y = B \cdot y + x \tag{3.1}$$

where y is the column vector of endogenous total receipts that each economic unit received. B denotes sub-matrices divided by the column sum of each account that is showing the income of endogenous account i received from endogenous account j as a proportion of expenditure of endogenous account j. x is the column vector of the exogenous export sector in this SAM.

M (the SAM multiplier matrix) can be derived when we solve Eq. (3.1) by y:

$$y = (I - B)^{-1}x = Mx. \tag{3.2}$$

As this SAM constructs two regions and one other sector, y, B, and x in Eq. (3.1) can be divided by three; then, the equation in the following matrix expression becomes

[2]Most of the SAM multiplier is based on Ito (2008).

$$\begin{bmatrix} y_1 \\ y_2 \\ y_3 \end{bmatrix} = \begin{bmatrix} B_{11} & \hat{b}_{12} & \hat{b}_{13} \\ \hat{b}_{21} & B_{22} & \hat{b}_{23} \\ \hat{b}_{31} & \hat{b}_{32} & B_{33} \end{bmatrix} \times \begin{bmatrix} y_1 \\ y_2 \\ y_3 \end{bmatrix} + \begin{bmatrix} x_1 \\ x_2 \\ x_3 \end{bmatrix}, \tag{3.3}$$

where y_1 is the set of endogenous variables for the disaster-affected region comprising four stricken prefectures, y_2 is the set of endogenous variables for the disaster-unaffected region, y_3 is the set of endogenous variables for other sectors, B_{11} represents the sub-matrix that economic units of the disaster-unaffected region pay each economic unit of the same area, B_{22} represents the sub-matrix that economic units of the disaster-unaffected region pay each economic unit of the same area, B_{33} represents the sub-matrix that economic units of other sectors pay each economic unit of the same sector, $\hat{b}_{ij}$ represents the sub-matrix that economic units among three sectors pay each other, and x_1, x_2, x_3 are the set of exogenous variables, where are the overseas sector.

Equation (3.4) in the following matrix expression can be obtained when we solve Eq. (3.3) by column vector y:

$$\begin{bmatrix} y_1 \\ y_2 \\ y_3 \end{bmatrix} = \begin{bmatrix} I & -D_{12} & -D_{13} \\ -D_{21} & I & -D_{23} \\ -D_{31} & -D_{32} & I \end{bmatrix}^{-1} \times \begin{bmatrix} (I - B_{11})^{-1} & 0 & 0 \\ 0 & (I - B_{22})^{-1} & 0 \\ 0 & 0 & (I - B_{33})^{-1} \end{bmatrix} \times \begin{bmatrix} x_1 \\ x_2 \\ x_3 \end{bmatrix}, \tag{3.4}$$

where $D_{ij} = (I - B_{ij})^{-1} \cdot \hat{b}_{ij}$

Thus, we rewrite the equation above:

$$y = M_r^{bt} M_r^{in} x, \tag{3.5}$$

where matrix M_r^{bt} represents the inter-regional linkage effect and the linkage effect between each region and other sectors, and matrix M_r^{in} represents the intra-regional linkage effect and the linkage effect in other sectors.

3.3 Magnitude of Multiplier Effects from the Damage to Agriculture and Fisheries

3.3.1 Estimation of Damage to Agriculture and Fisheries

Agricultural and fisheries production declined in the disaster-affected region, where floods from the tsunami damaged farmlands and facilities, such as nurseries, and the earthquake damaged the ground. To estimate such damages to production, we use surveys on the conditions to reopen businesses by the Ministry of Agriculture, Forestry and Fisheries that estimate cultivation acreage damaged by the tsunami for each of the disaster-affected prefectures. Regarding the damage to fisheries, with the exception of Fukushima Prefecture, we use surveys from the Ministry, which calculated the number of fishery businesses that were damaged on the basis of interviews. From these survey results, we calculate the damage ratio for each of the four disaster-affected prefectures and multiplied that by an agricultural and fisheries production value for each prefecture to obtain the monetary amount of damages to agriculture and fisheries due to the disaster. As depicted in Table 3.2, the damage to agriculture immediately following the disaster was approximately 75.6 billion yen, of which 23.0 billion was due to the tsunami and 52.6 billion was due to non-tsunami earthquake factors. The estimate for fisheries was 151.7 billion yen. These surveys were continued after the disaster, allowing us to estimate the damage 2 years later. Table 3.2 shows that after 2 years, agriculture recovered by 31.5% and the damage reduced to 51.8 billion yen. Fisheries recovered by 56.2% and the damage reduced to approximately 66.4 billion yen. This reduction in damage can be attributed to support subsidies amounting to 451.1 billion yen for the agricultural, forestry, and fisheries industries in the disaster-affected region in the reconstruction budgets for fiscal years 2011 and 2012 and to the launch of a variety of recovery and restoration works. Notwithstanding these factors, the recovery in Fukushima Prefecture lagged in comparison with that in other prefectures due to radioactive contamination arising from the Fukushima nuclear plant accident.

3.3.2 Production Loss Immediately After the Earthquake

In this section, we insert the immediate post-earthquake damage in production to agriculture and forestry into the agriculture and forestry production sectors in x_1 in Eq. (3.4), given in Sect. 3.2.2, to obtain y_1, y_2, and y_3. The results are shown in Tables 3.3a and 3.3b. Let us first consider the impact on production activities in the disaster-affected region. The production loss—the reduction in overall production in agriculture and fisheries in the disaster-affected region caused by the earthquake and tsunami—is estimated at 655.7 billion yen. Of this, 231.3 billion yen was for agriculture and 424.4 billion yen for fisheries. At the same time, Table 3.3a shows that the impact on production activity in the disaster-unaffected region, for all industry sectors, was a reduction of 2.69 trillion yen. Of this, 0.95 trillion yen was for agriculture and 1.74 trillion yen for fisheries. Notice that this production loss is a rough calculation assuming that the production damage to agriculture and fisheries persisted for 1 year following the earthquake. Analyzing monetary damage by industry in the disaster-affected region, the industry most affected was "other services," which was followed by finance/insurance/real estate, construction, and public services, in that order. Tertiary industries declined overall by 228.1 billion yen, accounting for 34.8% of the total damage. The greatest damage in the

Table 3.2 Estimation of damage to agriculture and fisheries

	Amount of damage on farm products							Amount of damage on fishery				
Unit: million yen	Immediately after the earthquake (A)	Damage to cultivation acreage by a tsunami (unit:ha)	Ratio of damage to cultivation acreage	Damage to cultivation acreage except a tsunami (unu:ha)	Ratio of damage to cultivation acreage	The two years period after the earthquake (B)	Ratio of restoration (=(A-B)/A)	Immediately after the earthquake (A)	Business unit being engaged in fishery (unit: number)	Ratio of damage	The two years period after the earthquake (B)	Ratio of restoration (=(A-B)/A)
Fukushima Prefecture	44,770	5,462	3.6%	27,502	18.3%	37,944	15.2%	20,256	740	99.6%	19,873	1.9%
Iwate Prefecture	10,459	725	0.5%	10,905	7.1%	6,707	35.9%	39,313	5,100	96.0%	11,008	72.0%
Miyagi Prefecture	17,518	14,341	10.5%	2,911	2.1%	7,174	59.0%	82,639	3,990	99.6%	35,537	57.0%
Ibaraki Prefecture	2,879	208	0.1%	1,444	0.8%	0	100.0%	9,518	210	43.8%	0	100.0%
Stricken Region	75,626	20,736	3.4%	42,762	6.9%	51,825	31.5%	151,725	10,040	95.2%	66,417	56.2%

Source Original data based on Survey by the Ministry of Agriculture, Forestry and Fisheries (2011a, b, 2015)

Table 3.3a Production losses of the disaster-affected and disaster-unaffected regions immediately after the earthquake

Unit:billion yen	Production loss caused by the earthquake and the tsunami			Unit:billion yen	Production loss caused by the earthquake and the tsunami		
The disaster-affected region	Total damage	The damage to farm products	The damage to fishery	The disaster-unaffected region	Total damage	The damage to farm products	The damage to fishery
Agriculture	−94.5	−88.9	−5.6	Agriculture	−35.0	−13.9	−21.0
Forestry	−1.1	−0.4	−0.7	Forestry	−3.9	−1.4	−2.5
Fishery	−156.7	−0.5	−156.2	Fishery	−7.2	−2.0	−5.3
Mining	−4.7	−1.6	−3.1	Mining	−37.9	−13.2	−24.7
Foods and tobacco	−35.6	−14.6	−21.0	Foods and tobacco	−118.9	−43.0	−75.8
Non-durable goods	−11.0	−3.9	−7.1	Non-durable goods	−63.3	−22.3	−41.0
Petroleum products and chemicals	−30.0	−10.2	−19.8	Petroleum products and chemicals	−160.0	−55.3	−104.7
Steel and metals	−18.4	−6.5	−11.9	Steel and metals	−124.6	−43.7	−80.9
General machinery	−9.5	−3.4	−6.1	General machinery	−82.9	−29.3	−53.6
Electronic parts and devices	−9.6	−3.4	−6.2	Electronic parts and devices	−72.8	−25.6	−47.2
Motor vehicle and Motor vehicle parts	−2.9	−1.0	−1.9	Motor vehicle and Motor vehicle parts	−71.2	−25.1	−46.2
Other durable goods	−6.4	−1.5	−4.9	Other durable goods	−46.5	−15.0	−31.5
Miscellaneous manufacturing goods	−5.0	−1.7	−3.3	Miscellaneous manufacturing goods	−45.5	−15.9	−29.6
Construction	−42.3	−15.4	−26.9	Construction	−176.1	−62.2	−113.9
Electricity power,gas and water supply	−21.2	−7.7	−13.5	Electricity power,gas and water supply	−74.2	−26.2	−48.1
Commerce	−31.4	−10.3	−21.0	Commerce	−271.1	−95.2	−176.0

Finance and insurance, Real estiate	−50.2	−16.5	−33.7	Finance and insurance, Real estiate	−323.1	−113.8	−209.3
Transport and Communication	−23.9	−8.7	−15.2	Transport and Communication	−132.1	−46.8	−85.3
Pulic services	−44.3	−15.1	−29.2	Pulic services	−354.3	−125.0	−229.3
Services	−57.1	−20.0	−37.0	Services	−488.5	−172.3	−316.2
Total industries	−655.7	−231.3	−424.4	Total industries	−2,689.4	−947.1	−1,742.3

Table 3.3b Impact on other sectors (except production activity) in the disaster-affected and disaster-unaffected region immediately after the earthquake

Unit:billion yen	Income loss caused by the earthquake and the tsunami			Unit:billion yen	Income loss caused by the earthquake and the tsunami			Unit:billion yen	Income loss caused by the earthquake and the tsunami		
The disaster-affected region	Total damage	The damage to farm products	The damage to fishery	The disaster-unaffected region	Total damage	The damage to farm products	The damage to fishery	Other sector	Total damage	The damage to farm products	The damage to fishery
Labor	−125.5	−39.0	−86.5	Labor	−688.2	−242.1	−446.1	Direct tax	−126.6	−44.7	−81.9
Capital	−163.3	−62.2	−101.0	Capital	−582.9	−205.6	−377.3	Indirect tax	−125.1	−44.7	−80.4
Household	−204.6	−68.2	−136.4	Household	−1,143.2	−402.9	−740.2	Import tax	−11.7	−3.9	−7.8
Income per house-hold(thousand yen)	−66.2	−22.1	−44.1	Income per house-hold(thousand yen)	−24.6	−8.7	−15.9	Tariff	−3.7	−1.1	−2.6
Non-profit corporation	−2.5	−0.9	−1.6	Non-profit corporation	−23.6	−8.3	−15.2	Total tax sector	−267.1	−94.4	−172.7
Non-financial corporaton	−103.8	−39.4	−64.4	Non-financial corporaton	−438.1	−154.6	−283.5	Property sector	−295.3	−105.4	−189.9
Financial institution	−8.3	−2.9	−5.4	Financial institution	−184.4	−65.7	−118.7	Other sector	−109.4	−38.5	−70.8
Branch office of Central governemnt	−5.7	−2.0	−3.7	Branch office of Central governemnt	−77.0	−27.2	−49.7	Central government	−126.2	−44.7	−81.5
Prefecture	−11.2	−4.0	−7.2	Prefecture	−88.5	−31.3	−57.3	GDP per-capita (thousand yen)	−13.3	−4.7	−8.6
Municipality	−9.8	−3.4	−6.4	Municipality	−93.2	−33.0	−60.3				
Social security fund	−32.7	−11.1	−21.7	Social security fund	−208.0	−73.4	−134.5				
Savings/investment	−73.0	−26.5	−46.5	Savings/investment	−346.2	−122.3	−223.9				

manufacturing sector was inflicted on the food and tobacco industry, being closely related to agriculture and fisheries, which declined by 35.6 billion yen.

Next, let us consider the impact of the damage caused by the earthquake and tsunami to agriculture and forestry with respect to various items in the institutional and all other sectors. The decline in an annual household income in the disaster-affected region was 66,000 yen per household and that in the disaster-unaffected region was 25,000 yen per household. Household budgets in the disaster-affected region suffered much greater damage than those in the disaster-unaffected region. The huge negative impact on the disaster-affected region is clearly visible in the 73.0 billion yen decline in the savings and investment section, 267.1 billion yen in the tax section; 126.2 billion yen in central government receipts, and 295.3 billion yen in the asset section.

3.3.3 Production Loss Two Years After the Earthquake

We obtain the production loss 2 years after the disaster from the production damage to agriculture and fisheries. The results are shown in Tables 3.4a and 3.4b. Let us first consider the impact on production activities in the disaster-affected region. The production loss 2 years after the earthquake —the reduction in overall production of agriculture and fisheries in the disaster-affected region caused by the earthquake and tsunami—was estimated at 344.3 billion yen. Of this, 158.5 billion yen was for agriculture and 185.5 billion yen for fisheries. When compared with the production loss immediately following the disaster, these amounts had shrunk by 311.4 billion yen, 72.8 billion yen, and 238.6 billion yen, respectively. Fisheries production exhibited greater recovery than agricultural production from the immediate production loss. We may also surmise that the 2 years' worth of recovery support for agriculture and fisheries mentioned above contributed greatly to this recovery; thus, the cost-effectiveness of this support is confirmed. At the same time, Table 3.4a shows that the impact on production activity in the disaster-unaffected region, for all industry sectors, was a reduction of 1.41 trillion yen, down by 1.28 trillion yen from immediate post-earthquake levels. Of the reduction, agriculture accounted for 0.65 trillion yen and fisheries for 0.76 trillion yen.

Next, let us consider the impact of damage caused by the earthquake and tsunami to agriculture and forestry after 2 years with respect to various items in the institutional and other sectors. The decline in an annual household income in the disaster-affected region had shrunk to 34,000 yen per household, about half of the immediate post-earthquake levels. The decline in household income in the disaster-unaffected region was 13,000 yen. The decrease in the savings and investment section was 38.5 billion yen, tax revenues was 140.3 billion yen, central government receipts was 66.2 billion, and the asset section was 155.3 billion yen, shrinking to about half of immediate post-earthquake levels.

Table 3.4a Production losses of the disaster-affected and disaster-unaffected regions two years after the earthquake

Unit:billion yen	Production loss caused by the earthquake and the tsunami			Unit:billion yen	Production loss caused by the earthquake and the tsunami		
The disaster-affected region	Total damage	The damage to farm products	The damage to fishery	The disaster-unaffected region	Total damage	The damage to farm products	The damage to fishery
Agriculture	−63.4	−60.9	−2.5	Agriculture	−18.8	−9.6	−9.2
Forestry	−0.6	−0.3	−0.3	Forestry	−2.1	−0.9	−1.1
Fishery	−68.7	−0.3	−68.4	Fishery	−3.7	−1.4	−2.3
Mining	−2.5	−1.1	−1.4	Mining	−19.9	−9.0	−10.8
Foods and tobacco	−19.2	−10.0	−9.2	Foods and tobacco	−62.7	−29.5	−33.2
Non-durable goods	−5.8	−2.7	−3.1	Non-durable goods	−33.2	−15.3	−17.9
Petroleum products and chemicals	−15.6	−7.0	−8.7	Petroleum products and chemicals	−83.7	−37.9	−45.9
Steel and metals	−9.7	−4.5	−5.2	Steel and metals	−65.4	−30.0	−35.4
General machinery	−5.0	−2.3	−2.7	General machinery	−43.6	−20.1	−23.5
Electronic parts and devices	−5.0	−2.3	−2.7	Electronic parts and devices	−38.2	−17.5	−20.7
Motor vehicle and Motor vehicle parts	−1.5	−0.7	−0.8	Motor vehicle and Motor vehicle parts	−37.4	−17.2	−20.2
Other durable goods	−3.2	−1.0	−2.1	Other durable goods	−24.1	−10.3	−13.8
Miscellaneous manufacturing goods	−2.6	−1.1	−1.5	Miscellaneous manufacturing goods	−23.9	−10.9	−13.0
Construction	−22.3	−10.6	−11.8	Construction	−92.5	−42.6	−49.9
Electricity power,gas and water supply	−11.2	−5.3	−5.9	Electricity power,gas and water supply	−39.0	−17.9	−21.0
Commerce	−16.3	−7.1	−9.2	Commerce	−142.2	−65.2	−77.0
Finance and insurance, Real estiate	−26.1	−11.3	−14.7	Finance and insurance, Real estiate	−169.6	−78.0	−91.6
Transport and Communication	−12.6	−6.0	−6.7	Transport and Communication	−69.4	−32.0	−37.3
Pulic services	−23.1	−10.4	−12.8	Pulic services	−186.1	−85.7	−100.4
Services	−29.9	−13.7	−16.2	Services	−256.5	−118.1	−138.4
Total industries	−344.3	−158.5	−185.8	Total industries	−1,411.7	−649.0	−762.7

Table 3.4b Impact on other sectors (except production activity) in the disaster-affected and disaster-unaffected regions two years after the earthquake

Unit:billion yen	Income loss caused by the earthquake and the tsunami			Unit:billion yen	Income loss caused by the earthquake and the tsunami			Unit:billion yen	Income loss caused by the earthquake and the tsunami		
The disaster-affected region	Total damage	The damage to farm products	The damage to fishery	The disaster-unaffected region	Total damage	The damage to farm products	The damage to fishery	Other sector	Total damage	The damage to farm products	The damage to fishery
Labor	−64.6	−26.7	−37.9	Labor	−361.2	−165.9	−195.3	Direct tax	−66.5	−30.6	−35.9
Capital	−86.9	−42.7	−44.2	Capital	−306.0	−140.9	−165.2	Indirect tax	−65.9	−30.6	−35.2
Household	−106.5	−46.8	−59.7	Household	−600.1	−276.1	−324.0	Import tax	−6.1	−2.7	−3.4
Income per household (thousand yen)	−34.4	−15.1	−19.3	Income per household (thousand yen)	−12.9	−5.9	−7.0	Tariff	−1.9	−0.7	−1.1
Non-profit corporation	−1.3	−0.6	−0.7	Non-profit corporation	−12.4	−5.7	−6.7	Total tax sector	−140.3	−64.7	−75.6
Non-financial corporaton	−55.2	−27.0	−28.2	Non-finan-cial corporaton	−230.0	−106.0	−124.1	Property sector	−155.3	−72.2	−83.1
Financial institution	−4.3	−2.0	−2.4	Financial institution	−97.0	−45.0	−52.0	Other sector	−57.4	−26.4	−31.0
Branch office of Central governemnt	−3.0	−1.3	−1.6	Branch office of Central governemnt	−40.4	−18.7	−21.8	Central government	−66.3	−30.6	−35.7

(continued)

Table 3.4b (continued)

Unit:billion yen	Income loss caused by the earthquake and the tsunami			Unit:billion yen	Income loss caused by the earthquake and the tsunami			Unit:billion yen	Income loss caused by the earthquake and the tsunami		
The disaster-affected region	Total damage	The damage to farm products	The damage to fishery	The disaster-unaffected region	Total damage	The damage to farm products	The damage to fishery	Other sector	Total damage	The damage to farm products	The damage to fishery
Prefecture	−5.9	−2.7	−3.2	Prefecture	−46.5	−21.4	−25.1	GDP per-capita (thousand yen)	−7.0	−3.2	−3.8
Municipality	−5.1	−2.3	−2.8	Municipality	−49.0	−22.6	−26.4				
Social security fund	−17.1	−7.6	−9.5	Social security fund	−109.2	−50.3	−58.9				
Savings/ investment	−38.5	−18.2	−20.4	Savings/ investment	−181.8	−83.8	−98.0				

3.4 Magnitude of Multiplier Effects from the Loss of Capital Stock to the Manufacturing Industry

3.4.1 *Overview of the Three-Regional SAM and SAM Multiplier Analysis*

In this section, using the three-regional SAM, we analyze the multiplier effects from the capital loss of the manufacturing industry in the disaster-affected region caused by the earthquake. The three-regional SAM comprises a 101 × 101 dimensional matrix, with production activities comprising 20 sectors for each of the three regions; production factors of labor and capital comprising two sectors for each of the three regions; and institution, savings, and investment comprising nine sectors for each of the three regions, an additional seven other sectors, and one overseas sector.

Accordingly, we must modify the following framework of the SAM multiplier because the SAM extends from two regions to three.

First, y, B, and x in Eq. (3.1) can be divided by four to generate the following matrix expression (3.6):

$$\begin{bmatrix} y_1 \\ y_2 \\ y_3 \\ y_4 \end{bmatrix} = \begin{bmatrix} B_{11} & \hat{b}_{12} & \hat{b}_{13} & \hat{b}_{14} \\ \hat{b}_{21} & B_{22} & \hat{b}_{23} & \hat{b}_{24} \\ \hat{b}_{31} & \hat{b}_{32} & B_{33} & \hat{b}_{34} \\ \hat{b}_{41} & \hat{b}_{42} & \hat{b}_{43} & B_{44} \end{bmatrix} \times \begin{bmatrix} y_1 \\ y_2 \\ y_3 \\ y_4 \end{bmatrix} + \begin{bmatrix} x_1 \\ x_2 \\ x_3 \\ x_4 \end{bmatrix} \tag{3.6}$$

where y_1 is the set of endogenous variables for Fukushima Prefecture, y_2 is the set of endogenous variables for the other three disaster-affected prefectures, y_3 is the set of endogenous variables for the disaster-unaffected region, y_4 is the set of endogenous variables for other sectors, B_{11} represents the sub-matrix that economic units of Fukushima Prefecture pay each economic unit of the same area, B_{22} represents the sub-matrix that economic units of the other three disaster-affected prefectures pay each economic unit of the same area, B_{33} represents the sub-matrix that economic units of other region pay each economic unit of the same area, B_{44} represents the sub-matrix that economic units of other sectors pay each economic unit of the same sector, $\hat{b}_{ij}$ represents the sub-matrix that economic units among four sectors pay each other, and x_1, x_2, x_3, x_4 are the set of exogenous variables, where are the overseas sector.

Equation (3.7) in the following matrix expression can be obtained when we solve Eq. (3.6) by column vector y:

$$\begin{bmatrix} y_1 \\ y_2 \\ y_3 \\ y_4 \end{bmatrix} = \begin{bmatrix} I & -D_{12} & -D_{13} & -D_{14} \\ -D_{21} & I & -D_{23} & -D_{24} \\ -D_{31} & -D_{32} & I & -D_{34} \\ -D_{41} & -D_{42} & -D_{43} & I \end{bmatrix}$$
$$\times \begin{bmatrix} (1-B_{11}) & 0 & 0 & 0 \\ 0 & (1-B_{22}) & 0 & 0 \\ 0 & 0 & (1-B_{33}) & 0 \\ 0 & 0 & 0 & (1-B_{44}) \end{bmatrix} \times \begin{bmatrix} x_1 \\ x_2 \\ x_3 \\ x_4 \end{bmatrix}, \quad (3.7)$$

where $D_{ij} = (I - B_{ij})^{-1} \cdot \hat{b}_{ij}$.

Thus, we rewrite the equation above such that

$$y = M_r^{bt} M_r^{in} x \quad (3.8)$$

3.4.2 *Magnitude of Multiplier Effects from the Loss of Capital Stock*

3.4.2.1 Estimated Direct Damage in Production

We now estimate the monetary value of the exogenous variables x_1, x_2 in Eq. (3.6) in Sect. 3.4.1 for each industry in the manufacturing sector in Fukushima Prefecture and the other three disaster-affected prefectures. We first use data from the 2010 Census of Manufacture for the four prefectures—Fukushima, Iwate, Miyagi, and Ibaraki— to estimate tangible fixed assets by industry in the manufacturing sector (in billions of yen). We also attempt to identify the disaster-affected region more accurately by using geospatial information instead of municipality information. For Fukushima Prefecture, we estimate tangible fixed capital amounts on the basis of the number of affected business locations damaged by the nuclear accident by industry in the manufacturing sector, considering business locations damaged by flooding from the tsunami as having been damaged by the nuclear accident. Assuming tangible fixed assets were destroyed in their entirety by the Great East Japan Earthquake, we solve for the proportion of tangible fixed assets damaged by the nuclear accident for each manufacturing industry. We multiply the ratio of destroyed value by production values for each manufacturing industry in Fukushima Prefecture in 2005 from the SAM and consider this the direct value of damages from the loss of capital stock in the prefecture. We treat this as the direct value of damages to production arising from the disruption of production accompanying the loss of capital stock in the manufacturing sector. Let us explain why this damage is x from Sect. 3.4.1. Assuming that production at some factories was shifted overseas, the closure of these factories constitutes the abandonment of its capital stock, which can be expressed as the loss of capital stock. Therefore, the direct damage estimated above can be synonymous with a reduction in the portion of production that was destined for export from that location. For the other three disaster-affected prefectures, we similarly estimate tangible fixed capital for each business location on the basis of the number of affected locations damaged by flooding from the tsunami to obtain the proportion of tangible fixed assets so damaged. We multiply this ratio by production values in 2005 from the SAM and consider this the value of direct damage to production from the disaster in the three prefectures. The resulting estimates are given in the "Impact on production value in terms of business locations damaged in the disaster" column of Table 3.5.

As shown in Table 3.5, the direct production damage to the manufacturing sector due to the nuclear accident in Fukushima Prefecture was estimated at 331.7 billion yen. Of this, the electronics industry accounted for 90.7 billion yen, 27.0% of the total, followed by non-durable manufacturing at 49.7 billion yen, or 15.0%, and then motor vehicle and motor vehicle parts at 35.8 billion yen, or 10.8% For the other three disaster-affected prefectures, the direct production damage to the manufacturing sector due to tsunami flooding was estimated at 1.078 trillion yen. Of this, the food and tobacco industry accounted for the largest share at 344.5 billion yen, or 32.0% of the whole, followed by petrochemicals at 22.9%, and equipment and infrastructure at 17.7%. These estimates confirm the considerable damage was incurred by tsunami flooding in the three disaster-affected prefectures in the food and tobacco industry, which is clustered in coastal areas. Note that Table 3.5 shows the production damage that could occur in the case of production activities being halted for 1 year due to the loss of capital stock in each sector.

3.4.2.2 Production Loss Immediately After the Earthquake

Using the direct production damage accompanying the loss of capital stock in the manufacturing sector by the disaster, as estimated in the previous section, let us consider the multiplier effects from the damage wreaked by the disaster on each industry (sector of economic activity) in Fukushima Prefecture, the other three disaster-affected prefectures, and the disaster-unaffected region (shown in Table 3.6). However, note that these multiplier effects are in each case dependent on the assumption that there is no recovery from the disaster and the halt in production activities caused by the damage persists for 1 year.

As shown in Table 3.6, production losses across all industries caused by direct production damage to the manufacturing sector in Fukushima Prefecture and the other three disaster-affected prefectures were 937.2 billion and 3.132 trillion yen, respectively. These losses reached 5.9% and 6.0% of gross production value in the economic activity sector in 2005 (in SAM terms, receipts), respectively. The production loss of the manufacturing sector in Fukushima Prefecture, including the impacts from within and outside, was 495.9 billion yen, exceeding direct production damage by 164.2 billion yen. The production losses to agriculture and construction/tertiary industries in Fukushima Prefecture were 23.2 billion and 418.1 billion yen, respectively. These production loss amounts accounted for by

Table 3.5 Estimation of production value in terms of business locations damaged and of changes in tangible fixed assets

		Businesses(unit:number)					
		2010	2012	The number of the stricken flooded area caused by a tsunami based on the spatical gergraphic information	(Ref.) The number of the stricken flooded area caused by a tsunami based on the administrative division	The number of the stricken area caused by a nuclear power plant disaster based on the administrative division	Variations of businesses from 2000 through 2012
		Number	Number	Number of stricken area	Number of stricken area	Number of stricken area	Number
Fukushima Prefecture	Total	4,186	3,893	48	868	301	−293
	Foods and tobacco	674	610	13	128	20	−64
	Non-durable goods	738	675	9	152	72	−63
	Petroleum products and chemicals	364	348	4	91	13	−16
	Steel and metals	512	489	6	132	30	−23
	General machinery	568	536	4	122	50	−32
	Electronic parts and devices	348	321	4	56	28	−27
	Motor vehicle and Motor vehicle parts	111	102	0	17	11	−9
	Other durable goods	199	192	2	46	22	−7
	Miscellaneous manufacturing goods	672	620	6	124	55	−52
The other three diasater-affected prefectures	Total	11,371	10,723	793	3,351	0	−648
	Foods and tobacco	2,540	2,198	386	938	0	−342
	Non-durable goods	1,499	1,379	95	374	0	−120
	Petroleum products and chemicals	1,056	1,038	42	264	0	−18
	Steel and metals	1,671	1,640	97	463	0	−31
	General machinery	1,301	1,306	53	331	0	5
	Electronic parts and devices	517	450	23	119	0	−67
	Motor vehicle and Motor vehicle parts	322	322	11	79	0	0
	Other durable goods	626	608	21	261	0	−18
	Miscellaneous manufacturing goods	1,839	1,782	65	522	0	−57

Source Original data based on Census of Manufacture 2010 by the Ministry of Economy, Trade and Industry (2013)

Estimated tangible fixed assets(unit:billion yen)						Estimated production value (unit:billion yen)		
2010 = A	2012 = B	The stricken flooded area caused by a tsunami based on the spatical gergraphic information	(Ref.)The stricken flooded area caused by a tsunami based on the administrative division	The stricken area caused by a nuclear power plant disaster based on the administrative division = D	Variations of tangible fixed assets from 2000 through 2012 = E	2005 = Z	Impact on production value in terms of business locations damaged in the disaster = D/A*Z	Impact on production value of changes in tangible fixed assets over the two years staring in 2010 = E/A*Z
Amount	Amount	Amount of damage	Amount of damage	Amount of damage	Amount	Amount	Amount	Amount
174,816	161,792	2,053	39,326	10,614	−13,024	5,639	−331.7	−464.6
18,254	17,098	248	2,303	381	−1,157	923	−19.3	−58.5
14,089	14,477	58	4,274	674	389	367	−17.6	10.1
29,315	25,832	626	10,439	1,038	−3,483	658	−23.3	−78.2
27,334	28,472	429	7,632	935	1,138	775	−26.5	32.3
13,490	12,649	87	2,562	1,019	−841	490	−37.0	−30.6
30,170	25,378	335	4,362	2,027	−4,792	1,350	−90.7	−214.4
11,233	9,505	0	1,485	1,071	−1,728	375	−35.8	−57.7
10,614	9,307	173	2,253	1,320	−1,306	400	−49.7	−49.2
20,318	19,074	97	4,016	2,148	−1,243	301	−31.8	−18.4
611,858	584,717	31,311	182,667	0	−27,141	20,809	−1,077.5	−1,011.4
91,661	90,888	9,599	33,495	0	−772	3,290	−344.5	−27.7
55,574	49,757	4,657	15,464	0	−5,816	1,269	−106.3	−132.8
97,906	93,131	5,302	33,844	0	−4,775	4,364	−236.4	−212.9
125,058	131,657	5,209	44,349	0	6,598	4,580	−190.8	241.6
86,474	83,184	1,163	20,193	0	−3,290	2,482	−33.4	−94.4
53,836	34,686	1,854	10,026	0	−19,150	2,201	−75.8	−783.0
26,761	32,671	1,187	4,023	0	5,910	874	−38.8	193.0
26,023	20,836	639	8,691	0	−5,187	924	−22.7	−184.1
48,565	47,906	1,700	12,582	0	−659	825	−28.9	−11.2

Table 3.6 Production loss caused by the loss of capital stock on the disaster-affected region

	Fukushima prefecture					The other three disater-affected prefectures					The disaster-unaffected region	
unit: billion yen	The amount of damage through the negative linkage effect (X = A + B)	Change % from the production value of 2005	Intra-regional negative linkage effect = A	Inter-regional negative linkage effect = B	Ratio of Intra-regional negative linkage effect = A/X	The amount of damage through the negative linkage effect (X = A + B)	Change % from the production value of 2005	Intra-regional negative linkage effect = A	Inter-regional negative linkage effect = B	Ratio of Intra-regional negative linkage effect = A/X	The amount of damage through the negative linkage effect	Change % from the production value of 2005
Agriculture	−13.5	−4.9%	−3.6	−10.0	26.2%	−94.3	−8.6%	−71.2	−23.1	75.5%	−210	−2.0%
Forestry	−1.7	−6.7%	−0.9	−0.8	51.7%	−7.5	−7.5%	−5.1	−2.4	68.1%	−25	−2.0%
Fishery	−1.1	−5.4%	−0.1	−0.9	13.0%	−15.2	−8.4%	−11.7	−3.5	76.9%	−37	−2.1%
Mining	−6.9	−4.0%	−3.4	−3.5	49.3%	−51.3	−6.3%	−34.1	−17.1	66.6%	−253	−1.9%
Foods and tobacco	−54.3	−5.9%	−26.9	−27.3	49.6%	−474.0	−14.4%	−407.0	−67.0	85.9%	−708	−1.9%
Non-durable goods	−30.0	−8.2%	−19.5	−10.5	65.0%	−174.2	−13.7%	−147.8	−26.4	84.8%	−403	−1.9%
Petroleum products and chemicals	−46.1	−7.0%	−27.8	−18.3	60.3%	−439.9	−10.1%	−357.8	−82.1	81.3%	−1,015	−1.8%
Steel and metals	−49.5	−6.4%	−33.6	−16.0	67.7%	−378.2	−8.3%	−306.1	−72.2	80.9%	−899	−1.8%
General machinery	−49.7	−10.1%	−39.0	−10.8	78.4%	−78.2	−3.2%	−43.0	−35.2	55.0%	−520	−1.4%
Electronic parts and devices	−123.8	−9.2%	−101.0	−22.8	81.6%	−142.2	−6.5%	−99.3	−42.9	69.8%	−496	−1.3%
Motor vehicle and Motor vehicle parts	−42.0	−11.2%	−36.7	−5.3	87.3%	−56.3	−6.4%	−47.2	−9.2	83.7%	−482	−1.0%
Other durable goods	−57.8	−14.5%	−50.4	−7.4	87.1%	−43.9	−4.7%	−31.3	−12.6	71.3%	−278	−1.4%
Miscellaneous manufacturing goods	−42.7	−14.2%	−34.0	−8.7	79.7%	−52.3	−6.3%	−37.5	−14.7	71.8%	−291	−1.7%
Construction	−61.9	−6.7%	−33.8	−28.1	54.6%	−152.1	−4.5%	−84.6	−67.5	55.6%	−1,097	−1.9%

Electricity power,gas and water supply	−47.3	−2.5%	−12.7	−34.6	26.9%	−77.6	−5.3%	−47.7	−29.9	61.5%	−470	−1.8%
Commerce	−45.2	−4.4%	−21.1	−24.1	46.7%	−136.3	−3.5%	−57.1	−79.3	41.9%	−1,691	−1.7%
Finance and insurance, Real estiate	−66.2	−4.8%	−36.3	−29.9	54.8%	−211.6	−4.2%	−115.3	−96.3	54.5%	−2,022	−1.7%
Transport and Communication	−32.1	−4.4%	−15.7	−16.3	49.0%	−95.0	−4.1%	−50.2	−44.8	52.9%	−822	−1.6%
Pulic services	−76.8	−3.6%	−33.3	−43.4	43.4%	−205.5	−3.1%	−77.7	−127.7	37.8%	−2,226	−1.8%
Services	−88.6	−5.2%	−49.5	−39.1	55.9%	−246.5	−3.7%	−113.6	−132.8	46.1%	−3,058	−1.8%
Total industries	−937.2	−5.9%	−579.1	−358.0	61.8%	−3,132.2	−6.0%	−2,145.5	−986.6	68.5%	−17,004	−1.7%

direct production damage to the manufacturing sector in Fukushima Prefecture was 8.0 billion yen and 202.4 billion yen, accounting for 34.3% and 48.4% of the total, respectively and the production loss with the remaining accounts for 40–50% stemming from the multiplier effects from the other three affected prefectures and the disaster-unaffected region. In contrast, the production losses of the manufacturing sector in the other three disaster-affected prefectures, omitting Fukushima Prefecture and including the impacts from within and outside, were 1.839 trillion yen, exceeding direct production damage by 761.8 billion yen. The production losses to the agriculture and construction/tertiary industries in the other three disaster-affected prefectures were 168.3 billion and 1.125 trillion billion yen, respectively. The portion of these losses accounted for by direct production damage to the manufacturing sector in the three disaster-affected prefectures was 122.2 billion yen and 546.3 billion yen, accounting for 72.6% and 48.6% of the total, respectively with the remaining portion stemming from the multiplier effects from Fukushima Prefecture and the disaster-unaffected region. The production loss in the disaster-unaffected region arising from the multiplier effects, excluding the four disaster-affected prefectures, and caused by direct production damage to the manufacturing sector in the disaster-affected region was in excess of 17 trillion yen. This amount represents 1.7% of the 2005 production value in the disaster-unaffected region. The highest production loss was incurred in construction/tertiary industries (11.386 trillion yen), which accounted for 67% of the production loss in the entire disaster-unaffected region. Meanwhile, production losses of the manufacturing sector were 5.092 trillion yen in the disaster-unaffected region and 525.5 billion yen for agriculture and fisheries.

3.4.3 Magnitude of Multiplier Effects from the Change of the Tangible Fixed Assets

3.4.3.1 Estimation of the Impact of the Change in Tangible Fixed Assets

We now estimate the variance in production values for Fukushima Prefecture and the other three disaster-affected prefectures caused by changes in tangible fixed assets in each industry in the manufacturing sector before and after the disaster. We assume an approach different from the one adopted in the previous section. We estimate tangible fixed assets by industry within the manufacturing sector in 2012 (after the disaster) and—from these estimates and values for 2010 (before the disaster)—derive the changes in tangible fixed assets from before to after the earthquake. We consider the tangible fixed assets at two points in time—2010 and 2012—to be capital stock at the beginning and end of the fiscal year. This is because we want to estimate the net increase in capital stock due to the disaster, including companies who withdrew in the wake of the earthquake, those who restored operations later because of fiscal aid, and those who seized the opportunity of the disaster to enter the market. We can now multiply the estimates of percentage change in tangible fixed assets by production value for each industry in the manufacturing sector in Fukushima Prefecture and the other three affected prefectures for 2005 from the SAM to calculate the impact on annual production value in the disaster-affected region caused by changes in tangible fixed assets in the manufacturing sector from before to after the disaster. The resulting estimates are given in the "Impact on production value of changes in tangible fixed assets over the two years starting in 2010" column of Table 3.5. Taking Fukushima Prefecture first, the change in tangible fixed assets over the 2-year period exhibited a decline of 13.024 trillion yen. Compared with the direct production damage estimated in Sect. 3.4.2.1, the impact of this decline on the production value across all manufacturing sectors was 464.6 billion yen, wider by 132.9 billion yen. In particular, it appears that tangible fixed assets in the food/tobacco and electronics industries declined over the 2-year period by 1.157 trillion yen and 4.792 trillion yen, respectively. On the other hand, tangible fixed assets in the non-durable manufacturing and equipment/infrastructure industries rose over the 2-year period, even as the number of business locations shrank.

Next, let us calculate the variance in production value in the other three disaster-affected prefectures. The drop in tangible fixed assets in the manufacturing sector in the other three prefectures over the 2-year period amounted to about 27.141 trillion yen. The resulting impact on production across the entire manufacturing sector was 1.011 trillion yen, a reduction of about 66.1 billion yen over the direct production damage from Sect. 3. 4.2.1. Meanwhile, tangible fixed assets in the steel/metals and automotive/automotive parts industries increased by 6.598 trillion yen and 59.1 trillion yen, respectively, over the 2 years, which led to respective increases of 241.6 billion yen and 193.0 billion yen, respectively, in production by the two industries. The electronics industry, on the other hand, witnessed a tangible fixed assets decline by 19.150 trillion yen over the 2 years, leading to a reduction in production value by 783.0 billion yen. This amount accounted for 77% of the total reduction in production value.

3.4.3.2 Production Loss from the Change in Tangible Fixed Assets

We now consider the multiplier effects on each industry (sector of economic activity) in Fukushima Prefecture, the other three affected prefectures, and the disaster-unaffected region using the changes in production value stemming from increases and decreases in tangible fixed assets in the manufacturing sector pre- and post-disaster. As shown in Table 3.7, the production losses across all industries in Fukushima and the other three disaster-affected prefectures were 1.131 trillion yen and 2.933 trillion yen, respectively. These figures represent 7.1% and 5.6% of the 2005 production value for the production activities sector (in SAM terms, receipts), respectively. Comparing these results with those in Sect. 3.4.2.2, Fukushima witnessed a further increase in production loss of 193.5 billion yen, whereas the other three disaster-affected prefectures experienced a reduction in production loss of 199.7 billion yen. The reduced production loss in the other three disaster-affected

Table 3.7 Production loss caused by the change in tangible fixed assets on the disaster-affected region

Unit: billion yen	Fukushima prefecture					The other three disater-affected prefectures					The disaster-unaffected region	
	The amount of damage through the negative linkage effect (X = A + B)	Change % from the production value of 2005	Intra-regional negative linkage effect = A	Inter-regional negative linkage effect = B	Ratio of Intra-regional negative linkage effect = A/X	The amount of damage through the negative linkage effect (X = A + B)	Change % from the production value of 2005	Intra-regional negative linkage effect = A	Inter-regional negative linkage effect = B	Ratio of Intra-regional negative linkage effect = A/X	The amount of damage through the negative linkage effect	Change % from the production value of 2005
Agriculture	−12.7	−4.6%	−6.6	−6.1	52.0%	−39.7	−3.6%	−16.2	−6.1	40.7%	−185	−1.8%
Forestry	−0.9	−3.8%	−0.1	−0.8	15.6%	−7.6	−7.5%	−5.5	−0.8	72.4%	−24	−1.8%
Fishery	−0.7	−3.6%	−0.3	−0.5	34.6%	−5.6	−3.1%	−2.2	−0.5	38.5%	−31	−1.8%
Mining	−3.1	−1.8%	0.1	−3.1	−1.8%	−44.1	−5.4%	−29.1	−3.1	65.9%	−234	−1.8%
Foods and tobacco	−89.6	−9.7%	−69.3	−20.3	77.3%	−131.7	−4.0%	−64.3	−20.3	48.8%	−639	−1.7%
Non-durable goods	−2.1	−0.6%	8.4	−10.5	−393.8%	−203.4	−16.0%	−178.2	−10.5	87.6%	−377	−1.8%
Petroleum products and chemicals	−103.4	−15.7%	−85.5	−17.9	82.7%	−408.8	−9.4%	−322.5	−17.9	78.9%	−973	−1.7%
Steel and metals	12.3	1.6%	25.4	−13.1	207.1%	171.3	3.7%	240.1	−13.1	140.2%	−731	−1.4%
General machinery	−43.4	−8.8%	−33.0	−10.4	76.1%	−144.5	−5.8%	−109.7	−10.4	75.9%	−500	−1.4%
Electronic parts and devices	−262.6	−19.5%	−234.4	−28.2	89.3%	−1,007.4	−45.8%	−941.9	−28.2	93.5%	−673	−1.8%
Motor vehicle and Motor vehicle parts	−62.0	−16.5%	−59.0	−3.1	95.0%	211.2	24.2%	219.7	−3.1	104.1%	−357	−0.8%
Other durable goods	−58.4	−14.6%	−50.1	−8.3	85.7%	−208.5	−22.6%	−195.4	−8.3	93.7%	−268	−1.3%
Miscellaneous manufacturing goods	−29.0	−9.6%	−21.1	−7.9	72.8%	−32.1	−3.9%	−17.6	−7.9	54.8%	−271	−1.6%
Construction	−70.4	−7.6%	−44.3	−26.1	63.0%	−137.6	−4.1%	−70.8	−26.1	51.5%	−1,026	−1.7%

Electricity power,gas and water supply	−48.6	−2.6%	−16.1	−32.5	33.2%	−63.6	−4.4%	−33.1	−32.5	52.1%	−439	−1.6%
Commerce	−50.3	−4.9%	−27.9	−22.4	55.4%	−127.1	−3.3%	−45.6	−22.4	35.9%	−1,581	−1.6%
Finance and insurance, Real estiate	−74.4	−5.4%	−46.4	−28.0	62.4%	−204.3	−4.0%	−108.4	−28.0	53.0%	−1,892	−1.6%
Transport and Communication	−34.3	−4.7%	−19.2	−15.0	56.1%	−82.8	−3.5%	−38.1	−15.0	46.0%	−764	−1.5%
Pulic services	−92.4	−4.3%	−50.8	−41.5	55.0%	−235.3	−3.5%	−111.3	−41.5	47.3%	−2,083	−1.7%
Services	−104.7	−6.1%	−68.0	−36.7	65.0%	−231.0	−3.5%	−97.6	−36.7	42.3%	−2,880	−1.7%
Total industries	−1,130.7	−7.1%	−798.3	−332.4	70.6%	−2,932.5	−5.6%	−1,927.5	−1,005.0	65.7%	−15,928	−1.6%

prefectures is most likely attributable, at least in part, to support subsidies of 515.5 billion yen for setting up locations vis-à-vis small- and medium-sized companies in the manufacturing and tertiary industries under the fiscal year 2011 and 2012 recovery budgets. We may conclude that while Fukushima Prefecture continues to be heavily impacted by the nuclear accident and notwithstanding differences in recovery from tsunami damage among the three other disaster-affected prefectures, the overall process of recovery is underway. In Fukushima Prefecture, the production loss from multiplier effects within and outside the prefecture was 798.3 billion yen, up by 219.2 billion yen compared with that in Sect. 3.4.2.2. In contrast, the multiplier effects from the disaster-unaffected region were 25.6 billion yen greater than in Sect. 3.4.2.2, with a loss of no more than 332.4 billion yen. Thus, we see a persistent negative impact in Fukushima but positive multiplier effects elsewhere. On the other hand, production loss from multiplier effects from within and outside the other three affected prefectures dropped by 218.0 billion yen compared with that in Sect. 3.4.2.2, limiting the loss to 1.928 trillion yen. However, multiplier effects from the disaster-unaffected region decreased about around 1.0 trillion yen, but the production loss increased by 18.4 billion yen over that in Sect. 3.4.2.2. By industry, whereas multiplier effects in the automotive industry increased 211.2 billion yen, there effects in electronics decreased 1.0 trillion yen accounting for 35% of the whole. Finally, the production loss in disaster-unaffected region was estimated at 15.928 trillion yen, of which the effect from construction/tertiary industries accounted for 67%. This, however, represented a loss 1.076 trillion lesser than that in Sect. 3.4.2.2.

3.5 Concluding Remarks

This chapter used a two-region and a three-regional SAM to analyze the multiplier effects caused by the decrease in economic activity (which was, in turn, induced by the direct external shocks resulting from the earthquake and tsunami).

First, we estimated the amount of production damage to agriculture and fisheries in the disaster-affected region immediately following the earthquake to be approximately 75.6 billion yen and 151.7 billion yen, respectively. The production loss—the magnitude of the negative effect on production values in all industries on the basis of the abovementioned production damage—was estimated at 655.7 billion yen in the disaster-affected region and at 2.69 trillion yen in the disaster-unaffected region. In addition, we also estimated the amount of production damage 2 years after the earthquake, when reconstruction and recovery works had restored production to some extent. After 2 years, agriculture recovered by 31.5% to reach a damage of 51.8 billion yen and fisheries recovered by 56.2%, with the decline shrinking by approximately 66.4 billion yen. The production overall loss was estimated at 344.3 billion yen. Compared with the production loss immediately following the disaster, this loss shrank by 311.4 billion yen; thus, the cost-effectiveness of the 451.1 billion yen support for recovery as part of the fiscal measures was confirmed.

Second, we estimated that the direct production damages to the manufacturing sector caused by the losses in manufacturing capital stock were at 331.7 billion yen from the nuclear accident in Fukushima Prefecture and 1.078 trillion yen from tsunami flooding for the other three disaster-affected prefectures. As a result, the production losses in Fukushima Prefecture and the other three disaster-affected prefectures were 937.2 billion and 3.132 trillion yen, respectively, and reached 5.9% and 6.0% of gross production value, respectively, in the economic activity sector in 2005. In addition, using the changes in production value stemming from increases and decreases in tangible fixed assets in the manufacturing sector pre- and post-disaster, the production losses in Fukushima and the other three disaster-affected prefectures were 1.131 trillion and 2.933 trillion yen, respectively. Comparing these above results, Fukushima saw a further increase in production loss of 193.5 billion yen, whereas the other three disaster-affected prefectures witnessed a reduction in production loss of 199.7 billion yen. The reduced production loss in the other three disaster-affected prefectures is most likely attributable, at least in part, to the 515.5 billion yen support subsidies.

Lastly, we should mention three limitations of our study. First, our estimations are based on multiplier effects on the assumption that the halt in production activities caused by the damage persists for 1 year. In addition, as mentioned in the section on the SAM multiplier analysis, this framework is lacking in the function of price mechanisms. As a result, it cannot be denied that production losses are pointed out in the overestimation. Second, our estimation of indirect damage depends on the speed of the recovery from the direct damage in the disaster-affected region. However, we surmise that the speed of recovery is faster in companies influenced by indirect damage because of the supply networks to the disaster-unaffected region; moreover, companies in the disaster-unaffected region will be faster to recover. Third, we estimated the effect of recovery support as the difference of production loss between periods immediately and 1 or 2 years after the earthquake. Tokunaga and Okiyama (eds) (2014) suggested that the production loss in the disaster-affected region could be eliminated owing to the fiscal measures in the reconstruction budgets for fiscal years 2011 and 2012 using the same methods and the same SAM employed in this study.

In any case, as the SAM multiplier analysis is limited in its application, we will overcome the above three limitations using the regional CGE model. For example, the second limitation will be overcome by offsetting up elasticity of substitution between final goods of different origins in the constant elasticity of substitution (CES) function. Please refer to Chap. 4 for the analysis on disrupted supply chains. The first and third limitations will be overcome by using the dynamic regional CGE model and including the loss of capital stock in all industries and the reconstruction budget for the intensive reconstruction period. Please refer to Chap. 5 for the analysis on fiscal measures for reconstructing the Tohoku region.

References

Hisatake M, Yamasaki K (2006) Development of a multi-region CGE model that incorporates monopolistic competition and other factors. RIETI Discussion Paper Series, 06-J-046 (in Japanese)

Ito H (2008) Interregional SAM model and structure path analysis focusing on institution sector. J Commer Sci Kwansei Gakuin Univ 56(1):33–70. (in Japanese)

Makino Y (1995) Features of the Japanese economy depicted by SAM. Input-Output Analysis-Innovation I-O Technique 7(3):62–80. (in Japanese)

Miller RE, Blair PD (2009) Input-output analysis: foundations and extensions, second edn. Cambridge University Press, Cambridge

Ministry of Agriculture, Forestry and Fisheries (2011a) Estimation of damage to cultivation acreage by the Great East Japan Earthquake (tsunami)(in Japanese).http://www.maff.go.jp/jtokei/saigai/pdf/shinsai.pdf

Ministry of Agriculture, Forestry and Fisheries (2011b) The current resuming situation about suffering and management of fisherman by the Great East Japan Earthquake(in Japanese). http://www.maff.go.jp/j/tokei/census/ fc/2008/pdf/jyokyo.html

Ministry of Agriculture, Forestry and Fisheries (2015) Survey result 2014 about management condition of farmer and fisherman in tsunami-disaster area by the Great East Japan Earthquake (in Japanese). http://www/maff.go.jp/j/tokei/keikou/pdf/fukkou-20150911.pdf

Ministry of Economic, Trade and Industry (2013) Census of Manufacture 2010. http://www.meti.go.jp/english/statistics/tyo/kougyo/index.html

Miyagi T (1998) A spatial computable general equilibrium approach for measuring multiregional impacts of large scale transportation projects. In: Lumdqvist L, Mallsson LG, Kim TJ (eds) Network infrastructure and the urban environment. Springer, Berlin, pp 224–241

Nidaira K (2000) A multiplier analysis of regional social accounting matrix (SAM) model for Indonesia. Stud Reg Sci 30(3):57–69. (in Japanese)

Pyatt G, Roe A (1977) Social accounting for development planning with special reference to Sri Lanka. Cambridge University Press, Cambridge

Pyatt G, Round JI (1979) Accounting and fixed price multipliers in a social accounting matrix framework. Econ J 89(356):850–873

Pyatt G, Round JI (2006) Multiplier effects and the reduction of poverty. In: de Janvry A, Kanbur R (eds) Poverty, inequality and development: essays in honor of Erik Thorbecke. Springer, New York, pp 233–259

Thorbecke E et al (1992) Adjustment and equity in Indonesia. OECD, Paris

Tokunaga S, Okiyama M (eds) (2014) Reconstruction the disaster-affected region of the great East Japan earthquake and recovery of the regional economy. Bunshindou, Japan. (in Japanese)

Chapter 4
Analysis of Supply Chain Disruptions from the Great East Japan Earthquake in the Automotive Industry and Electronic Parts/Devices

Suminori Tokunaga and Mitsuru Okiyama

Abstract This chapter constructs a two-region computable general equilibrium (2SCGE) model for the regions affected and not affected by the Great East Japan Earthquake of 2011. Using this model, we simulate negative supply shocks for both the upper-level sectors between regions and the lower-level sectors within the regions. Our results can be summarized as follows: (1) For negative supply shocks in the upper-level sectors across regions, production of automotive parts in the affected region could decline by as much as two times and still not have much of a negative effect on automotive assembly and production of automotive parts in other regions, if those parts are of commodity-grade. (2) For negative supply shocks in lower-level sectors within each region, to the extent that raw materials and intermediate goods produced by the manufacturing sector in the lower portion of the automotive production pyramid in the affected regions are difficult to source from non-affected regions; the more production in that industry drops, the greater is the negative effect on production of automotive parts and automobile production in the affected regions. From these results, we can derive the following implications: First, the automotive industry clusters in the affected region are inevitable, given the number of automotive parts. However, to construct a production pyramid structure, and from the standpoint of managing the risks from natural disasters, the pyramid must be completely formed while also being flexible across regions. This is another reason that it is preferable to avoid parts which can only be sourced from the same region and have even raw materials and intermediate goods be

S. Tokunaga (✉)
Faculty of Economics and Business Administration, Reitaku University, 2-1-1, Hikarigaoka, Kashiwa, Chiba 277-8686, Japan

University of Tsukuba, Tsukuba 305-8572, Japan
e-mail: tokunaga@reitaku-u.ac.jp

M. Okiyama
Reitaku Institute of Political Economics and Social Studies, Reitaku University, Kashiwa, Chiba 277-8686, Japan

S. Tokunaga, B.P. Resosudarmo (eds.), *Spatial Economic Modelling of Megathrust Earthquake in Japan*, New Frontiers in Regional Science: Asian Perspectives 11,
https://doi.org/10.1007/978-981-10-6493-7_4

commodity-grade and substitutable from other regions, supplied by manufacturers in the lower portion of the production pyramid in the same region. Second, in addition, industrial promotion measures are required to aid the formation of such automotive industry clusters. For this purpose, one should consider utilizing the fiscal measures currently being undertaken for the affected regions.

Keywords Great East Japan Earthquake • Supply chain disruptions • Disaster-affected region • Automotive industry and electronic parts/devices • A two region computable general equilibrium (2S CGE) model

4.1 Introduction

The Great East Japan Earthquake of 2011 had a massive economic impact on the four prefectures of Iwate, Miyagi, Fukushima, and Ibaraki (the "disaster affected regions"), and it also caused a negative supply shock in non-affected regions through supply chain disruptions. This impact was most prominent in the automotive industry. Renesas Electronics' Naka Plant located in Hitachi-Naka City in Ibaraki Prefecture produced microcontrollers for use in cars. The damage from the earthquake to this plant disrupted supplies of automotive parts such as microcontrollers used to control engines and chassis, forcing many domestic automotive assembly plants to shut down for a considerable period of time (the so-called "Renesas shock") as explained in Chap. 2. This situation also extended to overseas plants. Usually, the automotive assembly and auto parts industries have a pyramid structure with assembly companies positioned at the top. The structure is followed by primary and secondary parts manufacturers with parts becoming more commodity-grade toward the bottom. As shown on the left side of Fig. 4.1, this means that even if the supply of parts from region S (shown as a dotted line) were to be interrupted, parts from other regions can be substituted to avoid stopping production of the units and functional components positioned toward the top, let alone stopping the whole automotive production lines. However, it became clear that in the case of the Great East Japan Earthquake, one portion of the lower structure became a diamond-shaped production pyramid structure, as shown on the right side of Fig. 4.1. In other words, these were parts in the lower layers with low substitutability, which we shall call "core parts." Such parts were produced only in

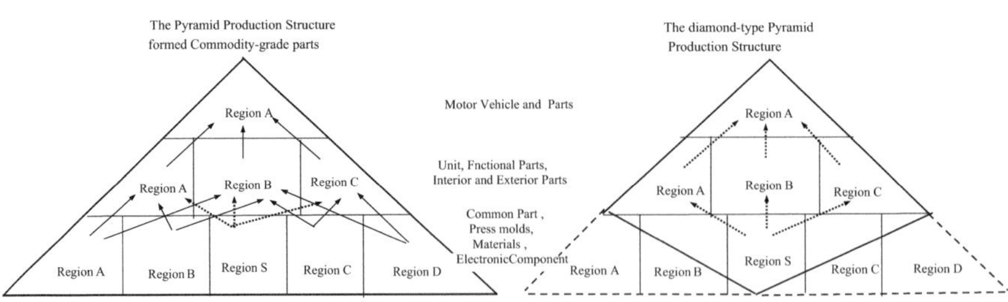

Fig. 4.1 The mechanism of a negative shock from "disrupted supply chains", Source: Tokunaga and Okiyama (eds) (2014, p. 93)

region S and hence could not be promptly supplied from other regions except region S. The destruction in region S made manufacturing there impossible, leading to an inability to produce automotive parts in the upper portion, for which the supply of parts had been coming from S, and eventually to a halt in automotive production lines in the top-most portion. Making the situation even more critical was the fact that because of economic globalization, the industrial structure was such that this caused a negative supply shock overseas too.

This chapter analyzes the effects of supply chain disruptions in both the disaster affected and non-disaster affected regions, focusing on the automotive industry and the electronics machinery industry, both of which are characterized by a wide range of supporting industries. To that end, we carry out simulations of the negative supply shock using a two-region computable general equilibrium (2SCGE) model for the affected and non-affected regions. In the following section, based on indices of industrial production for the affected regions before and after the disaster, we validate whether a negative supply shock was apparent outside the affected regions. In Sect. 4.3, we measure the impact of this phenomenon using the 2SCGE model. Finally, we present our conclusions and policy implications.

4.2 Supply Chain Disruptions from the Great East Japan Earthquake

4.2.1 The Nature of Supply Chain Disruption and Recovery in the Automotive Industry

Using the Great East Japan Earthquake as a case study, we now introduce previous research into supply chains. Tokui et al. (2012, 2015) examined the economic impact of supply chain disruptions caused by the earthquake using regional input–output tables. Hamaguchi (2012) studied the characteristics of the supply chain and the impact of the disaster based on a survey of manufacturing facilities located in areas affected by the earthquake. In addition, Saito (2012) and Todo et al. (2013) analyzed the nature of corporate networks in the supply chain, while Fujimoto (2011) and Otsuka and Ichikawa (2011) particularly assessed the supply chain in the automotive industry. In addition to the automotive industry, Nemoto (2012) discussed the reconstruction of local economies, including supply chains in the logistics and fishery industries.

This section examines the nature of supply chain disruptions in the automotive industry based on the indices of industrial production for Japan as a whole and for the four affected prefectures. Table 4.1 is based on these indices. We see that production in all manufacturing sectors in the four affected prefectures in the 3 months immediately following the quake dropped 25.5% compared to pre-quake levels (July 2010 through February 2011). The automotive industry (transportation equipment) fell 29.9%, a full ten points more than the electronics

Table 4.1 Change in amount of industrial products in the affected and other regions

Change amount of Average production value from September 2010 through February 2011 (unit: Billion yen, %)	From March 2011 through June 2011 (the 3 months period after the earthquake)	Change % from pre-the earthquake level	From July 2011 through September 2011 (from 3 months through 6 months period)	Change % from pre-the earthquake level	From October 2011 through January 2012 (from 6 months through 1 year period)	Change % from pre-the earthquake level	From February 2012 through January 2013 (from 1 year through 2 years period)	Change % from pre-the earthquake level
The stricken regions	−5,745	−25.5	−2,255	−10.0	−1348	−6.0	−502	−2.2
Foods and tobacco	−744	−22.3	−342	−10.2	−137	−4.1	−7	−0.2
Non-durable goods	−372	−32.2	−181	−15.7	−135	−11.7	−78	−6.8
Petroleum products and chemicals	−2,058	−44.2	−998	−21.4	−882	−18.9	−297	−6.4
Steel and metals	−1,155	−26.4	−364	−8.3	−195	−4.4	−172	−3.9
General machinery	−263	−9.7	228	8.4	319	11.8	157	5.8
Electronic parts and devices	−540	−19.2	−396	−14.0	−309	−11.0	−281	−10.0
Motor vehicle and motor vehicle parts	−342	−29.9	−43	−3.7	9	0.7	214	18.8
Other durable goods	−107	−7.6	−88	−6.3	−14	−1.0	−35	−2.5
Miscellaneous manufacturing goods	−164	−18.2	−71	−7.9	−4	−0.5	−3	−0.3

Other region	−16,587	−6.1	−22	0.0	1,958	0.7	−7,258	−2.7
Foods and tobacco	80	0.2	303	0.9	209	0.6	−172	−0.5
Non-durable goods	−20	−0.2	−192	−1.5	39	0.3	−446	−3.6
Petroleum products and chemicals	257	0.5	368	0.8	−196	−0.4	−836	−1.7
Steel and metals	−1,060	−2.5	−1,157	−2.7	−964	−2.3	−1,059	−2.5
General machinery	328	1.2	1,913	6.8	1,911	6.8	−1,384	−4.9
Electronic parts and devices	−3,554	−11.5	−3,482	−11.3	−5,550	−18.0	−6,429	−20.9
Motor vehicle and motor vehicle parts	−13,803	−32.7	−472	−1.1	4,177	9.9	1,536	3.6
Other durable goods	1,095	4.6	2,794	11.7	2,368	9.9	1,678	7.0
Miscellaneous manufacturing goods	90	0.8	−97	−0.9	−37	−0.3	−146	−1.3
Whole country	−22,332	−7.6	−2,277	−0.8	611	0.2	−7,760	−2.6
Foods and tobacco	−664	−1.8	−39	−0.1	72	0.2	−179	−0.5
Non-durable goods	−392	−2.9	−373	−2.7	−96	−0.7	−524	−3.9
Petroleum products and chemicals	−1,801	−3.4	−630	−1.2	−1,078	−2.0	−1,134	−2.1

(continued)

Table 4.1 (continued)

Change amount of Average production value from September 2010 through February 2011 (unit: Billion yen, %)	From March 2011 through June 2011 (the 3 months period after the earthquake)	Change % from pre-the earthquake level	From July 2011 through September 2011 (from 3 months through 6 months period)	Change % from pre-the earthquake level	From October 2011 through January 2012 (from 6 months through 1 year period)	Change % from pre-the earthquake level	From February 2012 through January 2013 (from 1 year through 2 years period)	Change % from pre-the earthquake level
Steel and metals	−2,216	−4.7	−1,521	−3.2	−1,158	−2.5	−1,230	−2.6
General machinery	65	0.2	2,141	6.9	2,230	7.2	−1,227	4.0
Electronic parts and devices	4,094	−12.2	−3,878	−11.5	−5,859	−17.4	−6,710	−19.9
Motor vehicle and motor vehicle parts	−14,145	−32.6	−515	−1.2	4,186	9.6	1,751	4.0
Other durable goods	989	3.9	2,706	10.7	2,354	9.3	1,643	6.5
Miscellaneous manufacturing goods	−74	−0.6	−168	−1.4	−41	−0.3	−148	−1.2

Source: Original database created based on index of industrial production published by METI and the statistics department of each prefecture government

equipment, which was down 19.2%. However, this fall in production was not substantially greater than in other sectors. Automotive production in other regions was down by 32.7%, a huge 21 points more than electronics; this decline was notably worse than in other sectors. This is the supply chain disruption, demonstrating the so-called negative supply shock. The degree of this negative supply shock is likely to be greater because of the high degree of concentration in Japan's automotive manufacturing industry, both within and across sectors (see Tokunaga et al. 2014).

In the subsequent period, as mentioned earlier, the post-quake response by automotive manufacturers brought the industry back up to a tiny 1.1% of pre-quake levels in the period from July to September 2011. In the affected regions as well, automotive industry production recovered to within 3.7% of pre-quake levels, which was better than the average 10% decline in those regions. In the subsequent one-year period starting in October 2011, the automotive industry in the affected regions bounced back to pre-quake levels, and in other regions moved into positive territory with a 9.9% gain. This trend continued through fiscal 2012 with the automotive industry in the affected regions experiencing a double-digit increase of 18.8%, even as the condition of the other sectors deteriorated. The automotive industry in other regions showed a 3.6% increase. On the other hand, the electronics equipment industry in the affected regions stood in stark contrast to the automotive industry. There were still no signs of recovery 6 months after the quake, and production in other regions too continued to deteriorate steadily.

The total for the four affected prefectures seems to indicate a rapid recovery in the automotive industry. However, production indices for the individual prefectures show that not all necessarily experienced a similar recovery. From Table 4.2, we see that it was only in Iwate that the automotive industry production recovered, growing to more than 60% over the pre-quake levels between 1 year and 2 years after the quake. The other three prefectures remained below pre-quake levels, with Miyagi remaining 7% down and Fukushima and Ibaraki 20% down, with their recovery lagging behind that of the electronics equipment industry. This may be attributed to the disparity between Iwate, where automotive assembly plants producing final goods are located, and Fukushima and Ibaraki, where automotive parts plants producing intermediate goods are located. Thus, we can infer that it was not the restoration and reconstruction of the automotive parts plants in the affected areas, which had been providing intermediate goods that led to the recovery in the automotive industry in other regions. The reason is that the automotive industry plants in the non-affected regions could have substituted supplies of automobile parts from non-affected areas, if the automotive industry plants in the affected regions produce commodity-grade products. After learning the lessons of supply chain disruptions, automobile manufacturers in the non-affected regions tried to switch to supplies from diversified production locations as quickly as possible. Another reason is that with the assembly plants in Iwate increasing the production of the final goods made under the domestic market revitalization plan based on eco-car subsidies, they could look for more suppliers of intermediate goods not only within the affected regions but also in the non-affected regions.

Table 4.2 Index of industrial product by industries in the affected and other regions

Index of Manufacturing and Mining production (2005 = 100)	Transportation equipment (motor vehicles and parts)					Electronic components/device				
	Average monthly index from 2010/7 to 2011/2	Average monthly index from 2011/3 to 2011/6	Average monthly index from 2011/7 to 2011/9	Average monthly index from 2011/10 to 2012/1	Average monthly index from 2012/2 to 2013/1	Average monthly index from 2010/7 to 2011/2	Average monthly index from 2011/3 to 2011/6	Average monthly index from 2011/7 to 2011/9	Average monthly index from 2011/10 to 2012/1	Average monthly index from 2012/2 to 2013/1
Iwate prefecture	86.7	61.9	82.3	86.6	69.7	117.3	96.7	113.0	114.9	105.9
Miyagi prefecture	117.2	76.8	118.5	123.9	195.5	97.5	84.2	69.9	65.3	69.0
Fukushima prefecture	94.7	75.4	88.2	85.8	88.5	93.8	68.4	66.7	76.6	99.7
Ibaraki prefecture	78.3	57.1	69.5	75.6	63.9	46.4	35.9	36.6	42.3	40.9
The whole country	91.6	61.7	90.5	100.4	95.3	112.6	98.9	99.6	93.0	90.1

Source: The same as Table 4.1

4.2.2 The Automotive Industry Production Pyramid in Affected Areas

Subsequently, we consider the input and output structure by focusing on the motor vehicles and motor vehicle parts industry using the 2005 inter-regional input–output table of the four affected prefectures and of other regions. Table 4.3 represents the input structure of motor vehicles and motor vehicle parts industry in the three affected prefectures, Ibaraki prefecture, and other regions. As of 2005, the intermediate inputs in other regions was valued at 35.4 trillion yen. Specifically, the value of motor vehicle parts bought from the three affected prefectures and Ibaraki prefecture was 333.1 billion yen and 116.9 billion yen, respectively. In addition, the electronic components and devices that contain the microprocessors were bought from Ibaraki prefecture for 7.9 billion yen. Thus, the value of intermediate goods in other regions of the motor vehicles and motor vehicle parts industry imported from

Table 4.3 Input structure of motor vehicles and motor vehicle parts industry

		Motor vehicles and parts		Motor vehicle parts		Motor vehicles and parts	
Billion Yen		The three stricken prefectures	Ratio	Ibaraki prefecture	Ratio	The other region	Ratio
The three stricken prefectures	Motor vehicle parts	112.2	14.7	3.6	1.8	333.1	0.9
	Electronic components/device	5.7	0.7	0.6	0.3	19.2	0.1
	Other industries	212.3	27.7	3.0	1.5	157.8	0.4
Ibaraki prefecture	Motor vehicle parts	1.2	0.2	9.8	4.8	116.9	0.3
	Electronic components/device	0.5	0.1	4.6	2.3	7.9	0.0
	Other industries	17.3	2.3	70.5	34.5	264.5	0.7
The other region	Motor vehicle parts	282.5	36.9	52.1	25.5	20,849.6	58.9
	Electronic components/device	4.0	0.5	9.2	4.5	675.9	1.9
	Other industries	130.0	17.0	50.8	24.9	12,972.4	36.6
Intermediate input		765.7	100	204.2	100	35,397.2	100
Production		919.7		265.6		44,173.0	

Source: Original database created based on a 2005 inter-regional input–output table of all prefectures created jointly by Ishikawa, Y. and the Mitsubishi Research Institute

the four affected prefectures was 899.2 billion yen, which represented a mere 2.5% of total intermediate input. The electronic components and devices made in Ibaraki prefecture accounted for less than 0.1% of among them.

On the other hand, as seen from Table 4.4, the ratio of motor vehicles and motor vehicle parts exported from a sector in one region to the same sector in another region was high compared with electronic components and devices and food and beverages. The vehicle parts produced by the affected regions were supplied in greater quantities to the automotive assembly industry in other regions. Country-wide, the automobile industry's pyramid structure was relatively robust, and this pyramid structure was found in regions where there were assembly plants. However, the vehicle parts industry in the region that was not a mass producer like the

Table 4.4 The yield rate for the same sector

		To Motor vehicles and parts in each region			
		The three stricken prefectures	Ibaraki prefecture	The other region	All region
From motor vehicle parts in each region	The three stricken prefectures	18.2	0.6	54.0	72.8
	Ibaraki prefecture	0.4	3.7	44.0	48.1
	The other region	1.1	0.2	78.5	79.8
		To electronic components/device in each region			
		The three stricken prefectures	Ibaraki prefecture	The other region	All region
From electronic components/device in each region	The three stricken prefectures	20.8	0.3	20.8	41.9
	Ibaraki prefecture	7.5	10.0	18.2	35.7
	The other region	1.3	0.4	32.4	34.0
		To beverages and food in each region			
		The three stricken prefectures	Ibaraki prefecture	The other region	All region
From beverages and food in each region	The three stricken prefectures	7.4	0.3	5.9	13.6
	Ibaraki prefecture	0.7	7.6	9.4	17.8
	The other region	0.2	0.5	14.3	15.1

Source: The same as Table 4.3

three affected regions and did not have an assembly plant as in the Ibaraki prefecture, supplied to assembly plants in the region with the pyramid structure rather than to assembly plants in their own region. The negative supply shock from "disrupted supply chains" appear to have been caused by the fact that the part is a core component and is supplied only from that region.

Based on our results, we will use the model explained in Chap. 2 to confirm the economic impact of the supply chain disruption caused by the Great East Japan Earthquake and also demonstrate the differences in economic spillover effects that occur in the event of such large-scale disasters when measures are in place to prevent such supply chain disruptions.

4.3 Measurement of Supply Chain Disruption

4.3.1 Two Simulations of Negative Supply Shocks

We shall now first measure the impact of the decrease in production of automotive parts and electronic parts and devices in the affected areas on the automotive and electronics industries in other regions. We will call this Simulation I or the "same-sector impact simulation." Second, we will measure the impact engendered by the decrease in production of raw materials and intermediate goods in the affected regions on the automotive parts and automobile manufacturing in the affected regions. We will call this Simulation II or the "cross-sector impact simulation." Figure 4.2 illustrates the relationship between the two simulations, taking the automotive industry as an example (see Fujita and Thisse 2013, Sect. 8.6; Tokunaga et al. 2014).

We will first describe Simulation I (same-sector impact simulation). We will measure the degree of negative supply shock affecting automotive and electronics production in other regions in the case of a decrease in production of automotive and electronic parts in a particular region owing to a natural disaster such as the 2011 earthquake. In addition, our measurement depends on whether the automotive parts and electronic parts are core to the automotive assembly and production of automotive parts and electronic devices in other regions or are commodity-grade parts substitutable from other regions. Specifically, we treat variables of automobile and automotive/electronic part production volumes in affected regions that are the endogenous variables in the 2SCGE model as exogenous variables. We also set up four cases where production volumes of each product decrease by 10%, 20%, 40%, and 60%.[1] The first is the case of a core product that is an almost non-substitutable

[1]Since exogenizing production volumes of automobiles and automotive parts and electronic parts within the region leaves an extra production function for that sector, we align the number of variables and equations by making the efficiency parmater, which is treated as an exogenous variable within that function, an endogenous variable.

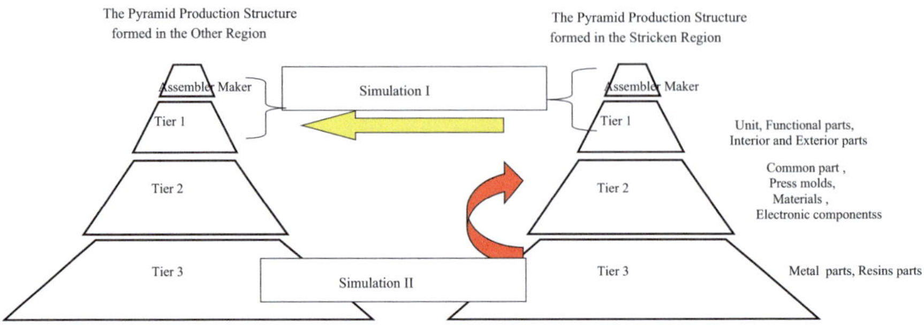

Fig. 4.2 Simulation contents using the 2SCGE model, Source: Tokunaga and Okiyama (eds) (2014, p. 106)

product (with the CES function's inter-regional substitution elasticity σ for deriving the integrated intermediate input in Table 4.5 set to 0.1). The second is the case of a semi-core product with low substitutability (σ = 0.5), the third is the case of a semi-commodity-grade, somewhat substitutable product (σ = 1.3), and the fourth is a commodity-grade, substitutable product (σ = 2.0). We use a 2SCGE model to measure a 4 × 4 matrix of these to find the ripple effect on the regional economics of the affected as well as the non-affected regions for each of the cases.

Next, in Simulation II (cross-sector impact simulation), for automobile and automotive parts production in a particular region, we focus on the automotive production pyramid and measure—in the case of a natural disaster which reduces production in the manufacturing industries located in the bottom portion of the production pyramid constructed in that region—the degree of negative supply shock on the automotive industry in the region in question. Our measurement depends on whether the products being supplied are core products or are commodity-grade products substitutable from other regions. Specifically, we treat as exogenous variables the production volumes for other manufacturing and mining sectors and electronic parts and devices in the affected regions, which are all the endogenous variables in the 2SCGE model. We also set up three cases where production volumes for the product declines by 10%, 20%, and 40%, and use the 2SCGE model to carry out the measurement.

4.3.2 Simulation I (Same-Sector Impact Simulation): Results

We will first present the results of Simulation I for the automobile and automotive parts industry, as shown in Tables 4.5 and 4.6. According to Section 2 (1), the earthquake caused an average 29.9% drop in automobile and automotive parts production in the affected regions in the 3 months following the disaster, and a 32.7% drop in other regions. Based on this and looking at the simulation results for Simulation I-B and Simulation I-C in Table 4.5, assuming these were core parts,

Table 4.5 Result 1 of simulation I (same-sector impact simulation) for the automotive industry

The production volume in the motor vehicles and motor vehicle parts industry of the other region		Substitution elasticities between commodities of different origin	The production volume in the motor vehicles and motor vehicle parts industry of the stricken region			
			Simulation I-A	Simulation I-B	Simulation I-C	Simulation I-D
			10% drop	20% drop	40% drop	60% drop
Changes in base value (2005) (%)	Core parts	$\sigma_m = 0.1$	▲ 0.830	▲ 2.209	▲ 5.375	▲ 9.469
	Semi-core parts	$\sigma_m = 0.5$	▲ 0.739	▲ 1.588	▲ 3.714	▲ 6.524
	Semi-commodity-grade parts	$\sigma_m = 1.3$	▲ 0.438	▲ 0.690	▲ 2.032	▲ 3.455
	Commodity-grade parts	$\sigma_m = 2.0$	▲ 0.286	▲ 0.591	▲ 1.279	▲ 2.129

Table 4.6 Result 2 of simulation I (same- sector impact simulation) for the automotive industry

Changes in base value (2005) (%)	Substitution elasticities between commodities of different origin	The production volume in the motor vehicles and motor vehicle parts industry of the stricken region							
		Simulation I-A: 10% drop		Simulation I-B: 20% drop		Simulation I-C: 40% drop		Simulation I-D: 60% drop	
		Stricken region	Other region	Stricken region	Other region	Stricken region	Other region	Stricken region	Other region
Equivalent variation (billion yen)	$\sigma_m = 0.1$	▲ 46.0	▲ 1.1	▲ 97.8	▲ 8.1	▲ 225.7	▲ 42.5	▲ 404.3	▲ 90.5
	$\sigma_m = 0.5$	▲ 38.8	0.8	▲ 81.2	▲ 2.5	▲ 182.9	▲ 23.2	▲ 324.3	▲ 59.8
	$\sigma_m = 1.3$	▲ 30.4	2.7	▲ 62.7	3.3	▲ 137.4	▲ 3.8	▲ 238.3	▲ 22.4
	$\sigma_m = 2.0$	▲ 26.1	3.7	▲ 53.6	5.9	▲ 115.9	4.7	▲ 199.0	▲ 4.9
Real GRP	$\sigma_m = 0.1$	▲ 0.330	▲ 0.002	▲ 0.713	▲ 0.005	▲ 1.703	▲ 0.017	▲ 2.976	▲ 0.037
	$\sigma_m = 0.5$	▲ 0.270	▲ 0.001	▲ 0.572	▲ 0.004	▲ 1.299	▲ 0.014	▲ 2.227	▲ 0.033
	$\sigma_m = 1.3$	▲ 0.201	0.000	▲ 0.415	▲ 0.002	▲ 0.896	▲ 0.009	▲ 1.474	▲ 0.025
	$\sigma_m = 2.0$	▲ 0.166	0.000	▲ 0.340	▲ 0.001	▲ 0.716	▲ 0.006	▲ 1.154	▲ 0.019
Total output	$\sigma_m = 0.1$	▲ 0.413	▲ 0.022	▲ 0.862	▲ 0.050	▲ 1.869	▲ 0.124	▲ 2.891	▲ 0.219
	$\sigma_m = 0.5$	▲ 3.570	▲ 0.016	▲ 0.732	▲ 0.035	▲ 1.544	▲ 0.084	▲ 2.392	▲ 0.149
	$\sigma_m = 1.3$	▲ 0.293	▲ 0.009	▲ 0.591	▲ 0.018	▲ 1.208	▲ 0.043	▲ 1.846	▲ 0.075
	$\sigma_m = 2.0$	▲ 0.261	▲ 0.005	▲ 0.523	▲ 0.011	▲ 1.055	▲ 0.024	▲ 1.600	▲ 0.043
Utility level for the household	$\sigma_m = 0.1$	▲ 0.256	▲ 0.001	▲ 0.520	▲ 0.006	▲ 1.118	▲ 0.027	▲ 1.971	▲ 0.054
	$\sigma_m = 0.5$	▲ 0.221	0.000	▲ 0.443	▲ 0.002	▲ 0.918	▲ 0.016	▲ 1.535	▲ 0.040
	$\sigma_m = 1.3$	▲ 0.180	0.001	▲ 0.358	0.001	▲ 0.718	▲ 0.005	▲ 1.128	▲ 0.021
	$\sigma_m = 2.0$	▲ 0.159	0.002	▲ 0.316	0.003	▲ 0.628	0.000	▲ 0.965	▲ 0.010
Unemployed person	$\sigma_m = 0.1$	1.841	0.001	3.751	0.023	8.128	0.136	14.574	0.287
	$\sigma_m = 0.5$	1.586	▲ 0.005	3.192	0.005	6.639	0.079	11.248	0.213
	$\sigma_m = 1.3$	1.290	▲ 0.011	2.571	▲ 0.013	5.171	0.017	8.177	0.104
	$\sigma_m = 2.0$	1.138	▲ 0.014	2.264	▲ 0.022	4.509	▲ 0.011	6.959	0.046

production volumes in other regions declined by 2.209% and 5.375%, respectively. The 2SCGE model estimates a reduction of around 3.8% as the impact on production volumes in the automotive industry in other regions caused by the negative supply shock that accompanied the disaster; however, the actual reduction was nearly ten times this number. The reason for this phenomenon is the huge impact of the supply of a single core part, out of a total of 20,000–30,000 total parts, being stopped by the earthquake. In the 2SCGE model, even Simulation I-D, in which the production of core parts in the affected areas goes down by nearly 60%, shows an impact on production volumes for automobiles and automotive parts of no more than 9.469%. This fact also tells us that the special factors mentioned above acted to reduce production in other regions by more than 30%. On the other hand, we also have a fascinating result from Simulation I, seen in Table 4.5: Even if production in affected regions declined by 60%, production volumes in other regions would decline by a mere 2.129% as long as those parts were commodity-grade. We also see that this result is about the same as in the simulation of the case in which production volumes were down 20% in affected areas and the parts were core ones. In other words, even if production declines in affected areas are tripled, as long as the degree of part substitutability goes up three ranks, the negative impact on production in other regions is nearly the same. Similarly, if production declines in affected areas double, as long as the degree of part substitutability goes up two ranks, the negative impact on production in other regions is nearly the same, as suggested by the matrix in Table 4.5. Even if production in the affected areas suffers greatly, as was the case with the earthquake, as long as the parts are highly commodity-grade and their supply can be switched to regions other than the affected ones, the impact of the negative supply shock can be largely avoided. Based on the results of this kind of simulation, supply of parts from other regions is feasible only if they become more commodity -grade. Moreover, if a supply of core parts must be procured from those other regions, then it is necessary to build mechanisms that can ensure multiple sources of supply.

Next, we consider the ripple effects caused by such differences in the commodity-grade automotive parts on economic welfare, local economies, and employment environments in the affected and other regions. If high commodity-grade automotive parts are being produced in the affected regions, then no matter how much production goes down, the decline in economic welfare (equivalent variation) in the affected regions can be cut to half of what it would be with core parts. In addition, the economic welfare in other regions remains positive. Based on the simulation results, even if automotive parts production in the affected areas declines by 40%, as long as the parts are highly commodity-grade, economic welfare in other regions increases by 4.7 billion yen, and the number of the unemployed decreases by 0.011%. In contrast, if the parts are core parts, we have the result that economic welfare in other regions decreases by 42.5 billion yen and the number of the unemployed goes up by 0.136%. It is critical that one create conditions such that commodity-grade parts can be procured across regions. This is because even if production of automotive parts is halted in a particular region owing to a natural disaster, one can mitigate the impact of the negative supply shock on automotive production in other regions, thereby holding the effect on the local economy in that region to a minimum.

Next, we discuss the results of Simulation I for electronics parts and devices, as shown in Tables 4.7 and 4.8. According to Section 2 (1), the earthquake caused an average 19.2% drop in electronic equipment production in the affected regions in the 3 months following the disaster, and an 11.5% drop in other regions. However, 3 months after the disaster and even 1 year after it, production in both sets of regions remained more than 10% below the pre-quake levels. We may conjecture that underlying this continued decline in production is the strong effect of factors such as offshoring of manufacturing sites and the decline in international competitiveness, rather than the impact of the catastrophe.

Let us now compare this with the automotive parts discussed to highlight the differences in the structures of the two industries in the affected regions. From Tables 4.5 and 4.7, we compare the degree of impact of the fall in production in both sectors on the same sector in other regions. We then see that the impact of the fall in production in the affected regions on production in other regions is almost half for electronic parts than for automotive parts, and this is true no matter what degree these parts are commodity-grade. For example, if the automotive parts and electronic parts are both core parts, a 40% drop in the production of electronic parts in the affected regions has the same impact on production in other regions as a 20% drop in the production of automotive parts. This tells us that even a large drop in production in electronics in a particular region will have a lower impact on the electronics sector in other regions than for the automotive sector. This is because the production structure of the electronics industry is less hierarchical than that of the automotive industry.

This is also indicated by the differences in ripple effects on economic welfare, local economies, and the employment environment in other regions as shown in Tables 4.6 and 4.8. Table 4.8 shows that regardless of the percentage decline in electronic parts in the affected regions, as the parts become more commodity-grade, the decline in economic welfare in the affected regions worsens, the local economies and gross regional product plummet, and the employment environment deteriorates as well. On the other hand, in other regions, economic welfare grows faster, the local economies and gross regional product expand, and the employment environment improves further. Compared with the case of automotive parts in Table 4.6, the affected regions show the same trend, while other regions show exactly the opposite trend. With electronics parts as well, the more core parts there are, if production in the affected regions deteriorates, the negative impact on economic welfare and the local economy is greater, similar to the situation with automotive parts. However, regarding the impact on economic welfare and the local economies of non-affected regions, as mentioned above, owing to differences in the degree of impact of the negative supply shock for automotive parts and electronic parts, automotive parts have a negative impact, whereas electronic parts have a positive ripple effect.

Table 4.7 Result 1 of simulation I (same- sector impact simulation) for electronic parts/devices

The production volume in the electronic components/device industry of the other region		Substitution elasticities between r commodities of different origin	The production volume in the electronic components/device industry of the stricken region			
			Simulation I-A	Simulation I-B	Simulation I-C	Simulation I-D
			10% drop	20% drop	40% drop	60% drop
Changes in base value (2005) (%)	Core parts	$\sigma_m = 0.1$	▲ 0.544	▲ 1.153	▲ 2.490	▲ 4.007
	Semi-core parts	$\sigma_m = 0.5$	▲ 0.454	▲ 0.945	▲ 2.043	▲ 3.299
	Semi-commodity-grade parts	$\sigma_m = 1.3$	▲ 0.291	▲ 0.605	▲ 1.314	▲ 2.140
	Commodity-grade parts	$\sigma_m = 2.0$	▲ 0.177	▲ 0.371	▲ 0.812	▲ 1.345

Table 4.8 Result 2 of simulation I (same-sector impact simulation) for electronic parts/devices

Changes in base value (2005) (%)	Substitution elasticities between commodities of different origin	The production volume in the electronic components/device industry of the stricken region							
		Simulation I-A: 10% drop		Simulation I-B: 20% drop		Simulation I-C: 40% drop		Simulation I-D: 60% drop	
		Stricken region	Other region	Stricken region	Other region	Stricken region	Other region	Stricken region	Other region
Equivalent variation (billion yen)	$\sigma_e = 0.1$	▲ 111.8	64.7	▲ 229.1	131.4	▲ 486.1	274.0	▲ 787.8	438.2
	$\sigma_e = 0.5$	▲ 105.2	60.8	▲ 215.1	123.2	▲ 454.7	255.5	▲ 735.7	406.5
	$\sigma_e = 1.3$	▲ 94.3	54.5	▲ 192.2	109.8	▲ 403.6	225.6	▲ 650.2	355.5
	$\sigma_e = 2.0$	▲ 86.8	50.0	▲ 176.4	100.6	▲ 368.5	205.3	▲ 591.5	321.3
Real GRP	$\sigma_e = 0.1$	▲ 0.595	0.008	▲ 1.225	0.016	▲ 2.601	0.030	▲ 4.152	0.041
	$\sigma_e = 0.5$	▲ 0.557	0.008	▲ 1.144	0.015	▲ 2.416	0.028	▲ 3.843	0.037
	$\sigma_e = 1.3$	▲ 0.495	0.007	▲ 1.011	0.013	▲ 2.116	0.024	▲ 3.345	0.031
	$\sigma_e = 2.0$	▲ 0.452	0.006	▲ 0.919	0.012	▲ 1.913	0.021	▲ 3.010	0.026
Total output	$\sigma_e = 0.1$	▲ 0.747	0.013	▲ 1.495	0.026	▲ 2.970	0.053	▲ 4.345	0.083
	$\sigma_e = 0.5$	▲ 0.710	0.013	▲ 1.419	0.025	▲ 2.816	0.052	▲ 4.124	0.081
	$\sigma_e = 1.3$	▲ 0.650	0.012	▲ 1.295	0.025	▲ 2.564	0.051	▲ 3.758	0.078
	$\sigma_e = 2.0$	▲ 0.607	0.012	▲ 1.209	0.025	▲ 2.389	0.050	▲ 3.505	0.076
Utility level for the household	$\sigma_e = 0.1$	▲ 0.881	0.039	▲ 1.793	0.078	▲ 3.743	0.161	▲ 5.953	0.257
	$\sigma_e = 0.5$	▲ 0.830	0.036	▲ 1.686	0.073	▲ 3.504	0.151	▲ 5.556	0.238
	$\sigma_e = 1.3$	▲ 0.747	0.033	▲ 1.511	0.066	▲ 3.116	0.133	▲ 4.910	0.207
	$\sigma_e = 2.0$	▲ 0.689	0.030	▲ 1.390	0.060	▲ 2.850	0.121	▲ 4.468	0.187
Unemployed person	$\sigma_e = 0.1$	6.435	▲ 0.237	13.159	▲ 0.478	27.795	▲ 0.989	44.837	▲ 1.568
	$\sigma_e = 0.5$	6.056	▲ 0.222	12.352	▲ 0.448	25.960	▲ 0.920	41.719	▲ 1.449
	$\sigma_e = 1.3$	5.436	▲ 0.199	11.040	▲ 0.399	22.994	▲ 0.810	36.662	▲ 1.258
	$\sigma_e = 2.0$	5.007	▲ 0.183	10.136	▲ 0.366	20.971	▲ 0.735	33.227	▲ 1.130

4.3.3 Results of Simulation II (Cross-Sector Impact Simulation)

Based on the results in the previous section, let us now examine the production pyramid for the automotive industry, which is relatively more hierarchical than that for the electronics industry. We then explore to what extent the production of automotive parts and automobiles, positioned in the upper portion of the pyramid, is affected by the commodity-grade of products in other manufacturing and mining sectors and electronic parts and devices, which are positioned in the lower portion of the pyramid.

We may conclude that the decline in production in other manufacturing and mining sectors in the affected regions does not have enough of a negative supply shock impact to decrease automotive industry production in those regions. In particular, Table 4.9 demonstrates that where commodity-grade automotive parts can be easily sourced across regions ($\sigma_m = 2.0$), to the extent that they are supplied by core parts from other manufacturing and mining sectors in the same region, there is no negative supply impact associated with the decline in production in other manufacturing and the mining sector. For example, in the case of Simulation II-B, we see that if the other manufacturing and mining sector products are core parts, then production volumes of automotive parts in the affected areas rises by 4.570%, a rise of 4.3 points compared with the 0.256% rise under circumstances where automotive parts cannot be easily sourced across regions ($\sigma_m = 0.1$). This is because production of automotive parts in the affected regions rises because of an increase (5.845%) in exports of commodity-grade automotive parts to other regions. Then the question is, why is it that the production of automotive parts in the affected areas increases in spite of the decrease in production of other manufacturing and mining sector production (20%), which makes sourcing impossible in the end, and the inability to substitute with imports from other regions owing to the core nature of the parts? The answer is that other manufacturing and mining sector products are substituted with foreign exports (increase of 9.404%). In this case, the negative impact on economic welfare and the local economies is 1093.6 billion yen and 7.640%, which is relatively greater than for other cases. Based on this fact, as long as both the automotive parts and other manufacturing and mining sector products being produced are commodity-grade in nature, even if a natural disaster strikes the region, reducing production in other manufacturing and the mining sectors, automotive part production can be maintained and increased by bringing in other manufacturing and mining sector products from non-affected regions and without depending on foreign imports. It also becomes possible to minimize the negative impact on economic welfare and the local economies of the region in question. For example, if all the products in Simulation II-B are high commodity -grade ($\sigma_o, \sigma_m, \sigma_e = 2.0$), a reduction of even 20% in other manufacturing and mining sector production volumes in the affected areas would result in production volumes and imports of automotive parts from the region of 3.703% and 3.854%, respectively. In addition, while production volumes in the affected areas

Table 4.9 Result of simulation II (cross-sector impact simulation) for other manufacturing and mining sectors

Changes in base value (2005) (%)	Region	Substitution elasticities between commodities of different origin	The production volume in the other industries of manufacturing and mining of the stricken region							
			Simulation II-A: 10% drop			Simulation II-B: 20% drop			Simulation II-C: 40% drop	
			$\sigma_o = 0.1$	$\sigma_o = 2.0$	$\sigma_o = 0.1$	$\sigma_o = 0.5$	$\sigma_o = 1.3$	$\sigma_o = 2.0$	$\sigma_o = 1.3$	$\sigma_o = 2.0$
The production volume in the motor vehicles and motor vehicle parts industry	Stricken region	$\sigma_m = 0.1$ $\sigma_e = 0.1$	0.178	0.402	0.256	0.393	0.615	0.767	0.892	1.295
		$\sigma_m = 2.0$ $\sigma_e = 2.0$	2.313	1.888	4.570	4.343	3.967	3.703	7.309	6.844
	Other region	$\sigma_m = 0.1$ $\sigma_e = 0.1$	0.266	0.167	0.513	0.462	0.377	0.316	0.662	0.545
		$\sigma_m = 2.0$ $\sigma_e = 2.0$	0.196	0.118	0.372	0.332	0.266	0.219	0.438	0.353
The export volume to other region in the motor vehicles and motor vehicle parts industry	Stricken region	$\sigma_m = 0.1$ $\sigma_e = 0.1$	0.888	0.544	1.775	1.587	1.276	1.058	2.424	1.961
		$\sigma_m = 2.0$ $\sigma_e = 2.0$	2.931	1.969	5.854	5.326	4.456	3.845	8.415	7.154
The production volume in the other industries of manufacturing and mining	Other region	$\sigma_m = 0.1$ $\sigma_e = 0.1$	▲ 0.173	0.126	▲ 0.362	▲ 0.196	0.077	0.269	0.204	0.625
		$\sigma_m = 2.0$ $\sigma_e = 2.0$	▲ 0.166	0.113	▲ 0.350	▲ 0.183	0.092	0.283	0.235	0.656
The import volume from other region in the other industries of manufacturing and mining	Stricken region	$\sigma_m = 0.1$ $\sigma_e = 0.1$	▲ 11.191	▲ 6.007	▲ 22.609	▲ 19.846	▲ 15.204	▲ 11.887	▲ 30.363	▲ 23.228
		$\sigma_m = 2.0$ $\sigma_e = 2.0$	▲ 11.105	▲ 5.923	▲ 22.350	▲ 19.582	▲ 14.949	▲ 11.650	▲ 29.494	▲ 22.454
The import volume from foreign countries in the other industries of manufacturing and mining	Stricken region	$\sigma_m = 0.1$ $\sigma_e = 0.1$	3.826	0.448	8.923	6.941	3.686	1.408	11.202	5.566
		$\sigma_m = 2.0$ $\sigma_e = 2.0$	3.954	0.501	9.404	7.320	3.928	1.582	12.256	6.261
The production volume in the electronic components/device industry	Stricken region	$\sigma_m = 0.1$ $\sigma_e = 0.1$	4.126	3.124	8.778	8.160	7.155	6.460	15.787	13.922
		$\sigma_m = 2.0$ $\sigma_e = 2.0$	7.104	5.159	15.620	14.318	12.248	10.850	28.186	24.137
Equivalent variation (billion yen)	Stricken region	$\sigma_m = 0.1$ $\sigma_e = 0.1$	▲ 546.1	▲ 381.4	▲ 1,128.2	▲ 1,036.3	▲ 883.9	▲ 776.4	▲ 1,865.0	▲ 1,622.3
	Stricken region	$\sigma_m = 2.0$ $\sigma_e = 2.0$	▲ 532.1	▲ 371.2	▲ 1,093.6	▲ 1,004.2	▲ 856.3	▲ 752.2	▲ 1,789.5	▲ 1,558.5

Real GRP	Stricken region	$\sigma_m = 0.1$ $\sigma_e = 0.1$	▲ 3.758	▲ 2.731	▲ 7.813	▲ 7.205	▲ 6.215	▲ 5.530	▲ 12.975	▲ 11.346
	Stricken region	$\sigma_m = 2.0$ $\sigma_e = 2.0$	▲ 3.688	▲ 2.679	▲ 7.640	▲ 7.044	▲ 6.076	▲ 5.408	▲ 12.600	▲ 11.031
Total output	Stricken region	$\sigma_m = 0.1$ $\sigma_e = 0.1$	▲ 4.402	▲ 3.393	▲ 8.955	▲ 8.374	▲ 7.424	▲ 6.765	▲ 14.918	▲ 13.414
	Stricken region	$\sigma_m = 2.0$ $\sigma_e = 2.0$	▲ 4.245	▲ 3.280	▲ 8.587	▲ 8.037	▲ 7.139	▲ 6.516	▲ 14.199	▲ 12.815
Unemployed person	Stricken region	$\sigma_m = 0.1$ $\sigma_e = 0.1$	27.871	18.889	57.954	52.823	44.367	38.447	93.905	80.158
	Stricken region	$\sigma_m = 2.0$ $\sigma_e = 2.0$	27.064	18.307	55.878	50.914	42.746	37.036	89.228	76.264

are down by 20%, the decline in imports of other manufacturing and mining sector products shrinks to 11.65% with foreign imports correspondingly held to a 1.582% increase. The negative impact on economic welfare and on the local economics of the affected region is 7.552 billion yen, i.e., a decline of 5.408%.

In addition, in the automotive industry production pyramid, the same result as for automotive parts is obtained for the impact on the production of electronic parts and devices which are above the other manufacturing and mining sectors in the pyramid. In other words, it is critical that commodity-grade electronic parts and devices are produced across regions is a way similar to products from other manufacturing and mining sectors. Even if a natural disaster should strike that region, it would not be affected by a negative supply shock although the other manufacturing sector and mining production in that region may shrink.

As with the other manufacturing and mining sectors, a decline in the production of electronic parts and devices in the disaster areas, as shown by Table 4.10, will not be impacted by a negative supply shock to the extent of decreasing automotive industry production in those areas. For example, we see in the case of Simulation II-B that when it is easy to source automotive parts in the affected areas and non-affected areas ($\sigma_m = 2.0$) for core electronic parts and devices, production volumes of automotive parts in the affected areas rose by 2.654%. This signifies 1.6 points more than 1.048% seen in non-affected regions when it is not easy to source automotive parts ($\sigma_m = 0.1$). Nevertheless, in this case, analogous to other manufacturing sectors and mining, the negative effect on economic welfare and on the local economy is relatively larger than in other cases. This means that if both automotive parts and electronic parts and devices being produced in the region are commodity-grade, the negative effect on economic welfare and on the local economy of the region can be held to a minimum. In other words, in Simulation II-B, if both products are commodity-grade across regions (or are core parts), even with a 20% reduction in the production volumes of electronic parts and devices in the affected region, production volumes and export volumes of automotive parts in the region would see an increase of 2.124% (1.048%) and an increase of 1.914% (0.779%), and imports of electronic parts and devices from non-affected regions and overseas would see a decrease of 7.123% (11.336%) and an increase of 15.53% (30.158%). The negative impact on economic welfare and on the local economy of the affected regions would be a reduction of 176.4 billion yen (217.5 billion yen) and a reduction of 0.919% (1.219%).

In any case, when products of the other manufacturing and mining sectors and of the electronics industry are supplied to the automotive industry in the same region, there is a dependence not only on the commodity-grade automotive parts in that region but also on the versatility of products from the other manufacturing and mining sectors and the electronics industry in that region. To the extent that they are highly commodity-grade, there would be no impact from a negative supply shock on the automotive parts and assembly at the top of the pyramid even if production from the other manufacturing and mining sectors and the electronics industry at the bottom of the automotive industry production pyramid were to be cut by a natural

Table 4.10 Result of simulation II (cross-sector impact simulation) for electronic parts/devices

Changes in base value (2005) (%)	Region	Substitution elasticities between commodities of different origin	The production volume in the electronic components/device industry of the stricken region							
			Simulation II-A: 10% drop		Simulation II-B: 20% drop				Simulation II-C: 40% drop	
			$\sigma_e = 0.1$	$\sigma_e = 2.0$	$\sigma_e = 0.1$	$\sigma_e = 0.5$	$\sigma_e = 1.3$	$\sigma_e = 2.0$	$\sigma_e = 1.3$	$\sigma_e = 2.0$
The production volume in the motor vehicles and motor vehicle parts industry	Stricken region	$\sigma_m = 0.1$ $\sigma_e = 0.1$	0.506	0.429	1.048	1.001	0.926	0.875	1.972	1.843
		$\sigma_m = 2.0$ $\sigma_e = 2.0$	1.285	1.041	2.654	2.511	2.281	2.124	4.844	4.463
	Other region	$\sigma_m = 0.1$ $\sigma_e = 0.1$	0.226	0.159	0.466	0.428	0.366	0.323	0.777	0.675
		$\sigma_m = 2.0$ $\sigma_e = 2.0$	0.205	0.143	0.423	0.388	0.330	0.291	0.695	0.607
The export volume to other region in the motor vehicles and motor vehicle parts industry	Stricken region	$\sigma_m = 0.1$ $\sigma_e = 0.1$	0.379	0.276	0.779	0.720	0.626	0.560	1.308	1.162
		$\sigma_m = 2.0$ σe = 2.0	1.225	0.945	2.514	2.353	2.093	1.914	4.373	3.965
The import volume from other region in the electronic components/device industry	Stricken region	$\sigma_m = 0.1$ $\sigma_e = 0.1$	▲ 5.645	▲ 3.361	▲ 11.336	▲ 10.153	▲ 8.186	▲ 6.797	▲ 16.770	▲ 14.006
		$\sigma_m = 2.0$ $\sigma_e = 2.0$	▲ 5.847	▲ 3.536	▲ 11.721	▲ 10.523	▲ 8.530	▲ 7.123	▲ 17.352	▲ 14.549
Equivalent variation (billion yen)	Stricken region stricken region	$\sigma_m = 0.1$ $\sigma_e = 0.1$	▲ 104.3	▲ 81.6	▲ 217.5	▲ 203.9	▲ 181.9	▲ 166.6	▲ 385.9	▲ 351.6
		$\sigma_m = 2.0$ $\sigma_e = 2.0$	▲ 111.8	▲ 86.8	▲ 229.1	▲ 215.1	▲ 192.2	▲ 176.4	▲ 403.6	▲ 368.5
Real GRP	Stricken region	$\sigma_m = 0.1$ $\sigma_e = 0.1$	▲ 0.591	▲ 0.447	▲ 1.219	▲ 1.137	▲ 1.004	▲ 0.912	▲ 2.110	▲ 1.905
	Stricken region	$\sigma_m = 2.0$ $\sigma_e = 2.0$	▲ 0.595	▲ 0.452	▲ 1.225	▲ 1.144	▲ 1.011	▲ 0.919	▲ 2.116	▲ 1.913
Total output	Stricken region	$\sigma_m = 0.1$ $\sigma_e = 0.1$	▲ 0.811	▲ 0.656	▲ 1.631	▲ 1.546	▲ 1.408	▲ 1.312	▲ 2.816	▲ 2.617
	Stricken region	$\sigma_m = 2.0$ $\sigma_e = 2.0$	▲ 0.747	▲ 0.607	▲ 1.495	▲ 1.419	▲ 1.295	▲ 1.209	▲ 2.564	▲ 2.389
Unemployed person	Stricken region	$\sigma_m = 0.1$ $\sigma_e = 0.1$	5.976	4.631	12.283	11.516	10.272	9.416	21.610	19.607
	Stricken region	$\sigma_m = 2.0$ $\sigma_e = 2.0$	6.435	5.007	13.159	12.352	11.040	10.136	22.994	20.971

disaster. The simulation results above show that the negative impact on economic welfare and on the local economy of that region can be minimized.

4.4 Conclusions and Policy Implications

In this chapter, we constructed a two-region computable general equilibrium (2SCGE) model for disaster affected regions and non-disaster affected regions in the wake of the Great East Japan Earthquake. Using this model, we simulated negative supply shocks for both the upper-level sectors between regions and the lower-level sectors within regions. The results can be summarized as follows:

1. Negative supply shocks in the upper-level same sectors across regions:

   If production of automotive parts in the disaster affected region declines by as much as two times, a negative effect on productions of automotive assembly and automotive parts in non-disaster regions is the same grade, furthermore if those parts are highly commodity-grade.

2. Negative supply shocks in lower-level sectors within each region:

   To the extent that raw materials and intermediate goods produced by the manufacturing sector in the lower portion of the automotive production pyramid in the disaster affected regions are difficult to import from the non-affected regions, if production in that industry drops, the negative effect on the production of automotive parts and automobile production in the affected regions is greater.

From these results, we can derive the following implications from our two simulations: (a) Automotive industry clusters in the affected region are inevitable, given the number of automotive parts. However, to construct a production pyramid structure, and from the standpoint of managing the risks from natural disasters, the pyramid must be completely formed while also being flexible across regions. This is another reason it is preferable to avoid parts which can only be sourced from the same region and have even raw materials and intermediate goods be commodity-grade and substitutable from other regions, supplied by manufacturers in the lower portion of the production pyramid in the same region. (b) In addition, industrial promotion measures are required to aid in the formation of such automotive industry clusters. For this purpose, one should consider utilizing the fiscal measures currently being undertaken for the affected regions.

References

Fujimoto T (2011) Supply chain competitiveness and robustness: a lesson from the 2011 Tohoku earthquake and 'virtual dualization.' MRC(Tokyo University Manufacturing Management Research Center) Discussion Paper Series, 354 (in Japanese)

Fujita M, Thisse J (2013) Economics of agglomeration: cities, industrial location, and globalization, 2nd edn. Cambridge University Press, Cambridge

Hamagushi N (2012) Impact of the great East Japan earthquake on local industries: synthesis and analysis from a survey. RIETI Discussion Policy Discussion Paper Series, 13-P-001 (in Japanese)

Nemoto T (2012) Vitalizing local economies through supply chain restoration. Plan Adm 35 (2):13–16. (in Japanese)

Otsuka T, Ichikawa Y (2011) How do you evaluate Japan-type supply chain? A collection of theses of Mizuho Research Institute 3 (in Japanese)

Saito(Umeno) Y (2012) Impact of East Japan earthquake disaster on companies in non-affected area: the structure of inter-company network of supply chain and its implication. J JSS 42 (1):135–144. (in Japanese)

Todo Y, Nakajima K, Matous P (2013) How do supply chain networks affect the resilience of firms to natural disasters? Evidence from the great East Japan earthquake. RIETI Policy Discussion Paper Series, 13-E-028

Tokui J, Arai N, Kawasaki K, Miyagawa T, Fukao K, Arai S, Edamura K, Kodama N, Noguchi N (2012) Economic impacts of the great East-Japan earthquake: comparison with the past great disasters, supply chain disruption, and effect of supply constraints in electric utility. RIETI Policy Discussion Paper Series, 12-P-004 (in Japanese)

Tokui J, Kawasaki K, Miyagawa T (2015) The economic impact of supply chain disruptions from the great East Japan Earthquake. RIETI Discussion Paper Series, 15-E-094

Tokunaga S, Okiyama M (eds) (2014) Reconstruction the disaster-affected region of the great East Japan earthquake and recovery of the regional economy. Bunshindou, Japan. (in Japanese)

Tokunaga S, Kageyama M, Akune Y, Nakamura R (2014) Empirical analyses of agglomeration economies in Japanese assembly-type manufacturing industry for 1985–2000: using agglomeration and coagglomeration indices. Rev Urban Reg Dev Stud 26(1):57–79

Part II
Reconstruction and Regional Renewal After the Great East Japan Earthquake

Chapter 5
Economic Analysis of Fiscal Measures for Reconstructing the Tohoku Region After the Great East Japan Earthquake: Using a Dynamic Two-Regional CGE Model

Mitsuru Okiyama and Suminori Tokunaga

Abstract This chapter measures the economic impact of the reconstruction budget and illustrates the transition of the regional economies in the disaster-affected region amid a declining population using a recursive dynamic regional CGE model. Our findings are as follows: Firstly, even if the Great East Japan Earthquake had not occurred, the disaster-affected region would have experienced weaker growth in its sub-regional economies owing to the impact of a declining population. Secondly, the fiscal measures in the intensive reconstruction period contribute toward recovering the real Gross Regional Product (GRP) of the disaster-affected region, which slumped following the Great East Japan Earthquake. But, the real GRP of the disaster-affected region begins to move downward after 2018. However, the fiscal measures would slow down the pace of curtailment of the regional economies in the disaster-affected region compared with the scenario without fiscal measures. Third, it is advisable that the government should provide fiscal measures to the disaster-affected region after the intensive reconstruction period by adopting a different form of reconstruction budget, that is, by revising the distribution ratios of the local allocation tax grants. However, this measure alone would not allow the regional economy in the disaster-affected region to recover and exceed the level prior to the earthquake. In order to realize sustained economic growth, it is necessary to take other measures after the intensive reconstruction period, such as productivity improvements in each industry located within the disaster-affected region, using fiscal transfers.

M. Okiyama (✉)
Reitaku Institute of Political Economics and Social Studies, Reitaku University, Kashiwa, Chiba 277-8686, Japan
e-mail: okiyama.m@gmail.com

S. Tokunaga
Faculty of Economics and Business Administration, Reitaku University, 2-1-1, Hikarigaoka, Kashiwa, Chiba 277-8686, Japan

University of Tsukuba, Tsukuba 305-8572, Japan

S. Tokunaga, B.P. Resosudarmo (eds.), *Spatial Economic Modelling of Megathrust Earthquake in Japan*, New Frontiers in Regional Science: Asian Perspectives 11,
https://doi.org/10.1007/978-981-10-6493-7_5

Keywords Great East Japan Earthquake • Disaster-affected region • Fiscal measures • Intensive reconstruction period • Recursive dynamic regional CGE model

5.1 Introduction

Four years after the Great East Japan Earthquake, the Reconstruction Agency issued a document on March 10, 2015, entitled "Current Status of Reconstruction and Challenges" in which the reconstruction progress and major challenges were concisely organized and compiled. The document summary of the current status was as follows: "The removal of disaster debris has almost been completed and the restoration of public infrastructure remains on track, the planning of housing reconstruction has finished and the construction work has accelerated. And we have grappled with the recovery of industries and livelihoods for the revival of a vibrant hometown and with the health and mental support of the evacuees. Concerning the revitalization and reconstruction of Fukushima, measures for the evacuees should respond to their intensions to either return home, remain where they are, or start a new life elsewhere." The document also described the basic principles and major initiatives undertaken, and it addressed the challenges of the Tohoku region by implementing innovative ideas and creating a "New Tohoku," as the region will not be returned to the way it was before the disaster. Yet, we need to question why the Reconstruction Agency reported the ongoing measures and policies without providing an assessment of the reconstruction budget at this point in time when a period of four out of the five-year designated reconstruction period has already passed since the earthquake and tsunami disaster. In contrast to the document issued by the Reconstruction Agency, which lacked such an assessment and summary, the Board of Audit conducted an audit from the point of view of the regularity, efficiency, and effectiveness of project implementations for the recovery and reconstruction of the region after the great East Japan Earthquake. The report of the Board of Audit (2015) mentioned one of the Board's opinions, as follows: "As it continues to cost a lot to carry out the various projects relating to restoration and reconstruction, the state government needs to endeavor not only to operate each project efficiently and effectively but also to secure the necessary finances. The government needs to limit the increased cost burden shouldered by the public, even though the finances have already been secured by a special tax for reconstruction."

On the other hand, Yomiuri Shimbun published the results of a public opinion poll concerning the Great East Japan Earthquake in its morning edition of March 8, 2015. Yomiuri Shimbun summarized the results of the poll as follows: "It has become evident that people's interest in reconstruction from the disaster is gradually weakening. While many demand an extension of the incentive reconstruction period following the Great East Japan Earthquake, a number of people are reluctant about incurring any new burden. The key issue toward supporting the disaster-affected area going forward lies in how to gain an understanding of the people's views about the reconstruction budget and the allocation of the associated burden." One particular view to point out regarding the results of the opinion poll is that in response to the question, "Has the country's reconstruction budget been properly used?" only 10% answered, "Yes, the budget has been properly used," whereas a large group (88%) responded, "I don't think so." We can infer that this result was caused by negligence on the part of the Reconstruction Agency to provide a thorough explanation of the use of the reconstruction budget and its outcome for each fiscal year. At the Reconstruction Promotion Council meeting held in January 2013, the size and source of funding for the recovery and reconstruction work during the intensive reconstruction period (FY 2011 to FY 2015) were revised. According to the revised plan, the total business expenditure for the period was set at 23.5 trillion yen and approximately 45% of the funding would be sourced by a special tax for reconstruction (roughly 10.5 trillion yen), including a special reconstruction income tax collected over a period of 25 years. In addition, a review is also under way to examine the budget and the associated burden beyond the intensive five-year construction period. This response also suggests that the Japanese people will continue to share the future burden in some way. Therefore, we believe that it is extremely important to form some sort of consensus regarding the reconstruction activities and subsequent funding among not only the residents in the disaster-affected region but also the people in the entire country. In regard to this consensus formation, the points addressed by Professor Satoru Masuda of Tohoku University, as published in the Nikkei Newspaper's "*Keizai Kyoshitsu* (economics class)" column on March 2, 2015, can serve as a guideline. That is, Professor Masuda pointed out that the success and failure of reconstruction work, and the increase and decrease in population in the disaster-affected communities (including the natural population increase and decrease, in some cases) are interrelated, particularly in the case of the Great East Japan Earthquake. He explained the importance of drawing up some future scenarios by performing simulations based on certain preconditions and checking from a wide-range perspective as to which scenario the actual status is moving toward as reconstruction progresses and then making the necessary adjustments. In consideration of this point by Professor Masuda, we first identify where the regional economies in the disaster-affected region are located and what is expected of these economies going forward. Then, we present other possibilities when certain measures and financial support are provided to these economies. By following this approach, we believe it will be easier to gain a consensus and allow people to deepen their understanding of the reconstruction activities and associated budget-related policies.

Thus, this chapter has two objectives. The first is to measure the economic impact of the reconstruction budget for the recovery and construction activities in the intensive five-year reconstruction period on the disaster-affected region and to evaluate the reconstruction budget. The second is to illustrate the transition of the regional economies of the disaster-affected region amid a declining population trend on the basis of four scenarios and to analyze the impact of the reconstruction-related budget on the regional economies beyond the intensive reconstruction period.

In order to analyze these points, we use a two-region social accounting matrix (SAM) for 2005, which comprises the region representing the four prefectures affected by the disaster and a region that was not affected by the disaster, as our database. As for the computable general equilibrium (CGE) model, we use an expanded and improved the dynamic two regional computable general equilibrium (D2SCGE) model to reflect the incomplete employment condition and an aging society. We then outline our simulation results for equivalent variation, real GRP, intra-regional gross production, and production and investment volumes by production activity for the disaster and non-disaster regions for the period between 2010 and 2035 on the basis of the D2SCGE model. In addition, we present the transition of capital stock, the number of unemployed persons, and the amount of savings for the four prefectures located in the disaster zone.

This chapter proceeds as follows: Section 5.2 provides the assumptions for building the scenarios for the simulations. Section 5.3 explains the dynamic process of the D2SCGE model. Section 5.4 discusses the establishment of four scenarios and the results of the simulations based on these scenarios. Finally, Sect. 5.5 presents our conclusions and discusses the policy implications.

5.2 Assumptions for Setting Scenarios

5.2.1 Assumptions for a Depopulating Society

The population forecasts frequently used in surveys and papers regarding a depopulating society are "Population Projections for Japan (January 2012 projections)" published by the National Institute of Population and Social Security Research (IPSS) in Japan (2012). These projections use data from the 2010 Population Census for Japan and the Vital Statistics of Japan, and they are based on three variants of the national population to 2060 incorporating the trend of social mobility in population of the inter-regional. According to these projections, as shown in Fig. 5.1, Japan's population is expected to decline by approximately 20.7 million people during the thirty-year period from 2010 to 2040 and to become the proportion of the elderly at approximately 36% by 2040. The National Institute of Population and Social Security Research (IPSS) in Japan (2013) has also released "Population Projections for Japan by Region" (March 2013 projections), in which it projects the population of prefectures and municipalities by gender and age group to 2040 on the basis of national medium-variant fertility and mortality assumptions. Figure 5.2 illustrates the information gathered for the region comprising the four disaster-affected prefectures and the non-disaster region in this study on the basis of these regional population projections.[1] As can be seen from the

[1]The population forecast by the IPSS does not consider the movement of population caused by the Great East Japan Earthquake.

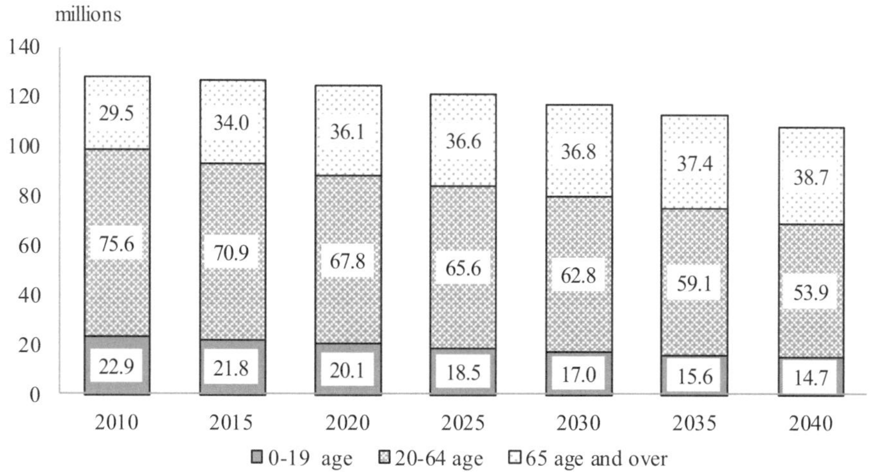

Fig. 5.1 Trends in the youth, working-age, and elderly populations in Japan (Source: Original data based on population projections for Japan (January 2012: IPSS))

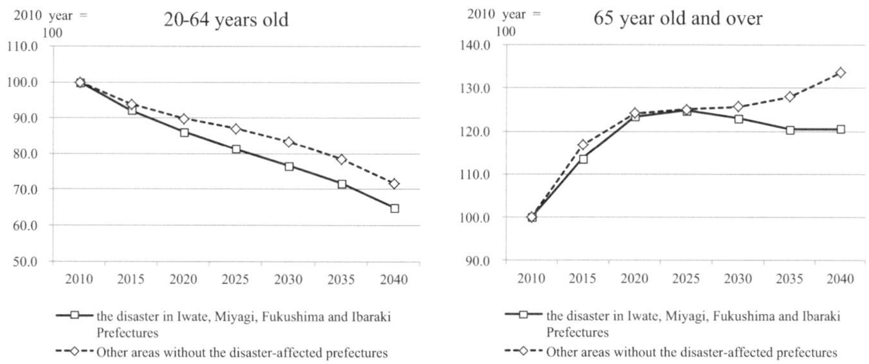

Fig. 5.2 Trends in population aged 20–64 and 65 years old and above for the two regions (Source: Original data based on population projections by regions (March 2013: IPSS))

trends for the population aged 20–64 and the population aged 65 and above by region (indexed to 2010 = 100), the former group is declining in both regions and the decline is becoming more pronounced year after year. This trend becomes particularly noticeable by 2040 in the region comprising the four disaster-affected prefectures, where the population aged 20–64 is projected to decline by 35% relative to 2010. And the population aged 65 and above in this region is projected to peak around 2025 before declining until 2035 and then stabilizing thereafter. On the other hand, although the population aged 65 and above in the non-disaster region is projected to rise, there is moderate growth between 2020 and 2030 that is then followed by a sharp increase.

This chapter uses the IPSS population projections until 2040 for each prefecture by gender and for five age groups spanning ages 20–64. The D2SCGE model incorporates the labor force population as given. We estimated this by taking the forecast labor force participation rate for each of the five age groups and then multiplying this rate by the total population. For the forecast labor force participation rate for each age group, we referenced the labor force participation rates for each of the five age groups and by gender through 2030 on the basis of the two projections—"the economic revival/progressive labor participation scenario" and the "zero growth/unchanged labor participation scenario"—given in the "Labor Supply and Demand Estimates" released by the Japan Institute for Labour Policy and Training (2014) in May 2014. This paper referred to the latter scenario, which features close to zero economic growth and labor force participation rates by gender and age group at the same levels as those in 2012.

5.2.2 The Estimated Loss of Capital Stock and the Reconstruction Budget

In regard to the formulation of the reconstruction budget, Saito (2015) states that the government officers considered that the estimated loss of capital stock was the important criterion for judging the reconstruction budget size and they used the data on the Hanshin–Awaji earthquake as a reference. In fact, the estimations of the direct damage (capital stock) from the Great East Japan Earthquake for the period up to 6 months following the earthquake were made by these respective parties: 16.9 trillion yen by the Cabinet Office (2011),[2] 17.8 trillion yen by Inada et al. (2011), 16.4 trillion yen by the Development Bank of Japan (2011), and 17.0 trillion yen by Hayashida et al. (2011). In response to these estimations, the Reconstruction Headquarters (in July 2011) set out a policy of securing a minimum budget worth approximately 19.0 trillion yen for the five-year intensive reconstruction period (2011–2015). Furthermore, as stated in the previous section, the budget was increased in January 2013 to 23.5 trillion yen. On the other hand, Saito (2015) also presents the results of the verification performed by the Cabinet Office on the estimations of the amount of damage for the building stock. According to these results, the total amount of damage in the three disaster-affected prefectures of Iwate, Miyagi, and Fukushima was estimated at 4 trillion yen, including the local municipalities that did not suffer from the tsunami, which represented about one-third of the total damage estimate by the Cabinet Office (2011).

In this chapter, we first estimate the amount of capital stock at the end of FY 2010 by industry using the data of private and public gross fixed capital formation for the four disaster-affected prefectures and the whole country in FY 2010, in addition to the

[2]Please refer to the Board of Audit (2015) concerning the results of the Cabinet Office's detailed estimations of the amount of damage to each institution and facility.

data for both the prefectural gross private capital stock of private enterprises and the social overhead capital stock in FY 2009, as estimated by the Cabinet Office. The results are shown in the two left-hand columns in Table 5.1. According to these results, the gross private and social overhead capital stocks in FY 2011 were 133.2 trillion yen and 58.1 trillion yen, respectively. Next, we estimate the capital stock that suffered large damages caused by the Great East Japan Earthquake in the four disaster-affected prefectures. Although Tokunaga and Okiyama (eds) (2014) estimated that the amount of damage on the capital stock for the four disaster-affected prefectures was 15.4 trillion yen, we again tried to estimate the amount of damage on the basis of the estimation results by Saito (2015) and the estimations released by each local government of the four disaster-affected prefectures. Using this approach, the amount of damage on the capital stock was estimated to be 9.4 trillion yen, a reduction of roughly 6 trillion yen. By comparing this amount with the total estimated value of capital stock, the loss rate for the industry located within the four prefectures was 4.9%, as shown in the right-hand column of Table 5.1. The figures in this table show that the loss rate was highest in the fisheries industry at 24.9%, followed by the foods and beverage industry at 7.8%, and the agriculture and forestry industry at 7.6%.

Table 5.1 The estimated loss of capital stock and loss ratio for each industry of the four disaster-affected prefectures

	Estimated amounts of gross capital stock (End of FY 2010)		The loss of capital stock by the Great East-Japan Earthquake	
Billon Yen, Base year = 2005 price	Private enterprises (A)	Social overhead (B)	Estimated amounts(C)	loss ratio immediately after Earthquake [= C/(A+B)]
Total	133,327	58,100	9,368.3	4.9%
Agriculture and forestry	7,596	0	579.2	7.6%
Fishery	1,232	0	307.2	24.9%
Foods and beverage	3,146	0	245.4	7.8%
Electrical devices and parts	1,332	0	69.8	5.2%
Motor vehicles and parts	1,614	0	62.6	3.9%
Other manufacturing products	27,719	0	1,363.7	4.9%
Construct	4,653	0	104.7	2.3%
Electricity,gas and heat-water supply	16,148	10,177	1,714.8	6.5%
Commerce	2,660	0	139.2	5.2%
Transport	7,938	0	180.1	2.3%
Financial and insurance, real estate, communcation, public adminstration, eduction and servics etc.	59,290	47,923	4,601.6	4.3%

Source: Estimated data based on statistics of the Cabinet Office

Table 5.2 provides the actual reconstruction budget for the intensive reconstruction period and the forward-looking budget based on the latest data by the Reconstruction Agency. The use of the reconstruction budget shown in Table 5.2 is based on reviewing the items disclosed. The figures in the table show that the largest amount of funding, worth 7.2 trillion yen, has been allocated to the restoration of public infrastructure and reconstruction grants, followed by 3.3 trillion yen each for the recovery and reconstruction from the nuclear disaster, the local allocation tax grants, and the reconstruction of industry. Together with other items, the total amount of expenditures budgeted for the intensive reconstruction period comes to 24.8 trillion yen. On the other hand, revenues disclosed up to FY 2013, excluding retained earnings from the previous fiscal year, amounted to 18.7 trillion yen, of which 76% is from the general account.

In this chapter, we evaluate the reconstruction budget to be allocated for purposes other than the following three items: "Removal and disposal of disaster waste," "Recovery and reconstruction from the nuclear disaster," and "Other expenditure." This is owing to the difficulty in incorporating these three items in the D2SCGE model. In particular, the analysis of the nuclear power disaster requires the preparation of three- regional SAM and separating the Fukushima prefecture from the disaster-affected region in order to accurately evaluate the reconstruction budget. In addition, it is not possible to properly reflect the fact that the nuclear disaster caused major changes in power transfers between the two regions in the D2SCGE model. We will re-examine these issues at a later date after we develop an inter-regional SAM for 2011 when the 2011 inter-regional input-output table for nine regions and the 2011 intra-prefectural input–output table for Fukushima, Miyagi, Iwate and Ibaraki prefectures are available. Therefore, although we consider the decrease in the power supply caused by the suspended operation of the Fukushima nuclear power station, this paper does not reflect the various effects of the nuclear disaster and the fiscal measures for Fukushima reconstruction.

Except for the three items described above, all other items allocated within the budget are incorporated into the D2SCGE model in accordance with the "Application to the D2SCGE model" column indicated in Table 5.2. Details will be explained in Sect. 5.4. Meanwhile, in regard to the source of funding for the expenditures relating to all items other than the above three mentioned items during the three-year period up to 2013, we recognize that the amount other than the tax revenues from the special reconstruction income tax and the special reconstruction corporation tax of 11.7 trillion yen was covered by issuing reconstruction government bonds. And we assume in the D2SCGE model that corporations in the non-disaster region purchased these government bonds, which the central government will redeem with no interest over a period of 25 years starting in 2014. We also assume that the source of funds for the redemption of government bonds and reconstruction in the remaining 2 years of the intensive reconstruction period will be the revenues from the special reconstruction income tax. As such, an increase of 0.14 percentage points in the D2SCGE model (derived through calibration)—the equivalent of a household income tax rate of 2.1% in the non-disaster region—will be in effect over a period of 25 years starting in 2012. Similarly, the corporate tax

Table 5.2 Actual and projected reconstruction budget balance for the intensive reconstruction period

Billion Yen		FY 2011	FY 2012	FY 2013	FY 2014	FY2015	Intensive reconstruction period
item	Application to the D2SCGE model	Settled expenditure	Settled expenditure	Settled expenditure	Budget	Budget	Accumlated expenditure
(1) Support for disaster-affected people	Expenditure for disater-affected local governments	1,224.4	352.3	139.8	111.7	128.7	1,956.9
(2) Removal and disposal of disaster waste		318.6	348.8	374.9	23.6	10.5	1,076.4
(3) Restoration of puiblic infrastructure and the Grant for Reconstruction	Savings for disater-affected local governments	936.5	2,249.4	1,426.9	1,305.4	1,337.4	7,255.6
(4) Rebuilding of Houses	Savings for disater-affected local governments	487.0	57.6	0.0	0.0	0.0	544.6
(5) Securing employment	Expenditure for disater-affected local governments	430.8	52.8	45.9	0.6	0.0	530.1
(6) Recovery and reconstruction from the nuclear disaster		977.5	369.0	580.5	660.0	780.7	3,367.7
(7) Local allocation tax grants	Expenditure for disater-affected local governments	2,140.8	670.4	577.1	0.0	0.0	3,388.3
(8) Reconstruction of industry							
(a) Support for Agriculture and fisheries industry	Subsidy to a primary industry	367.0	84.1	35.3	30.6	20.3	537.3
(b) Disaster-related loan	Savings for disater-affected local governments	1,305.6	153.1	125.2	22.1	30.7	1,636.7

(continued)

Table 5.2 (continued)

Billion Yen		FY 2011	FY 2012	FY 2013	FY 2014	FY2015	Intensive reconstruction period
item	Application to the D2SCGE model	Settled expenditure	Settled expenditure	Settled expenditure	Budget	Budget	Accumlated expenditure
(c) Support for small-and medium-sized companies and Creation of subsidies for the location of factories	Subsidy to Non-primary industry	387.0	128.5	228.8	58.1	80.5	882.9
(d) R & D and renewable sources of energy	Expenditure for disater-affected local governments	179.1	93.4	47.1	0.0	0.0	319.6
(9) Other expenditure		197.0	1,753.7	1,275.1	32.0	47.6	3,305.4
Total Expenditure		8,951.3	6,313.1	4,856.6	2,244.1	2,436.4	24,801.5
Total Expenditure except for (3), (6) and (9)		7,458.2	3,841.6	2,626.1	1,528.5	1,597.6	18,128.4
(A)Special-account for Reconstruction							
(1) Reconstruction bonds			2,303.2	0.0			
(2) From General -account		8,951.3	1,999.9	3,176.9			
(3) Tax revenues	(a) the special income tax for reconstruction		51.1	333.8			
	(b) the special corporate tax for reconstruction		649.3	1,204.3			
(B) Other revenues incluing excess cash reserve of previous FY			18.7	2,055.3			
Total Revenue			5,022.2	6,770.3			
Total Revenue except for (B)		8,951.3	5,003.5	4,715.0			

Source Original data based on materials from the Reconstruction Agency

rate in the non-disaster region will be increased by 0.69 percentage points in 2012 and 2013, which is equivalent to a 10% hike in the corporate tax rate, but the original rate will be restored in 2014 and thereafter.

5.2.3 The Recovery Phases for the Disaster-Affected and Non-disaster Regions

We can grasp the situation for the regional economic recovery in the disaster-affected region and the non-disaster region after the earthquake using both the National Accounts and the Prefectural Accounts of the four disaster-affected prefectures. Table 5.3 shows the real gross regional product (GRP) of FY 2011 and FY 2012 by production activity in the disaster-affected and non-disaster regions. According to Table 5.3, the real GRP of the four disaster-affected prefectures in FY 2011 fell to 99.3 from 100 in FY 2010, but in FY 2012 it recovered to 103.8. This is because the fiscal measures mentioned above were quite effective. In more detail, the rate of growth of the construction industry was 170.4 and it stood out compared with the other industries, which recovered to levels approximating or above those of FY 2010. On the contrary, the rates of growth of the fisheries industry, which suffered serious damage on its capital stock, and the electricity, gas, and heat-water supply industry, which was affected by the Fukushima nuclear power station accident, were sharply reduced to 67.6 and 65.0, respectively. Meanwhile, in terms of the effect to the regional economies in the non-disaster region, the real GDP in FY 2011 remained at the level of FY 2010 as a result of supply-chain disruptions; in particular, the rate of growth of the transportation equipment industry fell to 89.6. In addition, the rate of growth of the electricity, gas, and heat-water supply industry declined to 90.6 in FY 2011 and to 68.9 in FY 2012. According to the revised SNA figures of the Cabinet Office released on December 25, 2015, the GDP growth rate of Japan's economy (starting at 100 in 2010) stood at 99.5 in 2011, 101.3 in 2012, 102.7 in 2013, and 102.6 in 2014. In this paper, we have tried to adjust the efficiency parameter of production for each industry with the transition of the growth rate mentioned above in both regional economies under the D2SCGE model. This is because the new investment, which has increased in order to recover the previously existing capital stock by using the government's fiscal measures, is expected to contribute toward raising the productivity of each industry in the disaster-affected region. We will explain how to make the adjustments in Sect. 5.4.1 when we outline the establishment of each scenario.

Table 5.3 Actual real GRP after the Great East Japan earthquake

Billion yen ( ) is index of FY 2010 = 100	FY 2011						FY 2012					
	Whole country		The disaster in Iwate, Miyagi, Fukushima and Ibaraki Prefectures		Other areas without the disaster-affected prefectures		Whole country		The disaster in Iwate, Miyagi, Fukushima and Ibaraki Prefectures		Other areas without the disaster-affected prefectures	
Gross regional product	502,188	(99.7)	31,416	(99.3)	470,771	(99.7)	509,282	(101.1)	32,852	(103.8)	476,430	(100.9)
<production sector>												
Agriculture and forestry	5,543	(102.6)	598	(98.2)	4,944	(103.1)	5,558	(102.9)	582	(95.5)	4,976	(103.8)
Fisheries	730	(97.9)	48	(55.2)	682	(103.6)	748	(100.3)	59	(67.6)	689	(104.6)
Foods and beverage	11,865	(94.8)	998	(86.2)	10,866	(95.7)	12,349	(98.6)	1,212	(104.6)	11,137	(98.0)
Electrical machinery	28,301	(103.4)	2,111	(101.1)	26,190	(103.6)	27,153	(99.2)	2,126	(101.8)	25,028	(99.0)
Transportation equipment	12,887	(90.0)	383	(104.8)	12,504	(89.6)	14,200	(99.1)	385	(105.2)	13,815	(99.0)
Other manufacturing products and mining	54,162	(97.1)	3,798	(96.6)	50,363	(97.2)	55,936	(100.3)	3,892	(99.0)	52,044	(100.4)
Construct	25,823	(100.9)	2,270	(143.5)	23,553	(98.1)	26,233	(102.5)	2,695	(170.4)	23,537	(98.1)
Electricity,gas and heat-water supply	9,567	(86.9)	825	(60.6)	8,742	(90.6)	7,532	(68.4)	885	(65.0)	6,647	(68.9)
Commerce	65,943	(101.1)	3,293	(102.7)	62,650	(101.0)	67,461	(103.5)	3,373	(105.2)	64,088	(103.4)
Transport	23,407	(99.1)	1,206	(91.1)	22,201	(99.5)	24,128	(102.1)	1,337	(101.0)	22,791	(102.2)
Financial and insurance, real estate, communcation, public adminstration, eduction and servics etc.	263,962	(100.6)	15,886	(99.7)	248,076	(100.7)	267,985	(102.2)	16,307	(102.4)	251,678	(102.1)

Source: Original data based on SNA of the Cabinet Office

5.3 Dynamic Process of the Two-Regional Computable General Equilibrium Model

We constructed the recursive-dynamic two-regional computable general equilibrium (D2SCGE) model, this is, the two-regional computable general equilibrium (2SCGE) model with the addition of a recursive dynamic dimension.[3] Therefore, we explain the dynamic process in this section.[4]

First, the dynamic period in this study extends to 2035. We have taken 2009 as T = 0 and incrementally added a recursive dynamic dimension to the model for 26 periods (T = 26) until 2035. Then, the Great East Japan Earthquake, which occurred at the beginning of 2011, is T = 2. Since we have constructed the two regional SAM using the 2005 dataset, as noted above, the CGE model would normally use 2005 as the benchmark equilibrium year. In particular, since the simulation results using the CGE model are evaluated by the rate of change from the benchmark value, a simulation based on the benchmark year is preferred. Similarly, even in the dynamic CGE model, the value of capital stock in the initial period is calculated from the benchmark value noted below, with the simulation yielding the rate of change from the benchmark value for each period on the basis of a steady-state growth rate. Determination of the benchmark year is therefore important. Since 2005 is the most recent year for data at the present time, it should become the benchmark year when adding a recursive dynamic dimension to the model. However, even when adding a recursive dynamic dimension to the model with 2005 as the initial period, it is not possible to accurately trace the period from 2005 to 2010 because the so-called "Lehman shock" occurred during that time. Therefore, without replacing the benchmark year of 2005 with 2010, we carried out the simulation in order to reproduce each real GRP of 2010 in the disaster-affected region and the non-disaster region using the 2SCGE model, which is conditional on the rate of change in the labor and capital endowments from 2005 to 2010.[5] Then, we consider the benchmark year as the value for 2010, given such simulation results, and replace the benchmark year, 2005, with 2009. As a result, we can select 2010 to be the benchmark year in our study when adding a recursive dynamic dimension to the model. In addition, given the above simulation, we set up the rate of change in capital stock at the beginning of 2010 using the investment demand for 2010.

[3]Refer to Appendix 1 concerning the 2SCGE model.

[4]The method of adding a recursive dynamic dimension to the model is largely based on the GAMS code of Recursive Dynamics provided by EcoMod Modeling School (2012). We would like to note our appreciation.

[5]We estimate that each rate of change in labor endowment from 2005 to 2010 declines by 0.2% for the disaster-affected region and increases by 1.9% for the non-disaster region using the 2005 and 2010 population census of Japan. Though we cannot estimate the rates of change in capital endowment for two regions on the basis of any survey data, we can calculate those rates in the process of our simulations in order to reproduce the real GRP of 2010.

A simple description of the procedure for adding a recursive dynamic dimension to the model is that it consists of calculating the initial value of the total capital stock, KT_0, by dividing the steady-state growth rate (a 0.6% potential growth rate [*growth*] for both regions) by the initial value of total investment demand for the two-regional SAM, IT_0. We then use this total capital stock to derive the capital stock for each industry, K_0^a, from the percentage of capital in each industry according to the SAM. Multiplying the capital stock derived in this way for each industry by the steady-state growth rate (*growth*) results in the initial value of actual investment for each industry, INV_0^a. In addition, we derive the initial value of the return to capital for each industry, PK_0^a, by dividing the initial value of the capital demand (capital services) for each industry according to the SAM (referred to as "payment to capital," $KPAY_0^a$, in the dynamic model) by the capital stock for each industry calculated by the above process. An investment agent determines the total investment demand for each period, *IT*, by applying a certain percentage, *aIT*, of its own utility for each period, *UI*. We calculate the investment, INV_t^a, made by each industry in period *t* by multiplying the initial value of the actual investment for each industry, INV_0^a, by square root of the rate of the return to capital for each industry, PK_t^a, and the average return to capital, $PKAVG_t$[6]. We then add capital stock, K_{t-1}^a, in period $t-1$ to the actual investment, INV_t^a, in period *t* and deem the capital stock in period *t* to be K_t^a. In addition, the inter-regional intermediaries are the aggregated goods, IC_t^c, derived from the composite commodity according to the Armington assumption, XI_t^c, imported and exported from both regions and the intra-regional composite commodity according to the Armington assumption, XI_t^c, and the inter-regional current account balance that affects the savings, S_t, of both regions. The capital stock for each industry in each period derived this way is fixed in a similar manner as the labor supply, *LS*, and foreign savings, *SF*, for each period when multiplied by the steady-state growth rate (*growth*). These procedures add a recursive dynamic dimension to the 2SCGE model and the simulations are conducted with the wage rates of two regions incorporated into the formula for the endogenous variable, the labor supply, and the demand equation.[7]

5.4 Simulation Contents and Results

5.4.1 Setting the Simulation Scenarios

5.4.1.1 Labor Force and Capital Stock

In this chapter, we use the following projections for each scenario. As mentioned above in Sect. 5.2.1, Projection A assumes that the labor participation rate is

[6]Square root means the elasticity of the return to capital to investment demand.

[7]We took the wage rate for one of the two regions as the numeraire and checked whether Walras' law held true by excluding the labor supply and demand equation for that same region. However, Walras' law did not hold true, so we used this method as an alternative.

unchanged from the current level based on the rate of decline in the labor force by region. On the other hand, we estimated the size of the labor force in 2010 (T = 1) and 2011 (T = 2) using some labor statistics. As we have already mentioned in Note 5 concerning the estimation result for 2010, we will explain the estimation result for 2011. We determined the change in the labor force status for the three disaster-affected prefectures of Iwate, Miyagi, and Fukushima resulting from the Great East Japan Earthquake using the 2012 Employment Status Survey. Then, we estimated the rate of change in the labor force of the disaster-affected region using the survey results as follows—that is, whether or not the current type or number of jobs were affected by the earthquake. As a result, we estimated that each rate of change in the labor force population from 2005 to 2011 declined by 3.0% for the disaster-affected region and by 0.6% for the non-disaster region. We assume that the estimated labor force population in 2011 will be restored during the eight-year period to 2019 (T = 10), which is projected in Sect. 5.2.1. Table 5.4 shows the rates of change in the labor force population until 2035 for the two regions for projections A1 and A2, which are based on the above estimations and whether or not there was an earthquake. In addition, Table 5.5 shows the rates of change in the population aged 65 and over until 2035 for the two regions.

Next, we explain the setting of the capital stock. We assume that each rate of change in capital stock for all industries in 2010 declines by 5.44% for the disaster-affected region and increases by 0.6% for the non-disaster region. These numerical values are calculated by the rates of change in investment demand for the two regions that were obtained through the simulations that reproduced the real GRPs for 2010. Then, we use the figures in Table 5.1 in terms of the rates of change in capital stock in 2011 when the Great East Japan Earthquake occurred. In addition, the process to restore the capital stock after 2012 is determined by the actual investment in each industry, as mentioned above in Sect. 5.3.

5.4.1.2 Establishment of Each Scenario

Four scenarios, including the base scenario, are set for this purpose.

(a) Base Scenario

The base scenario, that is if the Great Earthquake had not occurred, assumes that the Great East Japan Earthquake did not occur on March 11, 2011 (hereinafter the "Base Scenario"). Taking into account a declining population trend under this scenario, changes in the labor force population are reflected in Projection A1, in which the labor force participation rate remains at the status quo based on the future population projections by the IPSS, as shown in Table 5.4. The scenario also takes into account the changes in the population for ages 65 and above. In terms of capital stock under this scenario, we apply the change ratio for 2010 in the preceding section.

Table 5.4 Projection of population in labor force

Population in labor force	Projection A1 (without earthquake, population decline, unchanged labor participation)				
(Annual rate, index of 2010 = 100)	2010–2015	2015–2020	2020–2025	2025–2035	2035/2010
Whole country	−0.84%	−1.30%	−0.74%	−1.20%	76.67
The disaster in Iwate, Miyagi, Fukushima and Ibaraki Prefectures	−1.20%	−1.55%	−1.06%	−1.34%	72.10
Other areas without the disaster-affected prefectures	−0.82%	−1.28%	−0.72%	−1.20%	76.96
Population in labor force	Projection A2 (with earthquake, population decline, unchanged labor participation)				
(Annual rate, index of 2010 = 100)	2010–2015	2015–2020	2020–2025	2025–2035	2035/2010
Whole country	−1.01%	−1.13%	−0.74%	−1.20%	76.67
The disaster in Iwate, Miyagi, Fukushima and Ibaraki Prefectures	−1.39%	−1.37%	−1.06%	−1.34%	72.10
Other areas without the disaster-affected prefectures	−0.98%	−1.12%	−0.72%	−1.20%	76.96
65 year and over	All scenario				
(Annual rate, index of 2010 = 100)	2010–2015	2015–2020	2020–2025	2025–2035	2035/2010
Whole country	2.50%	1.05%	0.23%	0.28%	123.92
The disaster in Iwate, Miyagi, Fukushima and Ibaraki Prefectures	2.07%	1.34%	0.24%	−0.15%	118.07
Other areas without the disaster-affected prefectures	2.53%	1.03%	0.23%	0.31%	124.32

Source Original data based on IPSS and JILPT

Table 5.5 Setting the change of productivity after the Earthquake for Scenario A

(Annual % point of increase within each period of time)	Fisheries				
	T = 1	T = 2	T = 3	T = 4	T = 5-6
The disaster in Iwate, Miyagi, Fukushima and Ibaraki Prefectures	0.0%	−13.0%	−13.0%	0.5%	0.5%
(Annual % point of increase within each period of time)	Foods and beverage				
	T = 1	T = 2	T = 3	T = 4	T = 5-6
Other areas without the disaster-affected prefectures	0.5%	−1.0%	−1.0%	0.3%	0.0%
(Annual % point of increase within each period of time)	Electrical devices and parts				
	T = 1	T = 2	T = 3	T = 4	T = 5-6
Other areas without the disaster-affected prefectures	0.5%	−1.0%	−1.0%	0.3%	0.0%
(Annual % point of increase within each period of time)	Motor Vehicles and parts				
	T = 1	T = 2	T = 3	T = 4	T = 5-6
Other areas without the disaster-affected prefectures	0.0%	−10.0%	10.0%	0.3%	0.0%
(Annual % point of increase within each period of time)	Other manufacturing products and mining				
	T = 1	T = 2	T = 3	T = 4	T = 5-6
Other areas without the disaster-affected prefectures	0.0%	−1.0%	−1.0%	0.3%	0.0%
(Annual % point of increase within each period of time)	Electricity, gas and heat-water supply				
	T = 1	T = 2	T = 3	T = 4	T = 5-6
The disaster in Iwate, Miyagi, Fukushima and Ibaraki Prefectures	0.0%	−20.0%	−20.0%	0.5%	0.5%
Other areas without the disaster-affected prefectures	0.0%	−20.0%	−20.0%	15.0%	2.5%
(Annual % point of increase within each period of time)	Transport				
	T = 1	T = 2	T = 3	T = 4	T = 5-6
Other areas without the disaster-affected prefectures	0.0%	−3.0%	3.0%	0.5%	0.5%

(b) Scenario A

Scenario A, that is after Great Earthquake, there were no fiscal measures and no change in labor force participation, assumes that, while the Great East Japan Earthquake occurred, no fiscal support was provided to the disaster-affected region (hereinafter "Scenario A"). The labor force population for this scenario is the same as that set for the Base Scenario, but it takes into account the changes in 2010 and

2011 outlined in the preceding section for deriving the future projections, as illustrated in Projection A2. The population for ages 65 and above is the same as that for the Base Scenario. The given conditions on the capital stock for Scenario A are the change ratio for 2010, as previously mentioned, and the rate of loss of capital stock caused by the Great Earthquake in 2011, as shown in Table 5.1.

In addition, we assume that, in order for the various changes in production activities caused by the earthquake to be reflected in the regional economy, the productivity for some industries declined, as shown in Table 5.5: (1) fisheries and electricity, gas, and heat-water supply in the disaster-affected region, which suffered large amounts of damage; and (2) manufacturing, transport, and electricity, gas, and heat-water supply in the non-disaster region, which suffered direct or indirect damages caused by supply-chain disruptions. So, we embody the change of productivity by the following method. It is that the efficiency parameter of the production function of some industries mentioned above, which was obtained through calibration of the D2SCGE model, is changed by the ratios indicated in Table 5.5.

(c) Scenario B

Scenario B, that is after Great Earthquake, there were fiscal measures implemented and no change in labor force participation, assumes that the Great East Japan Earthquake occurred and fiscal measures were implemented for the disaster-affected region, as shown in Table 5.2 (hereinafter "Scenario B"). The labor force population set under Scenario B is the same as that for Scenario A, whereas the population of ages 65 and above is the same as that for the Base Scenario. The capital stock for Scenario B is also the same as that for Scenario A. However, the difference in the setting from Scenario A is that, in conjunction with the fiscal measures provided by the central government to the local governments in the disaster-affected region, the subsidy rate for each industry and the savings rate of the local governments for each of the four disaster-affected prefectures have been set, as shown in Table 5.6 below. Table 5.6 (a) provides the subsidy rate by industry. According to this table, the amount of the subsidy for each industry in the intensive reconstruction period from T = 2 to T = 6 is made to correspond with the second item of the reconstruction budget for the industrial revitalization of each fiscal period shown in Table 5.2 by raising the subsidy rate, which then returns to the level of T = 1 at the completion of the intensive reconstruction period at T = 7. Table 5.6 (b) illustrates the savings rate for each local government in the disaster-affected region. The savings rate for the intensive reconstruction period from T = 2 to T = 6 is made to correspond with the reconstruction budget allocated for public works, residential housing investment, capital investments, and so on for each fiscal period shown in Table 5.2 by raising the rates in the range of 8.5–12.4 points above the rates in T = 1 as the source of funding for investments. The rise in savings rates will increase the amount of savings in the disaster-affected region and lead to promoting investments in construction and other industries. However, the savings rate in Scenario B returns to the level of T = 1 at the completion of the intensive reconstruction period at T = 7. As a result, the reconstruction budget allocated to the revenues of local governments in the

Table 5.6 Subsidy rate by production activity and projections on the savings rate of local governments in the four disaster-affected prefectures

(a) Setting the subsidy rate							
Scenario B and C Subsidy rate per industry	T = 1	T = 2	T = 3	T = 4	T = 5	T = 6	T = 7
Agriculture and forestry	1.4%	18.4%	4.8%	2.9%	2.7%	2.3%	1.4%
Fisheries	0.5%	21.5%	14.5%	7.0%	6.5%	4.5%	0.5%
Foods and beverage	0.5%	1.7%	0.9%	1.2%	0.7%	0.7%	0.5%
Electrical devices and parts	0.0%	0.5%	0.2%	0.2%	0.0%	0.1%	0.0%
Motor vehicles and parts	0.0%	1.0%	0.3%	0.5%	0.1%	0.2%	0.0%
Other manufacturing products and mining	0.0%	1.5%	0.5%	0.9%	0.3%	0.3%	0.0%
Commerce	0.0%	0.6%	0.2%	0.4%	0.2%	0.2%	0.0%
Transport	0.3%	1.4%	0.6%	0.9%	0.5%	0.5%	0.3%

(b) Setting the savings rate								
Savings rate of local government	T = 1	T = 2	T = 3	T = 4	T = 5	T = 6	T = 7-11	T = 12-26
Scenario B	2.6%	11.4%	15.0%	11.4%	11.0%	11.2%	2.6%	2.6%
Scenario C	2.6%	11.4%	15.0%	11.4%	11.0%	11.2%	10.2%	2.6%

disaster-affected region, as shown in Table 5.2, corresponds to the amount of public fiscal transfers from the central government minus the sum of the amount of the increase in subsidies to each industry and the amount of the increase in the savings of local governments in the disaster-affected region mentioned above. Meanwhile, the source of funding for the fiscal support from the central government to the local governments in the disaster-affected region is met by the issuance of reconstruction government bonds and an increase in direct taxes, as illustrated in Table 5.2.[8] Finally, Table 5.7 shows the rate of change in productivity for each industry located in both the disaster-affected region and the non-disaster region from T = 2 to T = 6 in order to trace the recovery phases for the regional economies after the earthquake, as mentioned in Sect. 5.5. In the same way as Scenario A, the efficiency parameters of the production function are changed by the ratios indicated in Table 5.7.

(d) Scenario C

Scenario C, that is after Great Earthquake, there were fiscal measures implemented, no change to labor force participation and continuation of fiscal measures, assumes the continuation of fiscal support for reconstruction based on a new approach in a five-year construction period from 2016 to 2020 following the

[8]As to how such funding sources are transferred from the central government to local governments in the disaster-affected region under the D2SCGE model, please refer to Section 2, Chapter 5 of Tokunaga and Okiyama (eds) (2014).

Table 5.7 Setting the productivity of each production activity by Scenario B and C

	Scenario B and C					Scenario B and C		
(Annual % point of increase within each period of time)	T = 1	T = 2	T = 3	T = 4	T = 5–6	T = 7–10	T = 11–20	T = 21–30
Agriculture and forestry								
The disaster in Iwate, Miyagi, Fukushima and Ibaraki prefectures	0.0%	1.3%	0.0%	0.0%	0.0%	0.0%	0.0%	0.0%
Other areas without the disaster-affected prefectures	0.0%	0.3%	0.0%	0.0%	0.0%	0.0%	0.0%	0.0%
Fisheries								
The disaster in Iwate, Miyagi, Fukushima and Ibaraki prefectures	0.0%	−13.0%	−13.0%	0.5%	0.5%	0.0%	0.0%	0.0%
Other areas without the disaster-affected prefectures	0.0%	0.0%	0.0%	0.0%	0.0%	0.0%	0.0%	0.0%
Foods and beverage								
The disaster in Iwate, Miyagi, Fukushima and Ibaraki prefectures	0.0%	11.0%	11.0%	2.0%	0.5%	0.0%	0.0%	0.0%
Other areas without the disaster-affected prefectures	0.5%	−1.0%	−1.0%	0.3%	0.0%	0.0%	0.0%	0.0%
Electrical devices and parts								
The disaster in Iwate, Miyagi, Fukushima and Ibaraki prefectures	0.0%	15.0%	15.0%	2.5%	0.5%	0.0%	0.0%	0.0%
Other areas without the disaster-affected prefectures	0.5%	−1.0%	−1.0%	0.3%	0.0%	0.0%	0.0%	0.0%
Motor vehicles and parts								
The disaster in Iwate, Miyagi, Fukushima and Ibaraki Prefectures	0.0%	17.0%	17.0%	3.5%	1.0%	0.0%	0.0%	0.0%
Other areas without the disaster-affected prefectures	0.0%	−10.0%	10.0%	0.5%	0.0%	0.0%	0.0%	0.0%
Other manufacturing products and mining								
The disaster in Iwate, Miyagi, Fukushima and Ibaraki prefectures	0.0%	6.5%	6.5%	1.5%	0.1%	0.0%	0.0%	0.0%
Other areas without the disaster-affected prefectures	0.0%	−1.0%	−1.0%	0.3%	0.0%	0.0%	0.0%	0.0%
Construct								

The disaster in Iwate, Miyagi, Fukushima and Ibaraki prefectures	0.0%	−3.0%	−3.0%	3.0%	3.0%	0.0%	0.0%	0.0%
Other areas without the disaster-affected prefectures	0.0%	15.0%	0.0%	0.0%	0.0%	0.0%	0.0%	0.0%
Electricity,gas and heat-water supply								
The disaster in Iwate, Miyagi, Fukushima and Ibaraki prefectures	0.0%	−20.0%	−20.0%	0.5%	0.5%	0.0%	0.0%	0.0%
Other areas without the disaster-affected prefectures	0.0%	−20.0%	−20.0%	15.0%	2.5%	0.0%	0.0%	0.0%
Commerce								
The disaster in Iwate, Miyagi, Fukushima and Ibaraki prefectures	0.0%	9.5%	9.5%	2.0%	1.0%	0.0%	0.0%	0.0%
Other areas without the disaster-affected prefectures	0.0%	0.0%	4.0%	1.5%	0.0%	0.0%	0.0%	0.0%
Transport								
The disaster in Iwate, Miyagi, Fukushima and Ibaraki Prefectures	0.0%	9.5%	9.5%	2.0%	1.0%	0.0%	0.0%	0.0%
Other areas without the disaster-affected prefectures	0.0%	−2.0%	2.0%	0.5%	0.5%	0.0%	0.0%	0.0%
Financial and insurance,Real estate, Communcation, Public adminstration,Eduction and Servics etc.								
The disaster in Iwate, Miyagi, Fukushima and Ibaraki Prefectures	0.0%	1.7%	6.0%	0.3%	0.05%	0.0%	0.0%	0.0%
Other areas without the disaster-affected prefectures	0.0%	0.3%	5.0%	1.0%	0.0%	0.0%	0.0%	0.0%

intensive reconstruction period, in addition to the parameters in Scenario B. The fiscal support under a new approach refers to the provision of 1.5 trillion yen by the central government to the disaster-affected region for the restoration of public infrastructure and reconstruction grants outlined in Table 5.2 until fiscal year 2020. The source of funding for this support will be the portion of the local allocation of tax grants distributed to the non-disaster region. As such, this does not call for new sources of funding for the reconstruction budget but rather for the alteration of the allocation ratios of the portions of the local tax grants. Under Scenario C, the savings rate does not return to the level of T = 1 at T = 7, as indicated for Scenario B, but it returns to the level of T = 1 at T = 12.

5.4.2 *Simulation Results*

5.4.2.1 Assessment of the Base Scenario

We examine the transition of real GRP by reviewing Table 5.8 and Fig. 5.3 for the four disaster-affected prefectures and the non-disaster region under the Base Scenario. Regional economies in the four disaster-affected prefectures are expected to weaken from about 2012 owing to the impact of population decline in the region. As a result, the real GRP in 2035 will fall to 83.65 (2010 = 100). From the transitions of the equivalent variations in Table 5.10 and Fig. 5.4, the equivalent variation will decrease roughly 810.7 billion yen after 2010 and the degree of decline will expand to 2.5 trillion yen by 2020. It will continue to move downward thereafter, reaching 6.2 trillion yen in 2035.

Meanwhile, the real GRP for the non-disaster region will show slower growth owing to a decrease in the labor force and further aging of the population. It is projected that the real GRP for the non-disaster region in 2035 will decline to 86.43 (2010 = 100). The equivalent variation for the non-disaster region will rise from 2010 and remain positive until 2013. However, from 2014 and thereafter, it will turn negative and decrease 69.9 trillion yen by 2035.

These simulation results indicate that in the four prefectures that were significantly affected by the Great East Japan Earthquake, the economic environment was prone to be negatively impacted by a declining trend in the overall population of the Tohoku region in the 2010s and thereafter, even when assuming that the Great Earthquake had not occurred.

5.4.2.2 Evaluation of Fiscal Measures for the Intensive Reconstruction Period

(a) Evaluation of fiscal measures from the standpoint of GRP

First, we look at Table 5.8 and Fig. 5.3 to examine the transition in Scenario A, which assumes no fiscal measures, and that in Scenario B, which assumes fiscal

Table 5.8 Simulation results on real GRP and intra-regional gross production by Scenario

		Real GRP					Total output				
(Annual rate, index of 2010 = 100)		2010–2015	2015–2020	2020–2025	2025–2035	2035/2010	2010–2015	2015–2020	2020–2025	2025–2035	2035/2010
The whole country	(1)Base Scenario	−0.35%	−0.64%	−0.41%	−0.77%	86.26	−0.40%	−0.67%	−0.47%	−0.79%	85.51
	(2)Scenario A	−0.72%	−0.56%	−0.41%	−0.77%	85.01	−1.06%	−0.59%	−0.47%	−0.80%	82.93
	(3)Scenario B	0.34%	−0.53%	−0.39%	−0.75%	90.03	0.67%	−0.53%	−0.43%	−0.75%	91.35
	(4)Scenario C	0.34%	−0.54%	−0.38%	−0.75%	90.04	0.67%	−0.51%	−0.40%	−0.75%	91.55
	(3)-(2)	1.06%	0.03%	0.02%	0.02%	5.02	1.73%	0.06%	0.05%	0.05%	8.42
The disaster in Iwate, Miyagi, Fukushima and Ibaraki prefectures	(1)Base Scenario	−0.52%	−0.75%	−0.58%	−0.85%	83.65	−0.50%	−0.74%	−0.57%	−0.85%	83.86
	(2)Scenario A	−1.71%	−0.69%	−0.60%	−0.88%	78.75	−1.67%	−0.67%	−0.60%	−0.87%	79.08
	(3)Scenario B	1.06%	−0.58%	−0.48%	−0.76%	92.55	1.33%	−0.53%	−0.49%	−0.76%	94.05
	(4)Scenario C	1.06%	−0.47%	−0.48%	−0.76%	93.13	1.33%	−0.48%	−0.42%	−0.76%	94.61

(continued)

Table 5.8 (continued)

		Real GRP					Total output				
(Annual rate, index of 2010 = 100)		2010–2015	2015–2020	2020–2025	2025–2035	2035/2010	2010–2015	2015–2020	2020–2025	2025–2035	2035/2010
	(3)–(2)	2.77%	0.10%	0.12%	0.11%	13.80	3.00%	0.14%	0.11%	0.11%	14.98
Other areas without the disaster-affected prefectures	(1)Base Scenario	−0.34%	−0.63%	−0.40%	−0.77%	86.43	−0.33%	−0.62%	−0.40%	−0.76%	86.62
	(2)Scenario A	−0.66%	−0.55%	−0.40%	−0.77%	85.42	−0.67%	−0.54%	−0.39%	−0.76%	85.49
	(3)Scenario B	0.29%	−0.53%	−0.38%	−0.75%	89.86	0.21%	−0.54%	−0.38%	−0.75%	89.55
	(4)Scenario C	0.29%	−0.54%	−0.38%	−0.75%	89.84	0.21%	−0.54%	−0.39%	−0.75%	89.52
	(3)–(2)	0.95%	0.02%	0.01%	0.01%	4.44	0.88%	0.01%	0.01%	0.01%	4.05

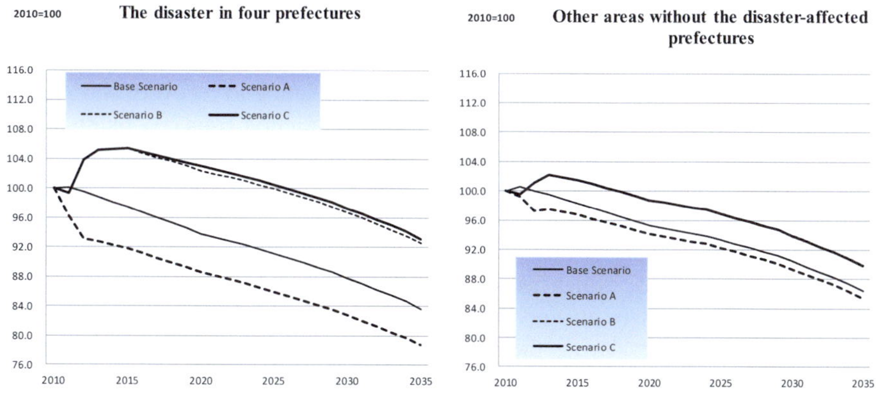

Fig. 5.3 Transition of real GRP by scenario

measures. We find that the fiscal measures in the intensive reconstruction period definitely contributed toward the recovery of the real GRP that had declined in the four disaster-affected prefectures after the earthquake. However, with the end of the intensive reconstruction period, the real GRP begins to move downward after 2018, though the degree of decline in the growth rate is slightly smaller than that under Scenario A. As a result, the real GRP for the four disaster-affected prefectures in 2025 reaches 99.93 (2010 = 100) and declines to 92.55 in 2035. Evaluating the effectiveness of the fiscal measures in 2035, we find a recovery equivalent to about 13.8 points compared with the case of Scenario A and 8.9 points compared with the Base Scenario under the assumption that the earthquake had not occurred. Meanwhile, in terms of the regional economies in the non-disaster region, the growth rate of real GRP begins to move upward to 2014 before turning slightly downward compared with the case under Scenario A. Although the non-disaster region must incur an incremental tax burden to fund the reconstruction budget, even after the intensive reconstruction period, the burden does not appear to be considerable when considering the transition of real GRP. Consequently, the real GRP of the non-disaster region in 2035 is expected to decline to 89.86 compared with 86.43 in the Base Scenario, and the decline is no more than 4.44 points from the level in Scenario A.

So, the combination of Scenario A and Scenario B in the disaster-affected region shows that, despite the declining trend in the population, which would curtail the growth of the regional economies, the fiscal measures during the intensive reconstruction period alone would enable the region to recover beyond the level of the Base Scenario. As a result, for example, the real GRP in Scenario B could reach the same level about 21 years late compared with the year when the real GRP in Scenario A reaches 92.0 (2010 = 100).[9]

[9] This model has no assumptions of labor movement and capital transfer between the disaster-affected region and the non-disaster region. Therefore, even if the return to capital in the disaster-affected region had increased by the equivalent of the damage to the capital stock, there is no capital transfer from the non-disaster region to the disaster-affected region. As a result, the real

(b) Evaluation of fiscal measures from the perspective of production and investment volumes by production segment

In this section, we evaluate the fiscal measures relating to recovery and reconstruction activities, such as public works and reconstruction grants. Looking at the transition of the indicators for production and investment volumes for each production sector under Scenario B, which assumes the implementation of fiscal measures as shown in Table 5.9, the construction business has a higher growth rate than any other industry during the intensive reconstruction period. In comparison with Scenario A, which assumes no fiscal measures, the disparity in the construction business is greater than that in the other industries. This suggests that financial support during the intensive reconstruction period has the most significant impact on the construction industry. In fact, the disparity between the two scenarios in terms of investment volume for the construction business is nearly 3.5 times in 2012 and 2013. Since the funds provided during the intensive reconstruction period were used mainly for such things as public works, we conclude that fiscal measures contributed to the high growth in investment and production volumes in the construction business. Meanwhile, the investment volume in fisheries also indicated high growth comparable to that in the construction business. However, the fact that this growth in investment did not lead to high production in the fisheries industry was due to the significant damage on the capital stock. And the production in electricity, gas, and heat-water supply during the intensive reconstruction period and thereafter remained at the lower 70 level (from 100 in 2010). This is because the Fukushima nuclear power station's operation had been suspended since the earthquake and this suspension is expected to continue in the future. In addition, the positive effects of the fiscal measures led to expediting the recovery of capital stock, as will be explained in part (3) of Subsect. 5.4.2, since investment showed higher growth than production in every production sector. This point agrees with Hamagushi (2015), who pointed out that the tangible fixed assets recovered quickly with the support of the post-earthquake recovery assistance measures; however, this recovery was not accompanied by the revitalization of production and employment. At any rate, the comparison of the scenarios indicates that, even after the completion of the reconstruction period, both the production and investment volumes of each production sector benefited from the effects and the effectiveness of the fiscal measures.

(c) Evaluation of fiscal measures from the standpoint of the equivalent variation

Next, we review Table 5.10 and Fig. 5.4, and examine the equivalent variations of the country, the four disaster-affected prefectures, and the non-disaster region for

GRP of the disaster-affected region continues to be influenced by the damage to its capital stock, even if the fiscal measures are implemented. In the future, we hope to conduct an improved CGE model that takes into account the movement of production factors between regions.

Table 5.9 Simulation results of production volume and investment volume by production sector

	The disaster in Iwate, Miyagi, Fukushima and Ibaraki Prefectures	Production(XD_a^o)					Investment(INV_a^o)				
	2010 year = 100	2012	2013	2015	2020	2030	2012	2013	2015	2020	2030
Scenario B	Total Industry	104.60	106.25	106.85	104.05	98.29	208.34	173.87	149.02	109.68	98.29
	Agriculture and forestry	96.00	96.97	98.54	101.05	103.41	225.89	176.83	152.73	110.37	103.41
	Fisheries	65.59	65.98	67.62	69.46	69.85	256.27	230.58	192.55	135.72	69.85
	Foods and beverage	104.68	107.13	108.11	108.76	106.25	198.59	157.35	134.57	99.80	106.25
	Electrical devices and parts	103.85	112.17	117.53	124.98	115.03	169.63	156.85	142.85	111.17	115.03
	Motor vehicles and parts	105.53	111.68	113.92	114.69	106.53	179.41	154.03	135.06	101.98	106.53
	Other manufacturing products and mining	99.69	104.25	105.60	106.00	100.56	186.91	162.96	144.20	108.31	100.56
	construct	167.50	156.10	154.25	115.31	106.51	271.10	215.88	172.73	114.79	106.51
	Electricity, gas and heat-water supply	73.17	72.81	73.57	74.70	72.82	221.56	221.81	185.99	136.48	72.82
	Commerce	105.39	108.52	111.06	109.68	101.26	195.25	165.55	144.07	107.53	101.26
	Transport	103.12	106.96	108.80	108.30	100.18	192.90	163.94	142.99	107.84	100.18
	Other services	102.36	102.82	101.90	99.63	94.20	212.20	170.36	145.32	106.61	94.20
Scenario A	Total industry	93.19	92.94	91.94	88.91	83.07	79.96	70.26	72.71	70.48	64.23
	Agriculture and forestry	94.21	94.47	94.89	95.59	95.91	82.53	71.13	73.55	69.61	59.36
	Fisheries	64.49	64.93	65.69	65.73	64.89	101.26	90.22	92.61	86.84	73.92
	Foods and beverage	94.97	95.00	94.79	93.63	90.33	81.80	71.22	74.39	72.67	66.21
	Electrical devices and parts	98.14	97.43	96.09	92.32	84.54	81.19	70.67	73.67	72.25	66.84
	Motor vehicles and parts	97.32	97.25	96.11	92.57	85.40	78.11	69.82	73.11	71.85	66.79
	Other manufacturing products and Mining	94.56	94.47	93.70	91.03	85.53	79.21	68.40	71.50	70.00	64.54
	Construct	81.75	84.11	83.49	79.83	71.97	72.35	60.75	64.60	63.92	59.75
	Electricity, gas and heat-water supply	75.78	73.37	72.76	71.48	68.68	93.21	92.42	90.97	86.67	76.99
	Commerce	95.57	95.39	94.26	90.67	82.91	79.07	69.01	72.19	70.85	65.44
	Transport	95.83	95.35	93.89	90.34	82.72	77.94	68.03	71.22	70.55	66.59
	Other services	94.88	94.39	93.06	89.44	83.51	78.20	67.79	70.38	68.08	62.13

Table 5.10 Simulation results of equivalent variation by Scenario

Trillion Yen		Equivalent variation								Accumulated equivalent variation				
		2011	2012	2013	2015	2020	2025	2030	2035	2013	2015	2020	2030	2035
The whole country	(1)Base Scenario	3.7	1.8	▲ 0.1	▲ 5.2	▲ 20.8	▲ 36.2	▲ 54.8	▲ 76.0	8.4	1.1	▲ 69.9	▲ 454.6	▲ 791.5
	(2)Scenario A	▲ 1.2	▲ 8.0	▲ 7.4	▲ 11.2	▲ 25.6	▲ 41.0	▲ 59.4	▲ 80.4	▲ 13.8	▲ 33.8	▲ 131.5	▲ 563.6	▲ 923.0
	(3)Scenario B	▲ 0.7	0.1	2.9	▲ 0.5	▲ 14.9	▲ 30.1	▲ 48.5	▲ 69.7	5.2	6.4	▲ 38.0	▲ 361.6	▲ 667.1
	(4)Scenario C	▲ 0.7	0.1	2.9	▲ 0.5	▲ 14.8	▲ 30.1	▲ 48.5	▲ 69.7	5.2	6.4	▲ 37.3	▲ 360.4	▲ 665.6
	(3)-(2)	0.5	8.1	10.3	10.7	10.7	10.9	10.9	10.7	18.9	40.2	93.5	202.0	255.9
	(4)-(3)	0.0	0.0	0.0	0.0	0.2	0.1	0.0	0.0	0.0	0.0	0.8	1.3	1.5
The disaster in Iwate, Miyagi, Fukushima and Ibaraki prefectures	(1)Base Scenario	▲ 0.9	▲ 1.1	▲ 1.2	▲ 1.5	▲ 2.5	▲ 3.5	▲ 4.8	▲ 6.2	▲ 4.0	▲ 6.9	▲ 17.4	▲ 54.3	▲ 82.3
	(2)Scenario A	▲ 1.8	▲ 2.4	▲ 2.5	▲ 2.8	▲ 3.6	▲ 4.7	▲ 5.9	▲ 7.3	▲ 7.5	▲ 12.8	▲ 29.1	▲ 77.3	▲ 110.8
	(3)Scenario B	0.5	0.2	0.1	▲ 0.2	▲ 1.3	▲ 2.2	▲ 3.4	▲ 4.7	▲ 0.1	▲ 0.4	▲ 5.3	▲ 29.2	▲ 50.0
	(4)Scenario C	0.5	0.2	0.1	▲ 0.2	▲ 0.8	▲ 2.1	▲ 3.2	▲ 4.5	▲ 0.1	▲ 0.4	▲ 3.0	▲ 25.0	▲ 44.9
	(3)-(2)	2.3	2.6	2.6	2.6	2.3	2.4	2.5	2.6	7.4	12.5	23.8	48.1	60.8

	(4)-(3)	0.0	0.0	0.0	0.0	0.5	0.2	0.2	0.2	0.0	0.0	2.3	4.2	5.1
Other areas without the disaster-affected prefectures	(1)Base Scenario	4.6	2.9	1.1	▲ 3.6	▲ 18.3	▲ 32.7	▲ 50.0	▲ 69.9	12.4	8.0	▲ 52.5	▲ 400.3	▲ 709.2
	(2)Scenario A	0.6	▲ 5.7	▲ 5.0	▲ 8.5	▲ 22.0	▲ 36.3	▲ 53.5	▲ 73.2	▲ 6.3	▲ 21.0	▲ 102.4	▲ 486.4	▲ 812.2
	(3)Scenario B	▲ 1.2	▲ 0.1	2.8	▲ 0.3	▲ 13.6	▲ 27.9	▲ 45.2	▲ 65.0	5.3	6.8	▲ 32.7	▲ 332.5	▲ 617.1
	(4)Scenario C	▲ 1.2	▲ 0.1	2.8	▲ 0.3	▲ 14.0	▲ 28.0	▲ 45.3	▲ 65.2	5.3	6.8	▲ 34.2	▲ 335.4	▲ 620.7
	(3)-(2)	▲ 1.8	5.6	7.8	8.1	8.4	8.4	8.4	8.1	11.5	27.8	69.7	153.9	195.1
	(4)-(3)	0.0	0.0	0.0	0.0	▲ 0.4	▲ 0.1	▲ 0.1	▲ 0.1	0.0	0.0	▲ 1.5	▲ 2.9	▲ 3.6

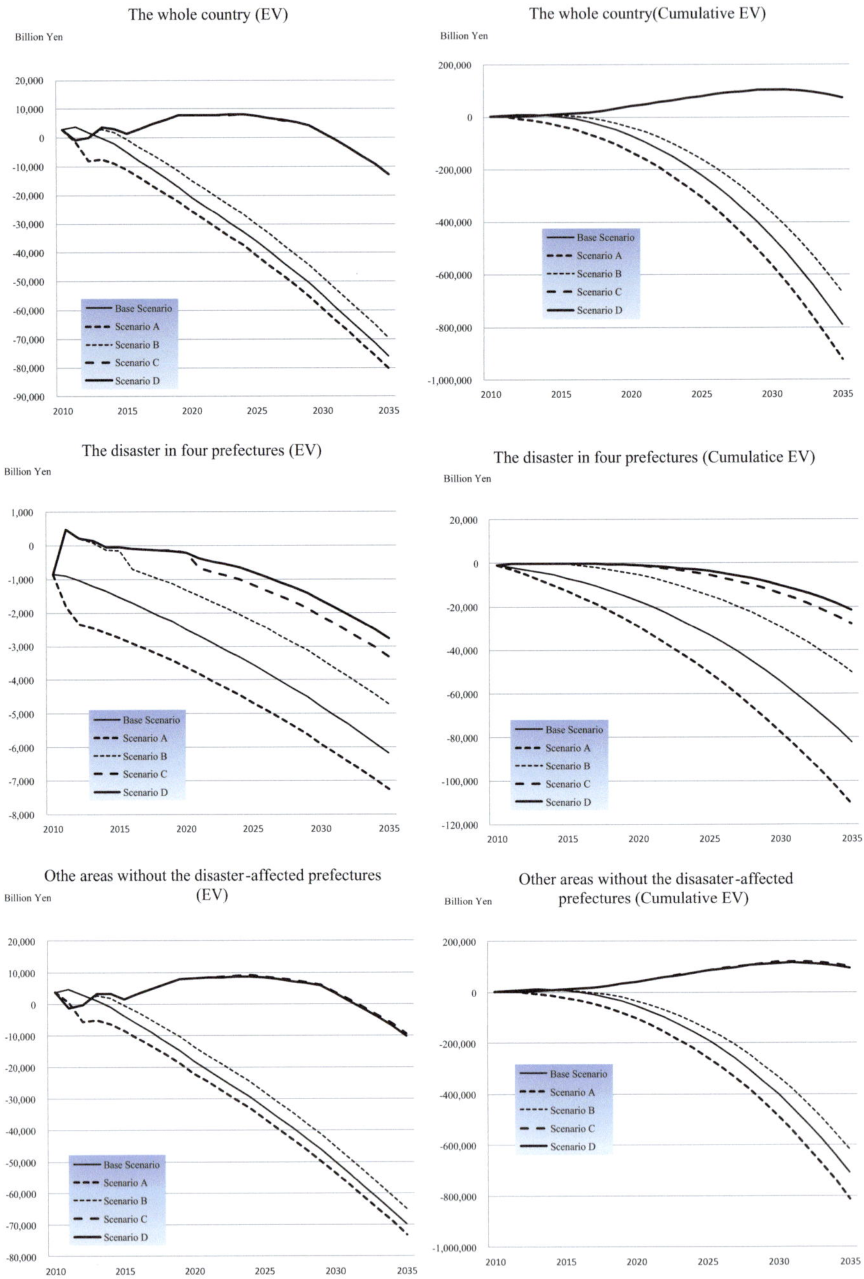

Fig. 5.4 Transition of equivalent variation of the country, four disaster-affected prefectures, and the non-disaster region

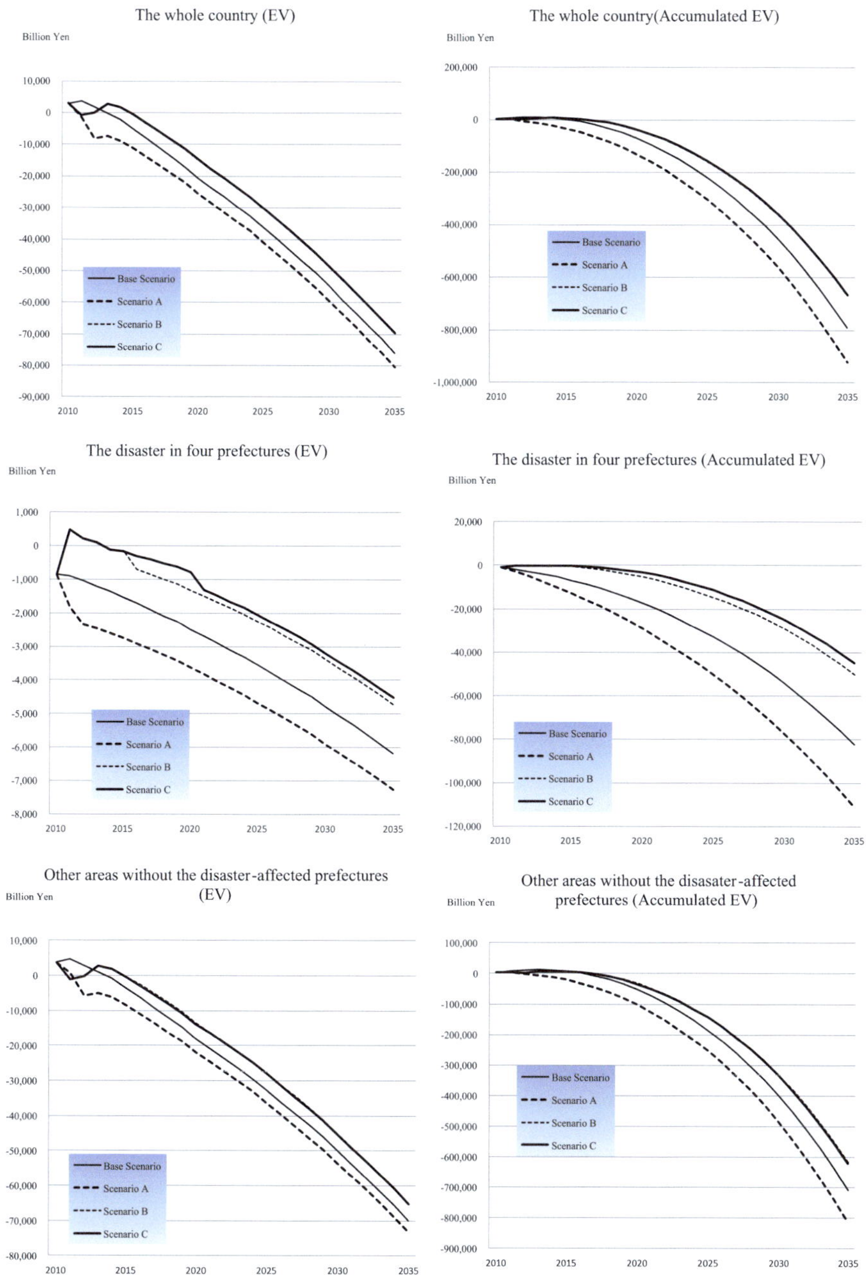

Fig. 5.4 (continued)

each year, as well as on a cumulative basis under each scenario.[10] During the three-year period from 2011 to 2013 in the intensive reconstruction period in Scenario B, the equivalent variation of the four disaster-affected prefectures was up by 0.5 trillion yen in 2011, 0.2 trillion yen in 2012, and 0.1 trillion yen in 2013. In the non-disaster region, it was down by 1.2 trillion yen in 2011 and 0.1 trillion yen in 2012, but it was up by 2.8 trillion yen in 2013. The mechanism that changes the equivalent variation of the disaster-affected region differs from that of the non-disaster region. Amid the tightening of labor demand in the disaster-affected region because of the increase in production for such industries as construction and the tertiary industry owing to the positive effects of the fiscal measures, the rate of increase of the wage rate exceeded that for consumer prices. As a result, the equivalent variation increased during the three-year period from 2011 to 2013. On the other hand, as the current account of the region comprising the non-disaster region turned from a deficit to a surplus because of the rapid increase in exports to the disaster-affected region, the income transfer from the non-disaster region to the disaster-affected region has increased, so the amount of investment has continued to decrease as a result of this income transfer. As a result, each industry in the non-disaster region substituted labor for capital stock, in which the rate of growth has been declining and the wage rate has been rising. However, as the rate of increase in wages was less than that for consumer prices in 2011 and 2012, the equivalent variation declined. Both rates reversed in 2013 and the equivalent variation continued to increase in 2014.

Taking the equivalent variations of Scenario A and Scenario B, and deriving the difference between the two scenarios from their cumulative values from 2010 to 2015, Scenario B exceeded Scenario A in the four disaster-affected prefectures by 12.5 trillion yen, which falls short of the cumulative public fiscal transfer amount in the intensive reconstruction period by 18.1 trillion yen. But the difference from the cumulative value from 2010 to 2020 is expected to be 23.8 trillion yen or an excess of 5.7 trillion yen.

Looking at the annual equivalent variation for Scenario B, the values turn downward for the disaster-affected region and the non-disaster region after 2014 and 2015, respectively, and they will continue to move downward thereafter, thus expanding the degree of decline. As a result, each value shows a decrease of 4.7 trillion yen and 65.0 trillion yen, respectively, by 2035. Meanwhile, replacing Scenario B with Scenario C and looking at the equivalent variation, the value of the disaster-affected region shows a decrease of 4.5 trillion yen and a reduction in the degree of decline of 0.2 trillion yen compared with Scenario B by 2035. The values for the non-disaster region show a decrease of 65.2 trillion yen and an increase in the degree of decline of 0.1 trillion yen compared with Scenario B by 2035.

[10]This chapter uses the equivalent variation as an index that measures the economic welfare. According to EcoMod Modeling School (2012), the equivalent variation is the difference between the consumption budget for the household of the "proposed change" deflated by the price index that represents the change in prices induced by the "proposed change" and the consumption budget of the "benchmark equilibrium." Please refer to Appendix in detail.

5.4.2.3 Evaluation of Fiscal Measures from the Perspective of Capital Stock and Other Major Economic Indicators

First, we examine the effectiveness of fiscal measures on the capital stock that was damaged in the earthquake and during the course of recovery by looking at the data by industry in Fig. 5.5. In agriculture and forestry, the fiscal measures contributed toward accelerating the recovery of capital stock to the levels prior to the earthquake by 2018, according to Scenario B. The recovery is accelerated by about 8 years compared with the results under Scenario A, which assume that no fiscal measures are implemented. On the other hand, the level of recovery in the fisheries industry remains at roughly 97, even in 2035, compared with a value of 100 prior to the earthquake. In the foods and beverage industry, its damaged capital stock recovers to the pre-earthquake level in 2019, which is almost comparable with the case for agriculture and forestry. The rate of recovery is quicker than that in Scenario A by about 7 years. The manufacturing and construction and utilities industries exceed 100 in 2015, reflecting a swifter recovery than that in Scenario A by 5 years. Meanwhile, the commerce, transport, and service industries recover and exceed the levels prior to the earthquake for the first time in 2014, which is the earliest year compared with the other industries. As a result, for the fisheries industry, in which the magnitude of the damage on its capital stock from the earthquake was significant, the recovery of capital stock falls behind that of the other industries, despite the fiscal measures. Therefore, there will be disparities in the recovery of capital stock between industries.

Next, we evaluate the fiscal measures according to the major economic indicators for the four disaster-affected prefectures in Fig. 5.6. Looking at the simulation results, the number of unemployed persons stays almost the same in the intensive reconstruction period, even though the number of unemployed persons was expected to decline owing to the current level of fiscal support. Moreover, this decline would be expected to decrease the number of unemployed persons to below that under the Base Scenario. Secondly, although the index of consumer prices in Scenario B and Scenario C has risen considerably owing to the fiscal measures in the intensive reconstruction period, the index for Scenario B remains at the Base Scenario level, whereas the index for Scenario C remains above that for Scenario A until the completion of the fiscal measures. Thirdly, the indicator for the consumption budget value of the households in Scenario B and Scenario C has risen considerably owing to the fiscal measures in the intensive reconstruction period. After the completion of the fiscal measures, this indicator in Scenario B and Scenario C falls by almost 10 points before continuing to slowly rise. Finally, looking at the transitions for three indicators including the revenue of local government, the level of savings, and the amounts of net imports, the size of changes in Scenario B and Scenario C obviously increases in the intensive reconstruction period because of fiscal transfers and imports from the non-disaster region. As a result, the savings value of the disaster-affected region largely increases compared with the Base Scenario and it becomes the source of local government funds for

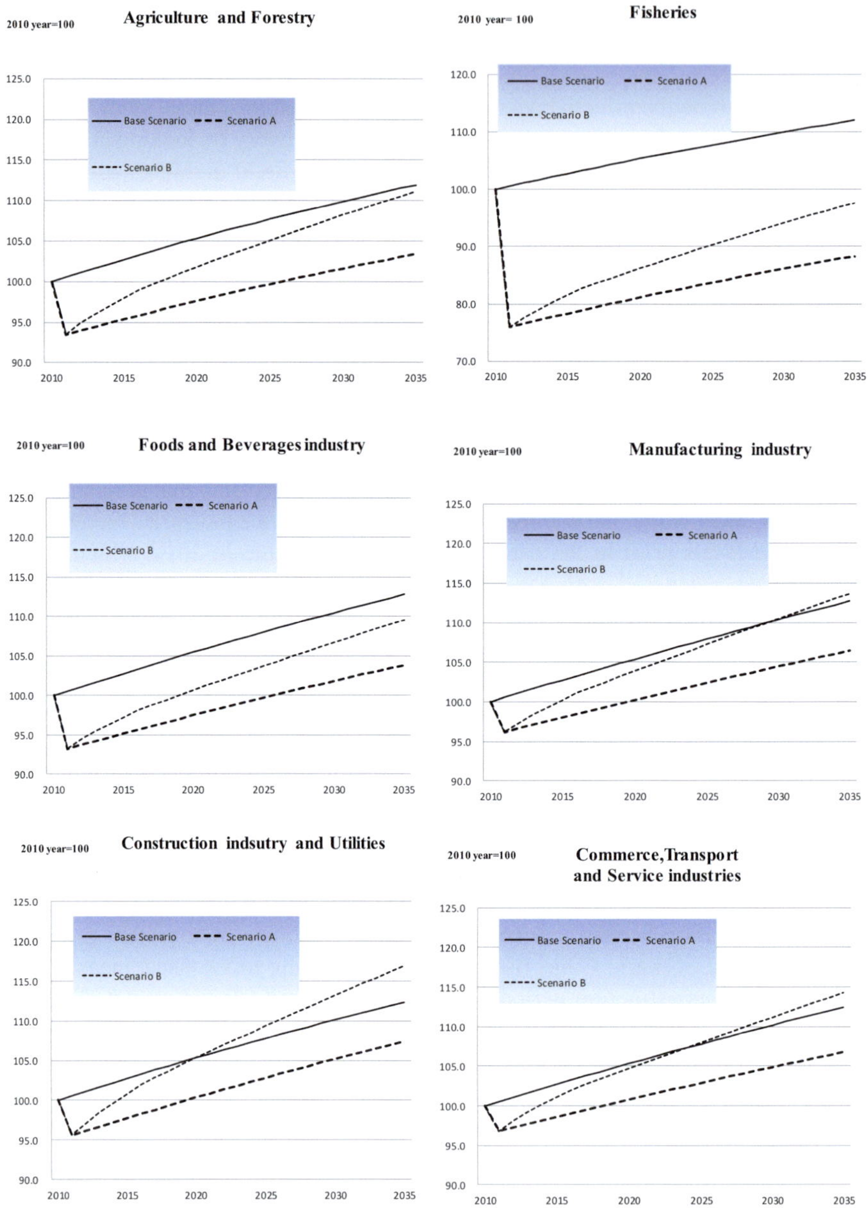

Fig. 5.5 Course of recovery of capital stock by industry in the four disaster-affected prefectures after the Great East Japan Earthquake

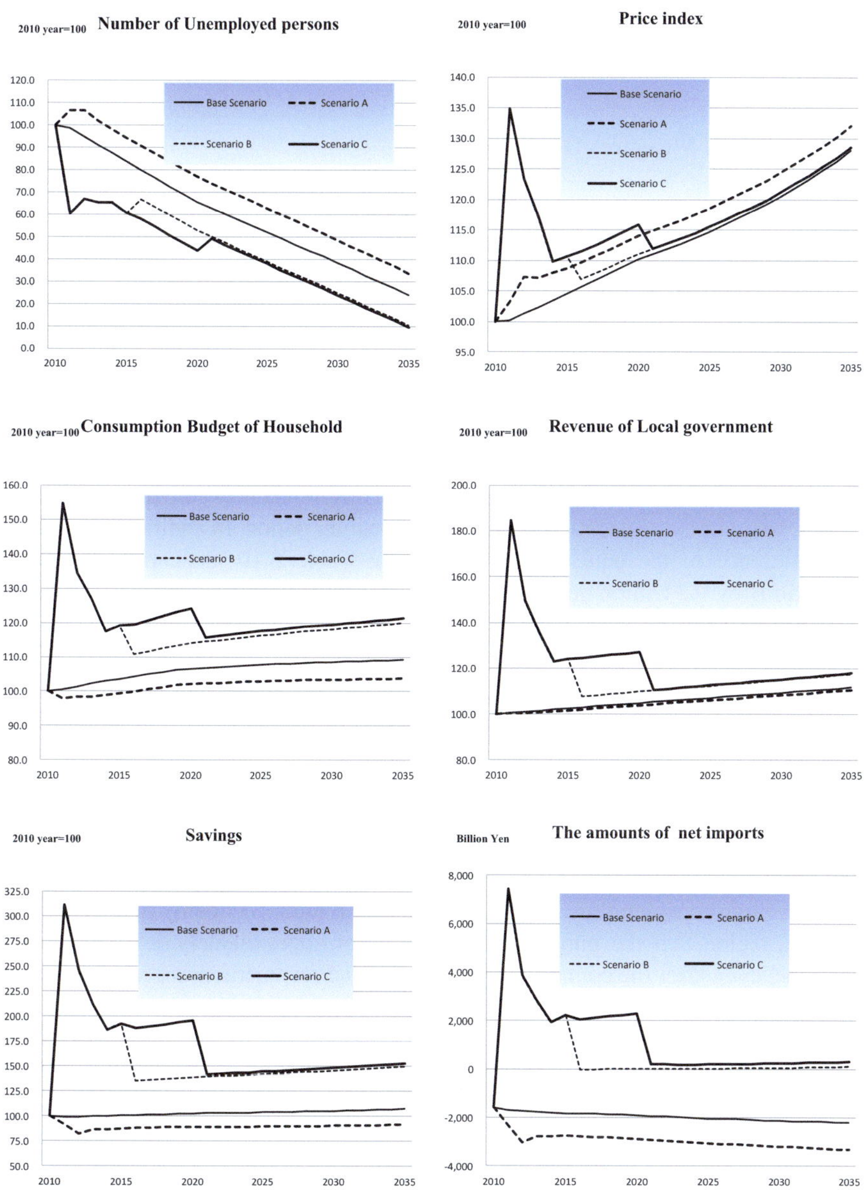

Fig. 5.6 Transition of major economic indicators in the four disaster-affected prefectures after the Great East Japan Earthquake

investment. In addition, Scenario C remains at a comparatively high level until the completion of fiscal measures before falling to the level of Scenario B thereafter.

5.4.2.4 Fiscal Measures Going Forward

There are key desired criteria in the implementation of any future fiscal measures. These criteria should ensure that the regional economies in the disaster-affected region exceed their levels prior to the Great East Japan Earthquake in spite of a declining population trend. We once again review the simulation results for Scenario B and Scenario C provided in Tables 5.8 and 5.10, and Figs. 5.3, 5.4, 5.5 and 5.6. The transition of real GRP for the disaster-affected region shown in Fig. 5.3 indicates that both scenarios do not meet the criteria mentioned above, which suggests that the strengthening of the declining population trend is relatively greater than the fiscal transfer of roughly 1.5 trillion yen after the intensive reconstruction period. This result is because there is not much difference between the scenarios in the transition of the real GRP after the reconstruction period. In addition, looking at the difference in the equivalent variations in Table 5.10, Scenario C exceeds Scenario B by 0.2 trillion yen after the 2020s, which equates to one-seventh times the fiscal transfer of 1.5 trillion yen.

As such, and looking at the transition of real GRP for the disaster-affected region, the positive effects of fiscal transfers after the reconstruction period cannot be recognized as a necessary factor for the sustained economic growth of the disaster-affected region. However, the transitions of the equivalent variation illustrated in Fig. 5.4 show that the value for Scenario C exceeds that for Scenario B, although they both move downward. Furthermore, looking at the major economic indicators of the disaster-affected region in Fig. 5.6, Scenario C shows more preferable results than the other scenarios during the construction period. Taking the above into consideration, it is desirable for the regional economy of the disaster-affected region that the public fiscal transfers of approximately 1.5 trillion yen continue, even after the completion of the intensive reconstruction period, and that these funds are procured by redistributing the local allocation tax grants to the disaster-affected region. In order to realize the conditions presented under Scenario C, it is necessary to take other measures such as productivity improvements in each industry by using fiscal transfers after the intensive reconstruction period ($T = 7$).

5.5 Conclusion and Policy Implications

In this chapter, we adopted the recursive dynamic regional CGE model with a database comprising an SAM between two regions; that is, the disaster-affected region comprising four prefectures and the non-disaster region. We first evaluated the economic impact of the reconstruction budget for the recovery of the disaster-affected region following the Great East Japan Earthquake during a five-year intensive reconstruction period, which is within the ten-year reconstruction period specified by the government. Then, looking ahead to the next 20 years after the intensive reconstruction period, we projected what the regional economies of the disaster-affected region would be in spite of Japan's population decline. At the same time, we clarified which fiscal measures would be necessary in the future.

Our simulation results revealed the following three key points. First of all, even if the Great East Japan Earthquake had not occurred, the four disaster-affected prefectures would still have experienced weaker growth in their regional economies starting in 2012 owing to the impact of a declining population, in which real GRP in the region is forecast to decrease to 83.65 by 2035 (2010 = 100). The equivalent variation of the disaster-affected region would be negative by 2.5 trillion yen in 2020 and further decline to negative 6.2 trillion yen by 2035. As such, we found that, in the four prefectures that were significantly affected by the Great East Japan Earthquake, the economic environment was prone to be negatively impacted by the declining trend in the overall population of the Tohoku region in the 2010s and thereafter, even assuming that the Great Earthquake had not occurred.

Second, we found that the fiscal measures in the intensive reconstruction period could contribute toward recovering the real GRP of the four disaster-affected prefectures, which slumped following the Great East Japan Earthquake. But the real GRP of the disaster-affected region would begin to move downward after 2018 with the end of the intensive reconstruction period. Hence, the fiscal measures alone during the intensive reconstruction period would not enable the region to return to a higher GRP level than that of the pre-earthquake period as the declining population trend would curtail the regional economies. However, the fiscal measures could slow down the pace of curtailment of the regional economies in the disaster-affected region compared with the scenario that assumed no fiscal measures.

We then proceeded to evaluate the fiscal measures for the intensive reconstruction period in terms of their cost effectiveness on the basis of the change in the size of the reconstruction budget and the economic welfare (equivalent variation) of the four disaster-affected prefectures. We found that the cumulative difference between the scenario with fiscal measures and the scenario without fiscal measures from 2010 to 2015 would be about 12.5 trillion yen, which falls short of the cumulative public fiscal transfer amount of 18.1 trillion yen in the intensive reconstruction period. But the cumulative difference from 2010 to 2020 would be 23.8 trillion yen, which exceeds the fiscal transfer amount by 5.7 trillion yen. Therefore, we found that the positive impact of the fiscal measures for the intensive reconstruction period would be realized after the first half of the 2020s.

Third, given these simulation results, it is advisable that the government provides fiscal measures to the disaster-affected region after the intensive reconstruction period by adopting a different form of reconstruction budget—that is, by revising the distribution ratios of the local allocation tax grants. This policy alone, however, cannot enable the regional economy in the disaster-affected region to recover to the level prior to the earthquake. In order to realize this goal, it will be necessary to take other measures after the intensive reconstruction period, such as productivity improvements in each industry located within the disaster-affected region, using fiscal transfers.

In this chapter, we do not consider the various negative effects and fiscal measures caused by the Fukushima Nuclear Power Station accident. This is because the reconstruction and rehabilitation of Fukushima cannot be analyzed accurately using the D2SCGE model. Therefore, we will construct three- regional SAM (the two nuclear disaster–affected prefectures as one region, Fukushima prefecture, and the non-disaster region) for 2011 when the 2011 inter-regional input–output tables for nine regions and the 2011 intra-prefecture input–output tables for Fukushima, Miyagi, Iwate, and Ibaraki prefectures are available. Then, by adopting the D3SCGE model with the three-regional SAM as our database, we will once again evaluate the fiscal measures for the intensive reconstruction period, which will include the funding for the reconstruction and rehabilitation of Fukushima, and present a few scenarios for the recovery and rehabilitation of the regional economies in the disaster-affected region.

References

Board of Audit (2015) About the result of audits concerning the situation of project implementation etc. for the reconstruction etc. form the Great East Japan earthquake Japan revitalization strategy as revised in 2014. The report following the provisions by article 30-3 of the Audit Laws (in Japanese)

Cabinet Office (2011) Estimation of the damage by the Great East Japan Earthquake (in Japanese)

Development Bank of Japan (2011) Estimation of damaged capital stock by the Great East Japan Earthquake -distinct areas such as inland and coast by prefecture(in Japanese) DBJ News

EcoMod Modeling School (2012) Advanced techniques in CGE modeling with GAMS global economic modeling network, Singapore, January 9–13

Hamagushi N (2015) On the recovery of the manufacturing industry from the great East Japan Earthquake. RIETI discussion paper series, 15-J-044, (in Japanese)

Hayashida M, Hamagata S, Nakano K, Hitomi K and Hoshino Y (2011) The impact on the macroeconomy of the great East Japan earthquake: calculation the impact based on CRIEPI Macroeconometric model. SERC Discussion paper, SERCI1024, Central research institute of electric power industry socio-economic research center (in Japanese)

Inada Y, Irie H, Shima A and Toizumi T (2011) The impact on the macro-economy suffered from the Great East Japan Earthquake: compound damage to earthquake, tsunami and nuclear power plant disaster. KISER report, Kansai institute for social and economic research (in Japanese).

Japan Institute for Labour Policy and Training (2014) Labor supply and demand estimates–policy simulations based on the labor supply and demand model (2013). Research Material Series129 (in Japanese)

National Institute of Population and Social Security Research in Japan (2012) Population projections for Japan (January 2012): 2011 to 2060. http://www.ipss.go.jp/site-ad/index_english/esuikei/gh2401e.asp

National Institute of Population and Social Security Research in Japan (2013) Regional Population projections for Japan: 2010–2040. .Population Research Series (in Japanese) http://www.ipss.go.jp/pp-shicyoson/j/shicyoson13/6houkoku/houkoku.pdf

Saito M (2015) Political economy for reconstruction of the earthquake disaster. Nippon Hyoron Sha, Japan. (in Japanese)

Tokunaga S, Okiyama M (eds) (2014) Reconstruction the disaster-affected region of the great East Japan Earthquake and recovery of the regional economy. Bunshindou, Japan. (in Japanese)

Chapter 6
Measuring Economic Gains from New Food and Automobile Industry Clusters with Coagglomeration in the Tohoku Region

Suminori Tokunaga and Mitsuru Okiyama

Abstract In this chapter, we will take the concepts of Porter's (1998, 2000) clusters and present an analysis of the feasibility of ongoing economic development in disaster regions, utilizing two new industry cluster models. The two clusters are the automobile industry, which targets the entire disaster region and where innovation comes from coagglomeration with different industries in megaregions, and the food industry, which, focusing on leveraging local resources, targets individual disaster-struck prefectures. This study applies a dynamic two-regional computable general equilibrium (D2SCGE) model, constructed in Chap. 5, assuming Scenario C (continuation of fiscal support for reconstruction on the basis of a new approach in a 5-year construction period from 2016 to 2020, following the intensive reconstruction period) as the base scenario. We constructed scenarios for each of these two clusters and evaluated the economic effects of each new industry cluster on disaster regions. A simulation analysis of scenarios for these two new clusters with positive and higher productivity in the coagglomerated industries reveals the following two effects: (i) economies of agglomeration from vertical and horizontal coagglomeration boost the real Gross regional Product (GRP) and productivity at the macro level when the two new industry clusters are formed jointly rather than separately and (ii) the clusters contribute to long-term, sustained growth in disaster region economies, thus reducing the gap between their growth and that of other regional economies. This can be interpreted to mean that the usual policies adopted, such as subsidies and corporate tax cuts, are unable to counteract economic stagnation resulting from sharp population decline in disaster regions. It also suggests that agglomeration externalities, evidenced by improved productivity in

S. Tokunaga (✉)
Faculty of Economics and Business Administration, Reitaku University, 2-1-1, Hikarigaoka, Kashiwa, Chiba 277-8686, Japan

University of Tsukuba, Tsukuba 305-8572, Japan
e-mail: tokunaga@reitaku-u.ac.jp

M. Okiyama
Reitaku Institute of Political Economics and Social Studies, Reitaku University, Kashiwa, Chiba 277-8686, Japan

S. Tokunaga, B.P. Resosudarmo (eds.), *Spatial Economic Modelling of Megathrust Earthquake in Japan*, New Frontiers in Regional Science: Asian Perspectives 11,
https://doi.org/10.1007/978-981-10-6493-7_6

the formation of new food/automobile industry clusters, can offer sustained economic development in disaster regions.

Keywords New food and automobile industry clusters • Coagglomeration • Economies of agglomeration • Higher productivity • Dynamic two regional computable general equilibrium (D2SCGE) model • Reconstruction from the Great East Japan Earthquake

6.1 Introduction

The analysis in Chap. 5 tells us that to recover and revitalize a region that has suffered a great deal of damage from an earthquake, a rudimentary shift of fiscal resources to the disaster region is not by itself sufficient to stimulate sustained growth in such regions' economies. Based on these conclusions from Chap. 5, this study considers Porter's concepts of clusters (Porter 1998, 2000) introduced in Chap. 2. We will analyze the potential for ongoing economic development in disaster regions using two new industry cluster models: the automobile industry cluster, which targets the entire disaster region and where innovation comes from the aggregation of different industries in megaregions, and the food industry cluster, which, focusing on leveraging regional resources, targets individual disaster-struck prefectures. Although these industry clusters are not new, policy makers and political officers have recently begun to consider industrial clusters in their formulation of industrial and regional policies in Japan.

This chapter will also adopt the dynamic two-regional computable general equilibrium (D2SCGE) model used in Chap. 5, assuming Scenario C (continuation of fiscal support for reconstruction based on a new approach in a 5-year construction period, from 2016 to 2020, following the intensive reconstruction period) as the base scenario. Furthermore, this chapter sets targets for revitalization derived from Japanese policy strategy as a base scenario, reflecting productivity improvement in each industry region and increased workforce participation across not only disaster regions but also other regions. With these ground rules set, we construct the "new food industry cluster formation scenario," which leverages local resources in each region, and the "new automobile industry cluster formation scenario," which applies more broadly to megaregions and innovation, as mentioned above. We will then assess the economic effects of the formation of each new industry cluster, particularly focusing on the ripple effects existing in disaster region economies, using a simulation based on these scenarios that shows the need to form two types of industry clusters within disaster regions.

6.2 Industry Agglomeration and Clusters

6.2.1 Concept of a New Food Industry Cluster

Before introducing the food industry cluster, let us begin by reviewing the state of agglomeration in Japan's food industry by utilizing the EG index, an agglomeration index proposed by Ellison and Glaeser (1997). Although the food industry ranks in a middling position when observing the two-digit industries[1] of in Japan's standard industry classification in 2010, by refining our scope to four-digit industries,[2] we see that "frozen seafood products," "salt-dried and salted products," and "canned or bottled seafood and seaweed" industries rank highly than median. We can also see that relatively highly-coagglomerated cross-industry pairs in 2010 include "frozen seafood products" and "salt-dried and salted products," as well as "frozen seafood products" and "sugar." This can be interpreted as meaning that food industries using regional resources such as natural resources (like agricultural and fishing resources) have an input-output structure in the same region. They are also often characterized by fairly high agglomeration and relatively high coagglomeration within their own industry class (Tokunaga et al. 2014, 2015).

Following our observations on the food industry, we propose the formation of a new food industry cluster that leverages resources of each individual disaster prefecture based on Porter's (1998, 2000) industry clusters concept, introduced in Chap. 2. According to the Japanese Ministry of Agriculture, Forestry, and Fisheries (2015, 2016), in the 21,480 ha of agricultural land damaged by tsunamis across six prefectures (Aomori, Iwate, Miyagi, Fukushima, Ibaraki, and Chiba), 74% of that land was capable of producing crops as of January 2016. In the three prefectures primarily hit in the disaster, 67% of the land in Iwate, 88% in Miyagi, and 33% in Fukushima had recovered as of January 2016. When referencing the recovery in fisheries, of the 319 fishing harbors damaged initially, 233 or 73% of harbors had restored landing functionality and 91% of the approximately 29,000 fishing boats affected were back to normal function by this same date. Furthermore, although all 22 market facilities across Iwate and Miyagi have returned to full operation, 11 of the 12 markets in Fukushima remain closed. Also, of the 816 seafood processing facilities that have sought to reopen after the disaster across the three primary disaster prefectures, 705 restarted operations as of the end of 2015. This hard data on recoveries in agriculture and fisheries denotes significant progress, excluding the portions of the Fukushima Prefecture that also were affected by the nuclear crisis. Separately, we can gain insight into the economic state of agriculture and fishery management entities that have reopened since suffering damage from the 2011 tsunami by using the "Economic Survey (results from 2014) of Agriculture/Fishery Management Entities," conducted yearly by the Ministry of Agriculture, Forestry

---

[1]The second industry tier as defined by the Japanese government (total: four tiers). For more (Japanese only), see http://www.soumu.go.jp/main_content/000317696.pdf

[2]The most refined industry tier as defined by the Japanese government.

and Fisheries (2016). According to this survey, while 133 management entities saw greater revenue from agricultural products after the disaster, 138 saw lower post-disaster levels, despite reopening, only reaching an index of 65 (with 100 as the pre-disaster base index). For fisheries, although 70 management entities in Iwate and Miyagi reached higher revenue levels than before the disaster, revenue in 84 have not reached pre-disaster levels despite reopening, only reaching an index of 64 (base index = 100). Though some of these agricultural and fishery management entities are expected to surpass pre-disaster levels in the future, management entities will be challenged in returning to or exceeding pre-disaster sales revenues. As selection continues to affect these reopened fishery and agricultural management entities, it is more likely that their activities will fail to lead to recovery in these increasingly sparsely populated disaster regions without forming a new industry cluster structure that generates new value with players from their industries, as well as from the food industry. In actuality, there are now new management organizations using advanced technology[3] and fisheries[4] seeking domestic and international sales channels for their products (Nikkei Newspaper, January 13, 2016, Morning Edition). Also, while the government conducts large-scale trial experiments utilizing advanced technology as part of public and private partnerships endeavoring to help new agriculture, forestry, and fishery businesses grow, there is also the "Ganbaru Fisheries and Aquaculture Restoration and Support Project" initiative, aimed at securing profitability at levels higher than before the disaster for fisheries and aquaculture companies. These cases, however, are only single dots; if they can be spread to cover an entire surface, economic recovery in the disaster region would become possible. In the next section, we will outline a scenario based on the formation of a new food industry cluster that leverages these sorts of regional resources.

6.2.2 *Concept of a New Automobile Industry Cluster*

Before beginning a discussion of the automobile industry cluster, let us first review the state of agglomeration/coagglomeration in manufacturing for the transportation

[3]As an example from the town of Yamamoto in the Watari District, Miyagi Prefecture, the Yamamoto Strawberry Farm reopened and incorporated very shortly after the disaster on 20 June, changing its soil cultivation methods to shelf cultivation and growing its sales in strawberry cultivation/sales/processing to 85 million yen (FY2014), according to onsite research by Tokunaga dated 26-Feb-2016.

[4]Eighteen local fishery processor companies in the city of Kessennuma formed the Kessennuma Shishiori Processing Co-op, running branding and sales activities based on their joint large-scale refrigeration facilities. Also launched was the Kesemo, an organization where other industry players beyond those in fisheries could participate, cooperating with research institutions, developing new products such as seasonings, working toward new industry generation and so on (Nikkei Newspaper, 29 Jun 2015, Morning Edition).

equipment industry, using the aforementioned EG index. In the two-digit industries of Japan's standard industry classification in 2010, this industry (which includes the automobile industry) ranks highly than median, with robust agglomeration. But we can also see, over time, that a downward trend has occurred. In four-digit industries, "forklift trucks and parts and accessories," as well as "motor vehicle parts and accessories," have relatively high levels of agglomeration in their manufacture of transportation equipment industry. In the automobile industry in 2010, we can also see a high coagglomeration relationship between "forklift trucks and parts and accessories" and "motor vehicle parts and accessories," or "motor vehicle parts and accessories" and "transportation machinery and equipment plastic products." In 2010, the motor vehicle parts and accessories manufacturing industry's upstream industries were "transportation machinery and equipment plastic products" and "electric switchgears and wiring devices." Its downstream industries were "forklift trucks and parts and accessories," "automobiles," and "the auto body industry," indicating the existence of both vertical coagglomeration and horizontal coagglomeration due to the relatively high agglomeration between industry groups in same industry. From this analysis of coagglomeration, we found that vertical and horizontal coagglomerations exist in the automobile industry, with automotive assemblers at the top, and a number of parts manufacturers below forming an automobile industry cluster as analyzed in Chap. 4 (Tokunaga et al. 2014; 2015).

Taking into account these empirical results, we will review current trends in the industries associated with automobile production in the Tohoku region and propose the formation of a new automobile industry cluster that brings innovation through agglomeration between different industry groups in a megaregion targeting the entire disaster region. Toyota Motors, one of Japan's leading automakers, has integrated three of its subsidiaries in the Tohoku region so far, establishing a fully owned subsidiary, Toyota Motor East Japan, Inc., in that region, after the disaster in July of 2012. This is its third major base, behind Aichi and northern Kyushu. This entity's 521,000-vehicle production capacity (as of 2015) is second to Toyota Auto Body Co., Ltd.'s 645,000-vehicle capacity (including Toyota Motor Kyushu, a wholly-owned Toyota subsidiary). This means that a new automotive industry cluster is being formed in the Tohoku region.

A number of issues must be negotiated to establish Tohoku as the third-largest domestic production site. The biggest of these is the difficulty obtaining local parts because of poor agglomeration of automobile production companies. Although the amount of local Tohoku parts in Toyota Motor East Japan's production is not known, reports suggest that a high proportion of auto and related parts are procured from the Mikawa region of Aichi (Nikkan Kogyo Newspaper, January 6, 2014). According to Orihashi (2013), use of local parts attracts Tier 1 manufacturers and also cultivates business with Tier 2 and Tier 3 suppliers. By fortunate happenstance, the Kanto region not only has assembly plants from Nissan, Honda, and Subaru, but automobile companies are currently shifting from the traditional integral architecture to a modular one. Furthermore, a shift to electric motors from conventional internal combustion engines is now occurring, with the spread of fuel cell powered and electric vehicles. To delay global warming, it is not necessary to build a traditional pyramid production structure for automotive manufacturing in the Tohoku region. This signals the need for a new automobile industry cluster, a shift from the traditional automobile industry cluster, which was weak in natural disaster risk management (see Fig. 6.1), to a new automobile industry cluster that takes globalization into account. It also requires a thriving regional economy that incorporates automotive parts factories in the region and its environs, and production facilities in other industries that exist lower in the production pyramid, such as electronics components and devices, and metal and materials products. In other

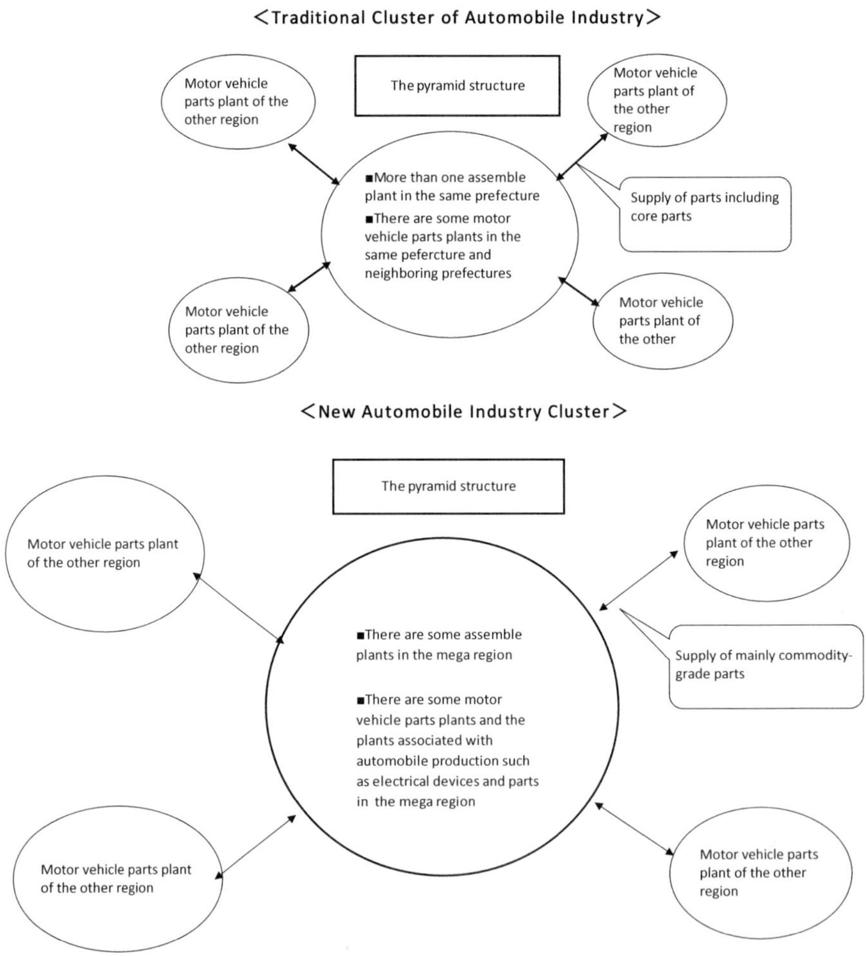

Fig. 6.1 Traditional and new automobile industry clusters
(Source: Reprinted Fig. 6-3 of ch.6 in Tokunaga, S. and Okiyama, M. eds, (2014) Reconstruction the disaster-affected Region of the Great East Japan Earthquake and Recovery of the Regional Economy)

words, the best system emphasizes general parts procurement across several regions, partnering with factories in other locations to better manage risk from natural disasters.[5]

If we reference new automobile industry policies, such as tax benefits adopted by Miyagi Prefecture and others, we notice local allocation tax grants (special disaster reconstruction grants), used to compensate for lost local revenue or to ease the burden on local regions for operations reconstruction, helping to form new industry clusters. These special disaster reconstruction grants totaled 3.8 trillion yen, from the FY2011 supplementary budget through FY2014, as part of the special account for reconstruction. The grants contributed to the recovery of regional disaster economies and industries. With this in mind, we will conduct a simulation that takes a portion of these fiscal resources and uses them as subsidies for industries forming new industry clusters, analyzing its effects on production and hiring in the disaster region's various industries, as well as economic ripple effects to the regional economy and household budgets.

6.3 Assumptions for Setting Scenarios

The assumptions for the scenarios in this chapter are as follows. Population forecasts use the same assumptions as in Chap. 5 (National Institute of Population and Social Security Research in Japan 2012, 2013), but the labor force participation rate is different. We adopt the economic revival/progressive labor participation scenario given in JILPT (2014). This scenario (hereinafter, "Projection B") assumes that the economic and employment policies indicated in the previously mentioned "Japan Revitalization Strategy" by the Cabinet Office (2014a, b), and elsewhere, are appropriately implemented. This scenario postulates 2% real growth and labor force participation by the young, women, and the elderly. Figure 6.2 illustrates the difference between Projection B and Projection A in Chap. 5. In the 2030 labor force, according to Projection B, the number of males and females in the labor force are projected to increase by 2.33 million and 3.69 million, respectively, compared to Projection A.

In addition, assumptions concerning the damage to capital stock in a disaster-affected region are the same in Chap. 5. This chapter also has the same settings for the rate of change of each industry's productivity, in both disaster-affected and unaffected regions, from $T = 2$ to $T = 6$, to trace the recovery phases for regional economies after an earthquake.

[5]There are already regional partnerships in place between auto industry players in the Tohoku region, including an Iwate/Miyagi regional partnership in 2005, the "Tohoku Automotive Industry Agglomeration Partnership Conference (Aomori, Akita, Iwate, Miyagi, Yamagata, Fukushima)" in 2007 and the "North Tohoku Three-Prefecture Automotive Technology Symposium (Iwate, Akita, Aomori)" in 2008 (Murayama 2013, Fig. 4-2 p. 69).

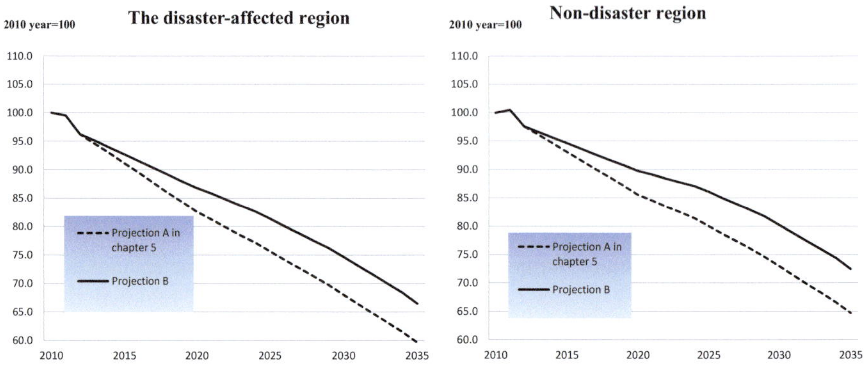

Fig. 6.2 Transition of the labor force through 2030, based on projections for two regions

6.4 Simulation Contents and Results

6.4.1 Establishment of Each Scenario

In this section, we can use the dynamic two-regional computable general equilibrium (D2SCGE) model, as in Chap. 5. The base scenario and two new industry cluster scenarios are set for these assumptions.

(a) Establishment of the Base Scenario

The base scenario of increasing productivity and labor participation for each industry assumes that after the Great East Japan Earthquake, fiscal measures are implemented for disaster-affected regions, while various industrial policies to revitalize Japan's economy are launched (hereafter, the "Base Scenario"). Under this Base Scenario, the labor force population in Projection B increases in line with overall Japanese economic growth, unlike the Base Scenario in Chap. 5. However, the same assumption is made for individuals 65 years of age and above, as in Chap. 5. Furthermore, the Base Scenario assumes that various industrial measures under the Japan Revitalization Strategy enhance productivity in each industry, not only in the four disaster-affected prefectures but also in the non-disaster regions from 2016 ($T = 7$) after the intensive reconstruction period ends. In addition, we referred to the medium-term forecast for the Japanese Economy, which the Japan Center for Economic Research (JCER) released in August 2015.[6] Under the D2SCGE Model, the efficiency parameter for each industry's production obtained

[6] JCER (2015) forecasted the Baseline Scenario that although the growth rate of Japanese economy can be achieved nearly 1% by 2020, the years 2021–2025 and 2026–2030 will likely see 0.3 growth and zero growth respectively.

Table 6.1 Changes in productivity for each simulation

Annual % point of increase within each period of time	Every simulation,					Base simulation and simulation III-b		Simulation I		Simulation II		Simulation III-a	
	T = 1	T = 2	T = 3	T = 4	T = 5–6	T = 7–10	T = 11–21	T = 7–10	T = 11–21	T = 7–10	T = 11–21	T = 7–10	T = 11–21
Agriculture and forestry													
The disaster in Iwate, Miyagi, Fukushima and Ibaraki Prefectures	0.0%	1.3%	0.0%	0.0%	0.0%	5.0%	6.0%	6.0%	8.0%	5.0%	6.0%	6.0%	8.0%
Other areas without the disaster-affected prefectures	0.0%	0.3%	0.0%	0.0%	0.0%	5.0%	6.0%	5.0%	6.0%	5.0%	6.0%	5.0%	6.0%
Fisheries													
The disaster in Iwate, Miyagi, Fukushima and Ibaraki Prefectures	0.0%	−13.0%	−13.0%	0.5%	0.5%	1.0%	5.0%	2.0%	7.0%	1.0%	5.0%	2.0%	7.0%
Other areas without the disaster-affected prefectures	0.0%	0.0%	0.0%	0.0%	0.0%	2.0%	5.0%	2.0%	5.0%	2.0%	5.0%	2.0%	5.0%
Foods and beverage													
The disaster in Iwate, Miyagi, Fukushima and Ibaraki Prefectures	0.0%	11.0%	11.0%	2.0%	0.5%	0.5%	1.0%	2.0%	4.0%	0.5%	1.0%	2.0%	4.0%
Other areas without the disaster-affected prefectures	0.5%	−1.0%	−1.0%	0.3%	0.0%	1.5%	2.5%	1.5%	2.5%	1.5%	2.5%	1.5%	2.5%
Electrical devices and parts													
The disaster in Iwate, Miyagi, Fukushima and Ibaraki Prefectures	0.0%	15.0%	15.0%	2.5%	0.5%	0.2%	0.2%	0.2%	0.2%	1.0%	1.5%	1.0%	1.5%
Other areas without the disaster-affected prefectures	0.5%	−1.0%	−1.0%	0.3%	0.0%	0.3%	0.25%	0.3%	0.25%	0.3%	0.25%	0.3%	0.25%

(continued)

Table 6.1 (continued)

Annual % point of increase within each period of time	Every simulation,					Base simulation and simulation III-b		Simulation I		Simulation II		Simulation III-a	
	T = 1	T = 2	T = 3	T = 4	T = 5–6	T = 7–10	T = 11–21	T = 7–10	T = 11–21	T = 7–10	T = 11–21	T = 7–10	T = 11–21
Motor vehicles and parts													
The disaster in Iwate, Miyagi, Fukushima and Ibaraki Prefectures	0.0%	17.0%	17.0%	3.5%	1.0%	0.3%	0.3%	0.3%	0.3%	1.5%	2.0%	1.5%	2.0%
Other areas without the disaster-affected prefectures	0.0%	−10.0%	10.0%	0.5%	0.0%	0.3%	0.3%	0.3%	0.3%	0.3%	0.3%	0.3%	0.3%
Other manufacturing products and mining													
The disaster in Iwate, Miyagi, Fukushima and Ibaraki Prefectures	0.0%	6.5%	6.5%	1.5%	0.1%	0.1%	0.1%	0.1%	0.1%	0.5%	1.0%	0.5%	1.0%
Other areas without the disaster-affected prefectures	0.0%	−1.0%	−1.0%	0.3%	0.0%	0.3%	0.3%	0.3%	0.3%	0.3%	0.3%	0.3%	0.3%
Construct													
The disaster in Iwate, Miyagi, Fukushima and Ibaraki Prefectures	0.0%	−3.0%	−3.0%	3.0%	3.0%	0.0%	0.0%	0.0%	0.0%	0.0%	0.0%	0.0%	0.0%
Other areas without the disaster-affected prefectures	0.0%	15.0%	0.0%	0.0%	0.0%	0.0%	0.0%	0.0%	0.0%	0.0%	0.0%	0.0%	0.0%
Electricity, gas and heat-water supply													
The disaster in Iwate, Miyagi, Fukushima and Ibaraki Prefectures	0.0%	−20.0%	−20.0%	0.5%	0.5%	0.0%	0.0%	0.0%	0.0%	0.0%	0.0%	0.0%	0.0%
Other areas without the disaster-affected prefectures	0.0%	−20.0%	−20.0%	15.0%	2.5%	0.0%	0.00%	0.0%	0.00%	0.0%	0.00%	0.0%	0.00%

Commerce													
The disaster in Iwate, Miyagi, Fukushima and Ibaraki Prefectures	0.0%	9.5%	9.5%	2.0%	1.0%	0.0%	0.0%	0.0%	0.0%	0.0%	0.0%	0.0%	0.0%
Other areas without the disaster-affected prefectures	0.0%	0.0%	4.0%	1.5%	0.0%	0.0%	0.0%	0.0%	0.0%	0.0%	0.0%	0.0%	0.0%
Transport													
The disaster in Iwate, Miyagi, Fukushima and Ibaraki Prefectures	0.0%	9.5%	9.5%	2.0%	1.0%	0.0%	0.0%	0.0%	0.0%	0.0%	0.0%	0.0%	0.0%
Other areas without the disaster-affected prefectures	0.0%	−2.0%	2.0%	0.5%	0.5%	1.5%	1.0%	1.5%	1.0%	1.5%	1.0%	1.5%	1.0%
Financial and insurance, real estate, communication, public administration, education and services													
The disaster in Iwate, Miyagi, Fukushima and Ibaraki Prefectures	0.0%	1.7%	6.0%	0.3%	0.1%	0.0%	0.0%	0.0%	0.0%	0.0%	0.0%	0.0%	0.0%
Other areas without the disaster-affected prefectures	0.0%	0.3%	5.0%	1.0%	0.0%	2.0%	1.0%	2.0%	1.0%	2.0%	1.0%	2.0%	1.0%

through calibration is raised by the growth ratios indicated in Table 6.1 (Base Simulation) to reflect productivity improvement in each industry.[7] The Base Scenario also assumes continuation of fiscal support for reconstruction based on a new approach in the construction period from 2016 to 2020, that is, the same as Scenario C in Chap. 5. Under the Base Scenario, after 2020, local governments in disaster-affected areas will get 1.5 trillion yen in additional annual revenue from the central government by applying some of the local allocation tax grants to non-disaster regions. These grants will partly restore the public infrastructure and also raise social security and pension benefit expenditures. In addition, we assume that special reconstruction income tax revenues will increase because of high growth rates compared with Scenario C in Chap. 5, although reduced by 0.05% points from $T = 11$.

(b) Formation Scenario for a New Food Industry Cluster

In this scenario (hereafter called the "new food industry cluster formation scenario"), we first incorporate industry recovery measures used to form new industry clusters that utilize local resources.[8] These include subsidizing the agricultural, forestry, fishing, and food and beverage industries, continuing the intensive reconstruction period through 2016 and thereafter. It also includes subsidizing industry rates above pre-disaster levels, as seen in Table 6.2, with significant jumps of 4.4%, 7.0%, and 4.5%. These funds will be set at 200 billion yen, partially sourced from local subsidy taxes paid to disaster regions. Furthermore, in the new food industry cluster formation scenario, it is expected that this sort of subsidized policy and cluster strategy will lead to new product development and new industry generation utilizing disaster region resources, improved product competitiveness, and stronger cost-competitiveness between regions. In the model, we change our elasticity of substitution from 2.0 to 3.0 for intra-regional products in the agricultural, forestry, fishery, and food and beverage industries. Finally, in the new food industry cluster formation scenario, these industry recovery measures lead to the formation of a new food industry cluster, where agriculture, forestry, and fishery industries utilize local resources. As a result, productivity in the agriculture, forestry, fishery, and foodstuff industries is expected to surpass that of other non-disaster regions by 2025.[9] In the D2SCGE model, raising the three production-function efficiency parameters, obtained by calibrating growth rates indicated in Simulation I (Table 6.1), yields results like those shown in Table 6.3.

---

[7]For this 2SCGE model, see Appendix 1. Furthermore, for the spatial CGE mode, see Tokunaga et al. (2003), Hosoe et al. (2010), EcoMod Modeling School (2012), and Okiyama et al. (2014).

[8]For the economics of Clusters, see Duranton et al. (2010), Fujita et al. (1999), Fujita and Thisse (2013), and Porter (1998, 2000).

[9]In Japanese food industry for 1985–2000, the output elasticities with respect to agglomeration is estimated at 0.023 (Kageyama and Tokunaga 2006).

Table 6.2 Subsidy rate by production activity and Savings rate of local government in the disaster-affected region

(a) Setting the subsidy rate

	Every simulation						Base simulation	Simulation I	Simulation II	Simulation III-a,b
Subsidy rate per industry	T = 1	T = 2	T = 3	T = 4	T = 5	T = 6	T = 7–21	T = 7–21	T = 7–21	T = 7–21
Agriculture and forestry	1.4%	18.4%	4.8%	2.9%	2.7%	2.3%	1.4%	4.4%	1.4%	4.4%
Fisheries	0.5%	21.5%	14.5%	7.0%	6.5%	4.5%	0.5%	7.0%	0.5%	7.0%
Foods and beverage	0.5%	1.7%	0.9%	1.2%	0.7%	0.7%	0.5%	4.5%	0.5%	4.5%
Electrical devices and parts	0.0%	0.5%	0.2%	0.2%	0.0%	0.1%	0.0%	0.0%	4.0%	4.0%
Motor vehicles and parts	0.0%	1.0%	0.3%	0.5%	0.1%	0.2%	0.0%	0.0%	6.0%	6.0%
Other manufacturing products and mining	0.0%	1.5%	0.5%	0.9%	0.3%	0.3%	0.0%	0.0%	2.0%	2.0%
Commerce	0.0%	0.6%	0.2%	0.4%	0.2%	0.2%	0.0%	0.0%	0.0%	0.0%
Transport	0.3%	1.4%	0.6%	0.9%	0.5%	0.5%	0.3%	0.3%	0.3%	0.3%

(b) Setting the savings rate

	Savings rate of local government							
	T = 1	T = 2	T = 3	T = 4	T = 5	T = 6	T = 7–11	T = 12–21
Every Simulation	2.6%	11.4%	15.0%	11.4%	11.0%	11.2%	10.2%	2.6%

Table 6.3 Change of the automobile industry's input structure using inter-regional IO tables

Billion Yen		Transportation equipment						Electrical machinery					
		1995	Input ratio	2000	Input ratio	2005	Input ratio	1995	Input ratio	2000	Input ratio	2005	Input ratio
Fukuoka prefecture	General machinery	2.4	0.2%	1.4	0.1%	3.4	0.2%	1.5	0.2%	1.2	0.1%	1.5	0.2%
	Electrical machinery	12.4	1.0%	12.2	0.9%	22.3	1.1%	36.9	4.6%	43.8	4.9%	74.9	12.0%
	Transportation equipment	107.3	8.6%	162.2	11.9%	371.8	19.2%	0.0	0.0%	0.0	0.0%	0.0	0.0%
	Other industries	226.7	18.1%	255.7	18.8%	426.8	22.0%	237.9	29.9%	308.5	34.7%	254.8	40.7%
	Total intermediate	348.7	27.9%	431.5	31.7%	824.3	42.5%	276.3	34.7%	353.4	39.8%	331.3	53.0%
Outside the prefecture	General machinery	9.8	0.8%	7.0	0.5%	7.7	0.4%	6.3	0.8%	6.1	0.7%	3.5	0.6%
	Electrical machinery	46.0	3.7%	50.2	3.7%	64.7	3.3%	137.4	17.3%	180.7	20.3%	82.9	13.2%
	Transportation equipment	527.6	42.1%	566.4	41.7%	680.4	35.1%	0.0	0.0%	0.0	0.0%	0.0	0.0%
	Other industries	88.4	7.1%	87.7	6.5%	123.8	6.4%	75.7	9.5%	85.5	9.6%	58.8	9.4%
	Total intermediate	671.9	53.7%	711.4	52.3%	876.6	45.2%	219.4	27.6%	272.3	30.7%	145.1	23.2%
Total of intermediate sectors		1020.6	81.5%	1142.9	84.0%	1700.9	87.7%	495.7	62.3%	625.7	70.5%	476.4	76.1%
Compensation of employees		120.9	9.7%	128.5	9.4%	140.7	7.3%	160.3	20.1%	160.2	18.0%	93.9	15.0%
Total of gorss values added sectores		231.4	18.5%	217.0	16.0%	238.5	12.3%	300.4	37.7%	262.3	29.5%	149.2	23.9%
Domestic production		1252.0	100.0%	1359.8	100.0%	1939.4	100.0%	796.1	100.0%	888.1	100.0%	625.6	100.0%

Source: Inter-regional input-output table of non-competitive imports for Fukuoka prefecture

(c) Formation Scenario for a New Automobile Industry Cluster

In this scenario (hereafter called the "new automobile industry cluster formation scenario"), we incorporated three factors for the formation of a new automobile industry cluster. The first is greatly increasing subsidies in 2016, and thereafter, for the automotive and auto parts industry, the electronic parts, devices and electronic circuits industry, and other manufacturing and mining industries that contribute to the new automobile industry cluster. Subsidy rates are set to 6.0%, 4.0%, and 2.0%, respectively, as per Table 6.2. These funds will be set at 600–700 billion yen, partially sourced from local allocation tax grants paid to disaster regions. Second, we incorporated the key terms of commodity-grade parts into our new automobile industry cluster formation scenario. Specifically, we altered the elasticity of substitution from 2.0 to 3.0 for inter-regional products in the above three categories, increasing inter-regional commodity-grade products. Third, we continued the round of policies and countermeasures mentioned above in our new automobile industry cluster formation scenario, incorporating productivity increases in the aforementioned three categories. At the end of our analysis, we incorporated perceived jumps in productivity, caused by changes in input-output structures in northern Kyushu, with the agglomeration that formed in the automotive industry beginning in the early 1990s. In 1975, Nissan Motor Company opened its Kyushu plant in Fukuoka Prefecture, adding the number two factory in 1992. In 2011, it also established Nissan Motor Kyushu Co., Ltd., as well as Nissan Shatai Kyushu in 2009, which began automotive assembly operations. Toyota Motors established its Miyata plant in 1992 in Fukuoka Prefecture, starting Toyota Motor Kyushu Inc. at the same time. It also added its Kanda engine plant in 2005 and its Kokura plant in 2008, totaling three plants in Fukuoka Prefecture. Daihatsu Motor also established Daihatsu Kyushu in 2004 in Nakatsu City, and Oita Prefecture added the Kurume engine plant in 2008 to its Oita assembly plant as part of its Kyushu fleet. Auto parts suppliers affiliated with Toyota Motor Corp., such as Toyoda Gosei Co., Ltd. and Toyota Boshoku Corporation, are also making inroads in the northern Kyushu region.[10] We observed economies of agglomeration in the automotive industry that resulted from the industrial concentration in the northern Kyushu region, as mentioned above, utilizing the inter-regional and intra-prefectural input-output tables in Fukuoka. The results are displayed in Tables 6.3 and 6.4. In Table 6.3, we can see the transportation and electric equipment input structure constructed from the Fukuoka Prefecture input-output tables and inter-regional input-output tables for 1995, 2000, and 2005. These tables reveal the following two points. The first is that in the automotive industry (more than half of transportation equipment is related to motor vehicles and their parts), procurement for automotive parts, from 1995 to 2005, dramatically shifted from extra-prefectural to intra-regional sources. We can conclude from this that agglomeration in the northern Kyushu automotive

[10]For more on the state of automotive industry agglomeration in Kyushu, see Mokudai and Ishiro (2013).

Table 6.4 Change of the automobile industry's output structure using intra-regional IO tables

Motor vehicles and motor vehicles parts	1990	1995	2000	2005	2011
Percentage of total demand: Unit:%					
Outflows and exports	64.5%	84.3%	79.5%	65.6%	92.3%
Percentage of total output: Unit %					
Outflows and exports	60.6%	83.3%	83.2%	73.2%	92.9%
Net outflows and exports	−11.3%	−6.3%	18.2%	22.1%	7.7%
Compensation of employees	7.7%	9.2%	9.0%	7.0%	8.1%
Total of gorss value added	16.7%	17.7%	15.3%	12.0%	10.0%

Source: Intra-regional input-output table of competitive imports for Fukuoka prefecture

industry greatly advanced after 2000. The second is that compensation per employee, and total value added to total transportation production, fell 2.1% and 3.7%, respectively, from 2000 to 2005. We can surmise from this that productivity increased in the automotive industry. Trends in the ratio of value-added, regional balance of payments to total production of motor vehicles and motor vehicle parts (in Table 6.4) between 1990 and 2011, derived from five examples in the intra-prefectural competitive import-type input-output tables. Table 6.4 reveals both higher agglomeration in the automotive industry and expanded transaction value between the automotive industry in Fukuoka Prefecture and other regions and nations. This case indicates that the advancement of industry agglomeration leads to increased intra-regional procurement rates for parts necessary for assembly, higher productivity rates in the industry, and more energized trade both inside and outside the region. Given this example of industry agglomeration in the Kyushu region, values seen in Table 6.5 show the higher efficiency parameters of D2SCGE model's three-production function, obtained by calibration, using growth rates from Simulation II, expressed in Table 6.1.[11]

(d) Corporate Tax Cuts for Firms Located in Disaster Regions

One further government policy that can promote the formation of new industry clusters is the reduction of corporate taxes for firms located in disaster regions.[12] More specifically, this refers to cutting corporate tax rates by 30% for companies in these regions. Although this policy decreases tax revenue for local governments in disaster regions, it strategically allows new industry clusters to form in these regions, with the goal that the tax cuts ultimately will contribute to increased total production volume for the industries in question.

[11]In Japanese assembly-type manufacturing industry for 1985–2000, the output elasticities with respect to agglomeration and coagglomeration are estimated at 0.06 and 0.013 (Tokunaga et al. 2014).

[12]Prefectures such as Miyagi have established special reconstruction zones (forgiving local taxes for 5 years as well as national tax, etc.) to attract new industry.

Table 6.5 Efficiency parameters for two scenarios

The disaster in Iwate, Miyagi, Fukushima and Ibaraki Prefectures		The efficiency parameter 2010	2015	2020	2025	2030
Agriculture and forestry	(1) Base scenario	0.036	0.037	0.047	0.063	0.085
	(2) Scenario for new food industry cluster	0.036	0.037	0.050	0.074	0.108
	(2)-(1)	0.000	0.000	0.003	0.011	0.023
Fisheries	(1) Base scenario	0.070	0.054	0.059	0.079	0.096
	(2) Scenario for new food industry cluster	0.070	0.054	0.063	0.088	0.123
	(2)-(1)	0.000	0.000	0.004	0.009	0.027
Foods and beverage	(1) Base scenario	0.611	0.776	0.799	0.840	0.883
	(2) Scenario for new food industry cluster	0.611	0.776	0.873	1.062	1.293
	(2)-(1)	0.000	0.000	0.074	0.222	0.410
Electrical devices and parts	(1) Base scenario	2.359	3.230	3.263	3.296	3.329
	(2) Scenario for new automobile industry cluster	2.359	3.230	3.412	3.676	3.960
	(2)-(1)	0.000	0.000	0.149	0.380	0.631
Motor vehicles and parts	(1) Base scenario	3.561	5.146	5.224	5.303	5.383
	(2) Scenario for new automobile industry cluster	3.561	5.146	5.571	6.151	6.792
	(2)-(1)	0.000	0.000	0.347	0.848	1.409
Other manufacturing products and mining	(1) Base scenario	1.352	1.560	1.564	1.568	1.571
	(2) Scenario for new automobile industry cluster	1.352	1.560	1.607	1.689	1.775
	(2)-(1)	0.000	0.000	0.043	0.121	0.204

6.4.2 *Simulation Designs and Results*

6.4.2.1 Simulation Designs

In this section, we implemented four simulations based on the above three scenarios and one measure. First, we implemented a simulation based on the Base Scenario. The following sections refer to this simulation as the "Base Simulation." Second, we implemented a simulation based on the scenario of a new food industry cluster. The following sections refer to this simulation as "Simulation I." Third, we implemented a simulation based on the scenario of a new automobile industry cluster. The following sections refer to this simulation as "Simulation II." Fourth, we implemented a simulation based on corporate tax cuts, while adding together scenarios for the new food industry and new automobile industry clusters. The following sections refer to this simulation as "Simulation III-a." Finally, we implemented a simulation based on the assumptions of a subsidy policy, the measure for corporate tax cuts, and commodity-grade products for two clusters, that is, without productivity improvement. The following sections refer to this simulation as "Simulation III-b." The contents of these simulations are summarized in Table 6.6. In this table, each factor marked with an "○" is incorporated in the above simulations.

6.4.2.2 Simulation Results

1. Assessment of the Base Simulation

We examine the transition of real GRP. Regional economies in disaster-affected regions in Table 6.7 and Fig. 6.3 would look different from Scenario C in Chap. 5. That is, even after an intensive reconstruction period, regional economies in disaster-affected regions will stay almost flat until 2024. Although regional economies in disaster-affected regions will see slower growth beginning in 2025, continuing thereafter with the declining population, real GRP in disaster-affected regions in 2030 reaches 104.94, above the 100 in 2010. As such, it is difficult for regional economies in disaster-affected regions to recover by simply implementing fiscal measures during the reconstruction period. Regional economies in disaster-affected regions can maintain pre-earthquake levels only if measures to invigorate the Japanese economy are launched in addition to continuing financial support, even after the reconstruction period. On the other hand, regional economies in non-affected regions will achieve 0.5–1.0% growth from the late 2010s to the 2020s. However, real GRP growth rates will turn negative after 2030. As a result, real GRP in non-affected regions in 2030 reaches 114.36, compared to 100 in 2010.[13]

Equivalent variations (EV) in the disaster-affected region, for each year of the Base Simulation in Table 6.8 and Fig. 6.4, turn negative after 2014 and continue to move downward thereafter with a greater degree of decline. As a result, a decrease of 1.6 trillion yen is shown in 2030. Meanwhile, an equivalent variation in the non-disaster region remains positive from 2013 to 2030, and although it turns downward after 2025, its value is a positive 4.8 trillion yen.

2. Economic Effects of the Formation of the Two New Industry Clusters in the Disaster Region

   (a) Assessing Economic Effects by GRP

If we look at Table 6.7 and Fig. 6.3, which express the change in real GRP over time in the disaster region as part of Simulations I to III-b, we see that in

[13]This model has no assumptions of labor movement and capital transfer between the disaster-affected region and the non-disaster region. Therefore, even if the return to capital in the disaster-affected region had increased by the equivalent of the damage to the capital stock, there is no the capital transfer from the non-disaster region to the disaster-affected region. As a result, the real GRP of the disaster-affected region continues to be influenced by the damage to the capital stock even if the fiscal measures were implemented. In the future, we hope to conduct an improved CGE model that takes into account the movement of production factors between regions.

Table 6.6 Contents of each simulation

	The factors are incorporated in each scenario	Base simulation	Simulation I	Simulation II	Simulation III-a	Simulation III-b
Base scenario	Progressive labor participation	○	○	○	○	○
	Enhancing productivity of each industry	○	○	○	○	○
Scenario for food-processing cluster in disaster-affected region(Industris targeted are agriculture-and -forestry, fisheries and foods-and-beverage)	Subsidy policy		○		○	○
	Commodity-grade products		○		○	○
	Productivity improvement		○		○	
Scenario for new automobile cluster in disaster-affected region(Industris targeted are electrical devices-and-parts, motor vehicles -and-parts and other manufacturing products-and-mining)	Subsidy policy			○	○	○
	Commodity-grade products			○	○	○
	Productivity improvement			○	○	
Measure of corporate tax cut in disaster-affected region					○	○

Table 6.7 Simulation results on real GRP and intra-regional gross production by scenario

		Real GRP					Total output				
(Annual rate, index of 2010 = 100)		2010–2015	2015–2020	2020–2025	2025–2030	2030/2010	2010–2015	2015–2020	2020–2025	2025–2030	2030/2010
The whole country	(1) Base simulation	0.51%	0.99%	0.67%	0.43%	113.78	0.83%	0.59%	0.39%	0.19%	110.52
	(2) Simulation I	0.51%	1.00%	0.68%	0.44%	113.96	0.84%	0.66%	0.49%	0.30%	112.04
	(3) Simulation II	0.51%	1.00%	0.68%	0.44%	113.94	0.84%	0.80%	0.54%	0.35%	113.41
	(4) Simulation III-a	0.51%	1.01%	0.70%	0.45%	114.15	0.84%	0.88%	0.64%	0.44%	114.95
	(5) Simulation III-b	0.51%	0.99%	0.67%	0.43%	113.82	0.84%	0.74%	0.40%	0.21%	111.52
	(2)-(1)	0.00%	0.01%	0.01%	0.01%	0.18	0.00%	0.07%	0.10%	0.10%	1.52
	(3)-(1)	0.00%	0.01%	0.01%	0.01%	0.17	0.01%	0.21%	0.15%	0.15%	2.89
	(4)-(1)	0.00%	0.02%	0.02%	0.02%	0.37	0.01%	0.28%	0.25%	0.25%	4.43
	(5)-(1)	0.00%	0.01%	0.00%	0.00%	0.04	0.01%	0.15%	0.01%	0.01%	1.01
The disaster-affected region (Iwate, Miyagi, Fukushima and Ibaraki Prefectures)	(1) Base simulation	1.23%	0.02%	−0.02%	−0.26%	104.94	1.50%	0.04%	−0.08%	−0.22%	106.32
	(2) Simulation I	1.24%	0.16%	0.21%	−0.03%	108.16	1.51%	0.20%	0.18%	0.05%	110.13
	(3) Simulation II	1.25%	0.23%	0.23%	0.00%	108.85	1.52%	0.57%	0.31%	0.19%	113.74
	(4) Simulation III-a	1.26%	0.37%	0.45%	0.21%	112.04	1.53%	0.74%	0.56%	0.43%	117.57
	(5) Simulation III-b	1.26%	0.13%	0.02%	−0.22%	106.07	1.53%	0.40%	−0.04%	−0.18%	108.90
The disaster-affected region (Iwate, Miyagi, Fukushima and Ibaraki Prefectures)	(2)-(1)	0.01%	0.14%	0.23%	0.23%	3.22	0.01%	0.16%	0.26%	0.27%	3.81
	(3)-(1)	0.02%	0.21%	0.25%	0.26%	3.91	0.02%	0.53%	0.40%	0.41%	7.42
	(4)-(1)	0.03%	0.35%	0.47%	0.47%	7.10	0.03%	0.71%	0.64%	0.64%	11.25
	(5)-(1)	0.03%	0.11%	0.04%	0.04%	1.13	0.03%	0.37%	0.04%	0.04%	2.58
Other areas without the disaster-affected region	(1) Base simulation	0.46%	1.05%	0.72%	0.47%	114.36	0.38%	0.98%	0.70%	0.46%	113.31
	(2) Simulation I	0.46%	1.05%	0.71%	0.47%	114.34	0.38%	0.98%	0.70%	0.46%	113.30
	(3) Simulation II	0.46%	1.05%	0.71%	0.46%	114.28	0.38%	0.97%	0.69%	0.45%	113.19
	(4) Simulation III-a	0.46%	1.05%	0.71%	0.46%	114.29	0.37%	0.97%	0.69%	0.45%	113.20
	(5) Simulation III-b	0.46%	1.05%	0.72%	0.47%	114.33	0.37%	0.97%	0.70%	0.46%	113.27
	(2)-(1)	0.00%	0.00%	0.00%	0.00%	▲ 0.01	0.00%	0.00%	0.00%	0.00%	▲ 0.01
	(3)-(1)	0.00%	0.00%	0.00%	0.00%	▲ 0.08	0.00%	−0.01%	−0.01%	−0.01%	▲ 0.12
	(4)-(1)	0.00%	0.00%	0.00%	0.00%	▲ 0.07	0.00%	−0.01%	−0.01%	−0.01%	▲ 0.11
	(5)-(1)	0.00%	0.00%	0.00%	0.00%	▲ 0.03	0.00%	0.00%	0.00%	0.00%	▲ 0.04

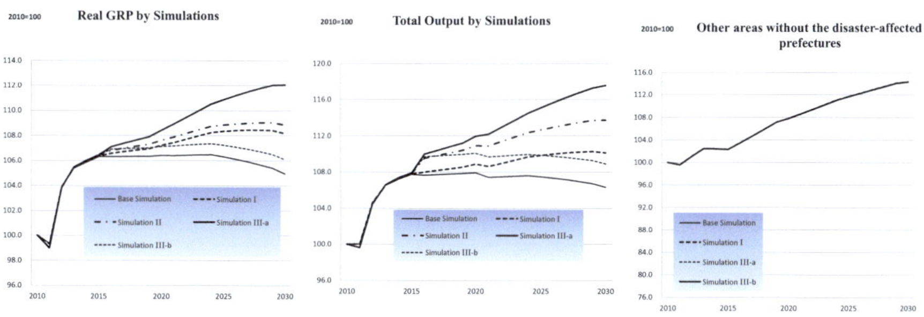

Fig. 6.3 Transition of real GRP and total output in the disaster-affected region

Simulation I, which was based on the formation of a new food industry cluster, that real GRP grows at a rate of 0.16% from the intensive reconstruction period to 2020. In 2020–2025, increased productivity from the agriculture, forestry, fisheries, and food and beverage industries, which form the new food industry cluster, contribute to a real GRP that rises 0.21%. However, this data suggests that in the latter half of the 2020s, population declines in the disaster region and regional economy's growth rates turn negative around 2030. However, the disaster area's real GRP in 2030 in this scenario hits an index of 108.16, up by 3.22 points over the Base Simulation, revealing positive effects on production in the disaster region. Specifically, we can see that production volumes per industry, for the aforementioned industries of the new food industry cluster, increase in the years following 2020 (see Fig. 6.5).

In the automobile industry, however, we can see from Fig. 6.3 that in Simulation II, based on the new automobile industry cluster formation scenario, real GRP is greater than in Simulation I over the entire period. Although growth rates trend upwards by 0.23% from the intensive reconstruction period to the early 2020s, downward population pressure in the disaster region during the latter half of the 2020s causes regional economic growth to flatten. As a result, the disaster region's real GRP index in 2030 is 108.85, 3.91 points above the Base Simulation, due to increased production. Specifically, we can see from Fig. 6.5 that the automotive and auto parts industry, as well as the electronic components and devices and other manufacturing and mining industries that make up this cluster, achieve stable, if modest, growth in production volumes.

By contrast, in Simulation III-a, where formation is simultaneous for both a new food industry cluster and a new automobile industry cluster, and where corporate taxes are cut in disaster regions, there is economic growth of 0.45% in the early 2020s, which is greater than the 0.37% after the intensive reconstruction period. Furthermore, although considerable population decline slows real GRP growth to 0.21% in the latter half of the 2020s, the disaster area's regional economy achieves continued growth through about 2030. As a result, real GRP in 2030 reaches an index of 112.04 versus 2010, 7.1 points above the Base Simulation, significantly boosting the disaster area's regional economy. If we look at the results of Simulation III-a in terms of total output in the disaster region, we see growth rates of 0.4%–

Table 6.8 Simulation results of equivalent variation by scenario

Trillion Yen		Equivalent variation							Accumulated equivalent variation			
		2011	2012	2013	2015	2020	2025	2030	2015	2020	2025	2030
Whole country	(l) Base simulation	▲ 0.7	0.1	3.6	1.6	7.8	7.6	3.1	10.6	40.4	80.0	106.8
	(2) Simulation I	▲ 0.7	0.1	3.6	1.6	8.0	8.0	3.7	10.6	40.8	81.9	110.9
	(3) Simulation II	▲ 0.7	0.1	3.6	1.6	8.1	8.1	3.7	10.7	41.7	82.7	110.9
	(4) Simulation III-a	▲ 0.6	0.1	3.6	1.6	8.3	8.5	4.3	10.7	42.3	85.5	117.5
	(5) Simulation III-b	▲ 0.6	0.1	3.6	1.6	8.1	7.9	3.4	10.7	41.7	82.7	110.9
	(2)-(l)	0.0	0.0	0.0	0.0	0.1	0.4	0.5	0.0	0.5	1.8	4.1
	(3)-(l)	0.1	0.0	0.0	0.0	0.3	0.4	0.6	0.1	1.4	2.7	4.1
	(4)-(l)	0.1	0.0	0.0	0.0	0.5	0.8	1.2	0.1	1.9	5.4	10.7
	(5)–(1)	0.1	0.0	0.0	0.0	0.3	0.3	0.3	0.1	1.4	2.7	4.1
The disaster-affected region (Iwate, Miyagi, Fukushima and Ibaraki Prefectures)	(l) Base simulation	0.5	0.2	0.1	▲ 0.0	▲ 0.2	▲ 0.8	▲ 1.6	▲ 0.1	▲ 0.9	▲ 3.7	▲ 10.0
	(2) Simulation I	0.5	0.2	0.2	▲ 0.0	▲ 0.1	▲ 0.5	▲ 1.2	0.0	▲ 0.4	▲ 2.2	▲ 6.9
	(3) Simulation II	0.6	0.3	0.2	0.0	0.2	▲ 0.1	▲ 0.6	0.3	1.2	0.1	▲ 4.1
	(4) Simulation III-a	0.6	0.3	0.2	0.0	0.3	0.2	▲ 0.2	0.3	1.7	2.8	2.6
	(5) Simulation III-b	0.6	0.3	0.2	0.0	0.1	▲ 0.4	▲ 1.2	0.3	1.2	0.1	▲ 4.1
	(2)–(1)	0.0	0.0	0.0	0.0	0.1	0.3	0.4	0.2	0.5	1.5	3.2
	(3)–(1)	0.1	0.1	0.0	0.0	0.5	0.7	1.0	0.4	2.0	3.8	5.9
	(4)–(1)	0.2	0.1	0.1	0.1	0.5	1.0	1.4	0.4	2.6	6.5	12.6
	(5)–(1)	0.2	0.1	0.1	0.1	0.3	0.4	0.4	0.4	2.0	3.8	5.9
Other areas without the disaster-affected region	(l) Base simulation	▲ 1.2	▲ 0.1	3.4	1.6	8.1	8.4	4.8	10.7	41.2	83.8	116.8
	(2) Simulation I	▲ 1.2	▲ 0.2	3.4	1.6	8.1	8.5	4.9	10.6	41.2	84.1	117.8
	(3) Simulation II	▲ 1.2	▲ 0.2	3.4	1.6	7.9	8.1	4.3	10.4	40.6	82.6	115.1
	(4) Simulation III-a	▲ 1.3	▲ 0.2	3.4	1.6	8.0	8.3	4.6	10.4	40.6	82.7	114.9
	(5) Simulation III-b	▲ 1.3	▲ 0.2	3.4	1.6	8.0	8.3	4.6	10.4	40.6	82.6	115.1
	(2)-(l)	▲ 0.0	▲ 0.0	▲ 0.0	▲0.0	0.0	0.1	0.1	▲ 0.1	▲ 0.0	0.4	0.9
	(3)-(l)	▲0.0	▲ 0.0	▲ 0.0	▲0.0	▲ 0.2	▲ 0.3	▲ 0.4	▲ 0.3	▲ 0.7	▲ 1.1	▲ 1.8
	(4)-(l)	▲ 0.1	▲ 0.1	▲ 0.1	▲ 0.1	▲0.1	▲ 0.1	▲ 0.2	▲ 0.3	▲ 0.6	▲ 1.1	▲ 1.9
	(5)-(l)	▲ 0.1	▲ 0.1	▲ 0.1	▲ 0.1	▲ 0.1	▲ 0.1	▲ 0.1	▲ 0.3	▲ 0.7	▲ 1.1	▲ 1.8

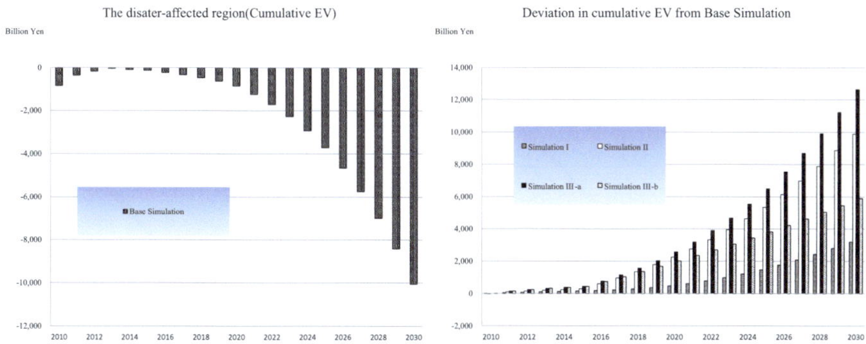

Fig. 6.4 Transition of equivalent variation in the disaster-affected region

0.5%, even into the 2020s, reaching an index of 117.57 in 2030 (versus 2010), which is 4.37 points above the index of 113.2 for other regions. If both new food industry and new automobile industry clusters formed in the disaster region, the economic disparity between the disaster region and other regions would shrink, and growth in total output would surpass that of other regions.

However, in Simulation III-b, where industrial productivity does not rise, despite the formation of the two new industry clusters and the enactment of a corporate tax cut, real GRP and total output in 2030 hit indexes of 106.07 and 108.09, respectively, versus 2010, nearly identical to the Base Simulation. In this case, achieving continued economic growth in the disaster region becomes a challenging proposition.

(b) Assessing Economic Welfare from Equivalent Variation

Table 6.8 expresses equivalent variation (EV) for each year in the disaster region, from Simulations I to III-b, with deviation from the Base Simulation expressed in Fig. 6.4. In Table 6.8, viewing data from Simulation I of the new food industry cluster, we see that the disaster region's EV trends are negative, similar to the Base Simulation, from 2014 onward. However, although it reached negative 1.2 trillion yen in 2030, Fig. 6.4 reveals that it has positive effects each year, compared with the Base Simulation.

Disaster region EV, from Simulation II, remains in the (positive) 200 billion yen range until 2020 and remains positive until 2024. Although it does trend downward after 2024, the drop is much gentler than in the Base Simulation scenario.

By contrast, in Simulation III-a, where formation advances for both clusters and corporate tax cuts are implemented, EV trends positive until 2027. Although it crosses into negative territory into 2028, the negative trend is small when measured in 2030. Divergence from the Base Simulation, as expressed in Fig. 6.4, reveals an increasingly greater EV after 2022. We can see from this that if policies are adopted that can simultaneously form new food industry and automobile industry clusters, reductions in economic welfare in the disaster region would be dramatically delayed. However, in Simulation III-b, where productivity in cluster industries

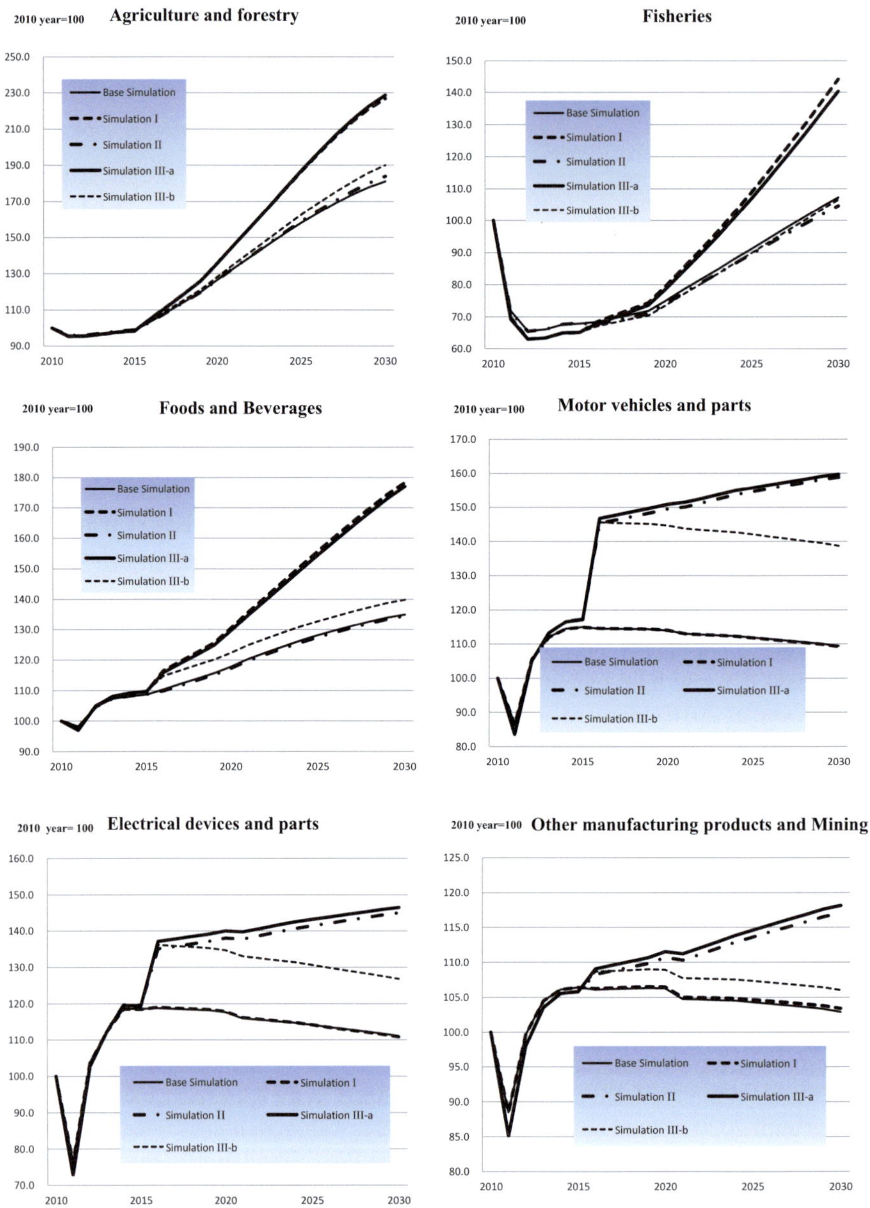

Fig. 6.5 Transition of the production volume by industries in the four disaster-affected prefectures

does not increase, despite the formation of these two clusters and the enactment of corporate tax cuts, the positive effects on EV are about half of those in Simulation III-a.

(c) Assessing Economic Effects from Production Volume per Industry

We will now use Fig. 6.5 to assess the effects on production volume in each industry. In Scenario I, production volumes for the agriculture and forestry, fishery, and food and beverage industries all rapidly increase from 2020 onward versus the Base Simulation, hitting indexes of 226 (agriculture and forestry), 144 (fisheries), and 178 (food and beverage) in 2030 versus 2010s base index of 100. Although these positive effects are not particularly strong compared with the real GRP values in Fig. 6.3, we can see that there are positive effects on production volumes for the above industries. However, when compared to the results of Simulation III-a, we see fiercer competition in III-a for both clusters in the capital markets; therefore, Simulation III-a's production volumes are slightly lower than those of Simulation I, where only a food industry cluster is formed.

On the other hand, looking at Simulation II of the automobile industry cluster, formed under a megaregion, the automotive and auto parts industry, the electronic components and devices industry, and other manufacturing industries all see modestly higher production volumes compared with the downward trends seen in the Base Simulation. By 2030, production volumes for these three industries reach indexes of 159, 145, and 117.8, respectively, versus 2010s base index of 100. By contrast, in Simulation III-a, where both new industry clusters form simultaneously, results are slightly positive versus those of Simulation II, where only the new automobile industry cluster is formed. However, if we compare these results to real GRP in the disaster region (see Fig. 6.3), the formation of two new megaregional industry clusters results in vertical and horizontal coagglomeration effects and economies of agglomeration, meaning that the joint formation of these two new industry clusters has more positive effects on real GRP and production values on the macro level than does the formation of an individual industry cluster.

(d) Assessing Economic Effects from the Formation of Two New Industry Clusters

We now assess economic effects in the event of the joint formation of two new industry clusters, using investment, capital stock, production volume, and economic welfare for each industry as barometers (Table 6.9). Let us begin by comparing economic effects from Simulation III-a with those of the Base Simulation, in terms of investment amounts and capital stock for each industry in the disaster region, resulting from the formation of two new industry clusters. In Simulation III-a, investment jumps 10–30 points for each industry through the 2020s versus the Base Simulation, and capital stock is half a point to three points greater for each industry as well. In the agriculture and forestry, fishery, and food and beverage industries that form the new food industry cluster, we see a 15–30 point jump in investment in the 2020s versus the Base Simulation and accelerated recovery of capital stock levels, which were impaired by the 2011 disaster. On the other hand, for the automotive and auto parts industry, the electronic components and devices industry,

Table 6.9 Transition of capital stock and investment by industry in disaster-affected region

	The disaster in Iwate, Miyagi, Fukushima and Ibaraki Prefectures	Capital stock(K_a^o)				Investment(INV_a^o)				Production(XD_a^o)			
	2010 year = 100	2015	2020	2025	2030	2015	2020	2025	2030	2015	2020	2025	2030
Simulation III-a	Total industry	100.76	105.74	110.34	114.92	152.43	159.02	144.74	147.01	107.88	111.95	115.09	117.57
	Agriculture and forestry	98.09	102.94	106.56	108.98	153.32	146.52	99.59	62.59	98.45	135.58	186.79	228.67
	Fisheries	81.29	87.10	92.04	96.30	180.41	184.53	147.05	128.31	65.18	78.58	107.05	140.29
	Foods and beverage	97.46	102.17	106.50	110.64	139.03	152.27	134.68	129.68	109.74	129.63	154.46	176.93
	Electrical devices and parts	99.95	105.05	109.79	114.48	146.88	165.54	148.30	150.71	119.43	140.01	143.15	146.41
	Motor vehicles and parts	101.25	106.30	110.98	115.58	139.76	164.50	146.22	146.82	117.18	150.86	155.75	159.63
	Other manufacturing products and mining	100.41	105.32	109.85	114.38	147.02	157.81	142.22	145.87	105.78	111.51	114.58	118.12
	Construct	104.53	110.41	115.74	121.15	177.65	188.89	167.62	175.93	157.97	157.33	142.53	141.82
	Electricity, gas and heat-water supply	100.03	106.19	111.90	117.62	190.37	197.13	179.81	184.18	74.06	76.05	77.86	78.84
	Commerce	100.24	105.10	109.62	114.20	147.45	156.10	142.80	148.32	112.15	112.01	110.23	108.50
	Transport	103.13	107.95	112.47	117.07	146.20	155.19	143.27	148.79	109.77	108.28	106.12	102.57
	Other services	101.32	106.15	110.66	115.22	148.85	153.55	142.90	146.98	102.86	100.52	102.60	101.29

Base simulation	Total industry	100.54	105.20	109.21	112.90	149.85	145.36	121.97	115.09	107.73	107.93	107.48	106.32
	Agriculture and forestry	97.98	102.43	105.46	107.41	153.94	129.37	80.88	50.17	98.76	126.40	157.99	181.01
	Fisheries	81.53	87.20	91.54	94.92	194.13	173.53	123.41	98.01	67.93	74.96	91.24	107.26
	Foods and beverage	97.25	101.54	105.39	109.03	135.28	136.67	118.77	114.05	108.69	117.82	128.09	134.99
	Electrical devices and parts	99.78	104.25	108.08	111.63	143.52	140.18	116.47	110.69	118.61	117.74	114.19	111.13
	Motor vehicles and parts	101.05	105.28	108.90	112.26	135.69	132.20	110.39	105.04	114.95	113.94	111.85	109.40
	Other manufacturing products and mining	100.26	104.78	108.65	112.24	144.92	141.68	117.81	112.42	106.41	106.27	104.30	102.89
	Construct	104.13	109.50	113.92	117.95	173.42	167.30	132.76	125.83	155.73	145.87	123.54	115.89
	Electricity,gas and heat-water supply	99.79	105.56	110.53	115.14	187.18	180.78	151.51	143.65	74.04	76.08	77.47	77.95
	Commerce	100.02	104.55	108.47	112.14	144.75	142.48	119.70	114.93	112.08	112.06	109.81	107.63
	Transport	102.91	107.41	111.32	114.98	143.52	141.47	119.71	114.65	109.79	108.16	105.10	100.73
	Other services	101.10	105.65	109.63	113.33	146.15	142.23	121.66	115.68	102.78	102.29	104.10	102.30

and other manufacturing industries that form the new automobile industry cluster, we see that investment jumps 30–40 points above that of the Base Simulation in and after the 2020s, and capital stock also increases between two to three points over the Base Simulation. We can therefore confirm the accelerating recovery of capital stock in other industries, which had slowed recovery since the disaster, such as agriculture and forestry and fisheries. As can be seen in Fig. 6.5, these expansionary effects on capital stock, from Simulation III-a, can help increase production volumes in various industries. In Simulation III-b, where production fails to increase despite the formation of two new industry clusters, we see a slight increase in production volumes versus the Base Simulation and a modest upward trend in agriculture, forestry, fisheries, and foodstuffs. By contrast, the automotive, electronic parts and devices, and other industries trend downward year-by-year. We can therefore conclude that subsidy policies, corporate tax cuts, and commodity-grade products alone cannot stamp out decreased production volumes that result from the sharp decline of population in disaster regions. We can also conclude that industry agglomeration, from the formation of the new food and automobile industry clusters, will improve production volumes over time barometers.

6.5 Conclusions and Policy Implications

In this chapter, we applied the concepts of clusters espoused by Porter (Porter 1998, 2000) and presented an analysis of the ripple effects on the economic welfare and regional economy of disaster-affected regions, resulting from the formation of two new industry clusters. The first is a new food industry cluster that leverages regional resources, and the second is a new megaregional automobile industry cluster that innovates based on an agglomeration of different industries. This analysis was composed of simulations, utilizing the dynamic two-regional computable general equilibrium (D2SCGE) model used in Chap. 5. The results of this analysis revealed the following four implications for government policy, which also serve as the conclusions reached by this chapter.

First, although Japan's Revitalization Strategy will surely maintain fiscal measures through the mid-2020s and will just barely keep a disaster area's regional economy out of negative growth territory, there will be stronger downward economic pressure thereafter from a shrinking population, and the disaster area's regional economy likely will see negative growth on its current course. On the other hand, non-disaster regions will enjoy the effects of Japan's Revitalization Strategy, seeing economic growth of 0.5–1.0%, even into the 2020s, widening the gap between non-disaster and disaster regions.

Second, if a portion of the special disaster reconstruction grants are diverted as production subsidies to support industries that form the two new industry clusters, and if a corporate tax cut for a disaster region is enacted, positive effects on regional economies could appear in disaster regions. But we now understand that these actions cannot ensure long-term, sustained economic growth in these regions. Our simulations, which form a new food industry cluster and a new automobile industry cluster, increasing productivity in related industries, make clear that (i) vertical and horizontal coagglomeration leads to economies of agglomeration, meaning that joint formation of these two industry clusters has a positive effect on real GRP and productivity on a macro level, versus formation of individual industry clusters, and (ii) these actions stimulate sustained, long-term growth in disaster region economies, mitigating the growth disparities between economies in disaster regions and other regions.

Third, formation of a new food industry cluster boosts how much investment is made in the agriculture, forestry, fisheries, and food and beverage industries and accelerates the recovery of capital stock for the agriculture, forestry, and fisheries industries, which were stunted by the disaster. Also, although the formation of a new food industry cluster does not greatly change the economic landscape, brought about by Japan's Revitalization Strategy, the formation of a new automobile industry cluster does have the effect of increasing both investment and capital stock.

Finally, formation of these two new industry clusters in a disaster region also has a positive effect on economic welfare in these regions, closing the economic welfare gap between disaster and non-disaster regions. Therefore, formation of two new industry clusters, which will boost productivity in related industries, is essential for sustained economic development in the disaster region. We conclude, therefore, that the standard subsidies and corporate tax cut policies usually adopted in such situations, by themselves, will be unable to hold off the stagnation caused by decreasing populations in disaster regions. Agglomeration externalities, in the formation of new food industry and automobile industry clusters, will increase productivity in relevant industries and finally stimulate sustained economic development in disaster regions.

References

Cabinet Office (2014a) Basic policies for the economic and fiscal management and reform 2014-from deflation to an expanded economic virtuous cycle (in Japanese). http://www5.cao.go.jp/keizai-shimon/kaigi/minutes/2014/0624/shiryo_01.pdf

Cabinet Office (2014b) Japan revitalization strategy as revised in 2014 (in Japanese). http://www5.cao.go.jp/keizai-shimon/kaigi/minutes/2014/0624/shiryo_02_1.pdf

Duranton G, Martin P, Mayer T, Mayneris F (2010) The economics of clusters. Oxford University Press, Oxford

EcoMod Modeling School (2012) Advanced techniques in cge modeling with gams global economic modeling network, Singapore, January 9–13

Ellison G, Glaeser EL (1997) Geographic concentration in U.S. manufacturing industries: a dartboard approach. J Polit Econ 105(5):898–927

Fujita M, Thisse JF (2013) Economics of agglomeration, 2nd edn. Cambridge University Press, New York

Fujita M, Krugman P, Venables AJ (1999) The spatial economy: cities, regions, and international trade. MIT Press, Cambridge

Hosoe N, Gasawa K, Hashimoto H (2010) Textbook of computable general equilibrium modeling. Palgrave Macmillan, New York

Japan Center for Economic Research (2015) Potential economic collapse in the late 2020s-adverse effects of the declining population after the Tokyo olympics. 42st Medium-term forecast for the Japanese economy: FY2035-FY2030 (Baseline scenario)

Japan Institute for Labor Policy and Training (2014) Labor supply and demand estimates: policy simulations based on the labor supply and demand model (2013). Research Material Series129 (in Japanese)

Kageyama M, Tokunaga S (2006) An empirical analysis of co-agglomeration effects in the Japanese food industry. J Rural Econ Spec Issue 2006:132–138 (in Japanese)

Ministry of Agriculture, Forestry and Fisheries (2015) Survey result 2014 about management condition of farmworker and fisherman in Tsunami-disaster area by the Great East Japan Earthquake (in Japanese). http://www.maff.go.jp/j/tokei/keikou/pdf/fukkou-20150911.pdf

Ministry of Agriculture, Forestry and Fisheries (2016) Commitment for reconstruction support of agriculture, forestry and fisheries after the Great East Japan Earthquake (in Japanese). http://www.maff.go.jp/j/kanbo/kihyo02/fukkou/pdf/1603-zen.pdf

Mokudai T, Ishiro K (2013) The current situation and challenges facing automobile industry in Kyushu. In: Orihashi S, Mokudai T, Murayama T (eds) Tohoku region and automobile industry. Soseisha, Tokyo, pp 168–186 (in Japanese)

Murayama T (2013) Automobile industry promotion by industry-academia-government collaboration in Iwate. In: Orihashi S, Mokudai T, Murayama T (eds) Tohoku region and automobile industry. Soseisha, Tokyo, pp 62–107 (in Japanese)

National Institute of Population and Social Security Research in Japan (2012) Population Projections for Japan (January 2012): 2011 to 2060. http://www.ipss.go.jp/site-ad/index_english/esuikei/gh2401e.asp

National Institute of Population and Social Security Research in Japan (2013) Regional Population Projections for Japan: 2010–2040. Population Research Series (in Japanese). http://www.ipss.go.jp/pp-shicyoson/j/shicyoson13/6houkoku/houkoku.pdf

Okiyama M, Tokunaga S, Akune Y (2014) Analysis of the effective source of revenue to reconstruct the disaster- affected region of the great East Japan earthquake: utilizing the two–regional CGE model. J Appl Reg Sci 18:1–16 (in Japanese)

Orihashi S (2013) The problem for the development of automobile industry in Tohoku. In: Orihashi S, Mokudai T, Murayama T (eds) Tohoku region and automobile industry. Soseisha, Tokyo, pp 248–258 (in Japanese)

Porter M (1998) On competition. Harvard Business School Press, Cambridge, MA

Porter M (2000) Location, competition, and economic development: local clusters in a global economy. Econ Dev Q 14(1):15–34

Tokunaga S, Okiyama M (eds) (2014) Reconstruction the disaster-affected region of the Great East Japan Earthquake and recovery of the regional economy. Bunshindou, Tokyo (in Japanese)

Tokunaga S, Resosudarmo BP, Wuryanto LE, Dung NT (2003) An inter-regional CGE model to assess the impacts of tariff reduction and fiscal decentralization on regional economy. Stud Reg Sci 33(2):1–25

Tokunaga S, Kageyama M, Akune Y, Nakamura R (2014) Empirical analyses of agglomeration economies in Japanese assembly-type manufacturing industry for 1985–2000: using agglomeration and coagglomeration indices. Rev Urban Reg Dev Stud 26(1):57–79

Tokunaga S, Akune Y, Ikegawa M, Okiyama, M (2015) An empirical analysis of japanese industrial agglomeration and factors of supply chain internationalization in East Asia: a case of Japanese food, electronics, and automobile firms. RIETI Policy Discussion paper series, 15-P-021: 1–69 (in Japanese)

Chapter 7
Production Recovery of Fishery and Seafood Manufacturing After the Disaster in Japan: Economic Evaluation Using Dynamic CGE Model

Yuko Akune

Abstract The purpose of this chapter is to evaluate the economic effects of the recovery of fishery and seafood manufacturing after the disaster in Japan under depopulation by employing a dynamic Computable General Equilibrium (CGE) model. We confirmed a clear contrast between the production recovery of seafood manufacturers in Iwate and Miyagi Prefectures after the Great East Japanese Earthquake (GEJE) of 2011. The economic evaluation involves setting up simulation scenarios based on actual recovery evidence after GEJE. The results of five simulations indicated the following five points. (1) The disaster accelerated problems caused by depopulation. (2) Prompt capital restoration and production recovery contributed to shortening the period of output loss. (3) The degree of output recovery of the downstream industry was faster than that of the upstream industry. (4) No scenario based on GEJE evidence was sufficient to reach base scenario without disaster. (5) Stepwise production recoveries contributed to increases not only in specific industries but also economic welfare in the long term.

Keywords Production recovery • Fish food system • Geographic concentration • Great East Japan Earthquake • Dynamic CGE model

7.1 Introduction

The purpose of this chapter is to evaluate the economic effects of recovery at fishery and seafood manufacturing after a disaster under depopulation by employing a dynamic Computable General Equilibrium (CGE) model. It involves setting up a simulation scenario based on actual recovery evidence of these industries after the Great East Japanese Earthquake (GEJE) of 2011. This purpose is stimulated by the

Y. Akune (✉)
Faculty of Economics and Business Administration, Reitaku University, 2-1-1, Hikarigaoka, Kashiwa, Chiba 277-8686, Japan
e-mail: yakune@reitaku-u.ac.jp

S. Tokunaga, B.P. Resosudarmo (eds.), *Spatial Economic Modelling of Megathrust Earthquake in Japan*, New Frontiers in Regional Science: Asian Perspectives 11,
https://doi.org/10.1007/978-981-10-6493-7_7

following five motivations: threats of natural disaster, grave damage to fishery and seafood manufacturing after the GEJE, differences in recovery of these industries, structural problems, and lack of existing studies.

Nature not only provides numerous benefits to humankind but also thoroughly despoils life and property through natural disasters. Japanese communities endure a significant number of earthquakes and have endeavored to decrease damage through the development of technology. However, the destructive power of the GEJE in 2011 exceeded the efforts of human technology. Furthermore, huge earthquakes are forecasted to recur in the near future, such as the Nankai earthquakes, Tokai earthquakes, and their combined earthquakes. It is predicted that these earthquakes will simultaneously cause huge tsunamis and severe damage to major coastal regions of the Pacific, such as Kochi, Wakayama, Mie, and Shizuoka Prefectures.

Coastal areas have numerous economic benefits from their natural resources. At the same time, strong supply linkages between fishery and seafood manufacturing develop in these regions; as a result, they support these regional economies. These industries' major productive regions suffered severe damage in the GEJE, namely Iwate, Miyagi, and Fukushima Prefectures. Vast amounts of aid and support have been allocated in public and private ways to restore these regional economies. However, there is a strong contrast between recoveries of industries in coastal areas and in inland areas. Recovery of fishers and seafood manufacturers has not reached the level of production before the disaster. In contrast, inland manufacturers achieved a complete recovery in six months.

In addition, we can confirm a clear contrast between the recovery of fishery and seafood manufacturing in Iwate and Miyagi Prefectures. A distinct difference, especially, exists between the restoration of seafood product manufacturers in both prefectures. The recovery speed in Iwate was faster than that of Miyagi. At the same time, the disparity between the each sector's recovery in Iwate is greater than that in Miyagi. As a result, the production composition of seafood manufacturing has changed since before the disaster. In Iwate Prefecture, many producers in these industries prioritized recovery as fast as possible, because they had felt that they were at risk of unsustainability due to depopulation before the disaster. On the other hand, producers in Miyagi prioritized coordination between related parties more than those in Iwate because the production scale of seafood manufacturing in Miyagi is four times that of Iwate. It took, naturally, a long time to communicate with each producer. These details of recovery in these prefectures may provide some essential insights for comprehensive consideration about sustainable industries and economies in regions forecasted to suffer significant damage due to earthquakes and tsunamis.

A disaster leads us to address structural social problems such as a decline in population and aging society. It is a frequent tendency to defer these problems because it is hard to envisage them in our everyday lives. However, in the GEJE, the affected regions had faced already problems from the severe decline in population and aging society before the catastrophe. These problems caused Iwate locals to feel that sustainability of their regional economies was at risk before the disaster;

many older fishers and producers who work in coastal areas had retired because of their age. Hence, we should consider these structural problems when discussing economic damage caused by natural disasters.

Hallegtte and Prsyluski (2010) introduced five methods to analyze the economic evaluation of natural disasters. The CGE model employed in this chapter is one of them. It has the remarkable feature of being able to evaluate not only direct losses caused by destroyed capital stock but also indirect losses by seen throughout the supply-chain. Numerous studies have employed the CGE model to evaluate economic losses caused by an earthquake (e.g. Rose and Liao (2005), Tsuchiya et al. (2007), Shibusawa et al. (2009), Tokunaga et al. (2013), Akune et al. (2013), and Huang and Hosoe (2016)).

Several recent papers have focused on economic losses in national economies caused by specific damaged industries, while many early studies focused on economic losses in national and regional economies caused by destroyed infrastructure, such as aqueducts and transportation networks. For instance, Tokunaga et al. (2013) focused on the auto industry; Akune et al. (2013) examined fishers and seafood manufacturers, as does this chapter. Huang and Hosoe (2016) focused on the electronics industry in Taiwan. Particularly, evidence of damage by the GEJE was utilized by Tokunaga et al. (2013) and Akune et al. (2013). However, they did not involve actual recovery situations because of the short time of analyses since the disaster. Hence, the purpose of this chapter is to evaluate economic effects based on confirmed recovery evidence from fishery and seafood manufacturing after the GEJE by employing a Dynamic CGE model.

The structure of this chapter is as follows. Section 7.2 describes the recovery of fishery and seafood manufacturing in Iwate and Miyagi Prefectures with a focus on recovery of production, labor, productivity, and geographic proximity. Section 7.3 indicates activities in the model and the model structure. Section 7.4 shows the simulation results with scenarios. Finally, Sect. 7.5 discusses results and policy implications, along with remaining tasks.

7.2 Different Recovery Patterns After the Great East Japan Earthquake

This section indicates the difference between recoveries experienced at fishery and seafood product manufacturing in Iwate and Miyagi Prefectures. Iwate, Miyagi, and Fukushima Prefectures, which are major Japanese fishing and seafood production regions, were tremendously affected by the tsunami in the GEJE. A distinct contrast has emerged between the recovery of Iwate and Miyagi 5 years after the disaster. Evidence regarding Fukushima Prefecture is excluded in this chapter since it is facing the different problem of the nuclear accident at the Fukushima Daiichi Nuclear Power Plant. The contrasting recoveries describe regarding production, labor, productivity, and geographic proximity.

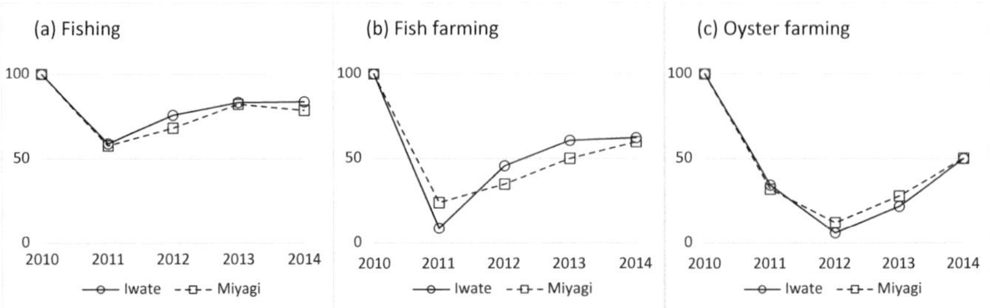

Fig. 7.1 Recovery of production in fishing and fish/oyster farming (2010 = 100) (Source: Fisheries Agency "Annual Statistical Report of Fishing and Aquaculture")

7.2.1 Production of Fishery and Seafood Product Manufacturing

Figure 7.1 shows the recovery process of fishing, fish farming, and oyster farming. The figures are standardized indices of 100 in 2010, which was the year before the disaster. Fishing production indices in both prefectures fell below 60 at the time of the catastrophe in 2011. The recovery pace in Iwate was slightly faster than that in Miyagi in 2012; Iwate was at 76 whereas Miyagi was at 68. The damage to fish/ oyster farming was much more grievous than the damage to fishing. Fish farming production dramatically fell to 9 in Iwate and 24 in Miyagi. The damage in Iwate was more extensive than in Miyagi; nevertheless, the production recovery degree in Iwate was higher than that of Miyagi in 2012 and 2013. The return ratio was at the same level in 2014 since the restoration of fish farming in Miyagi was consistent. Oyster farm production had declined to around 30 in both prefectures by 2012 since it takes longer to grow oysters to the size required for shipping. We can find a similar pattern in fish farming; the recovery of oyster production reached 50 in 2014 in both prefectures, while Miyagi recovered at a faster rate than Iwate from 2012 to 2013. Thus, the recovery levels at fisheries in both prefectures were at the same level in 2014, despite their different processes.

In contrast to the recovery at the fishery, there is a noticeable difference between Iwate and Miyagi Prefectures regarding seafood product manufacturing recovery. Figure 7.2 shows the recovery of all seafood manufacturing, which is categorized as 920 in the Japan Standard Industry Classification (JSIC). In 2011, the levels of production dropped to 32 in Iwate and 53 in Miyagi due to the disaster. Production in Iwate has steadily recovered since 2012. In comparison with the processes of recovery, the degree of recovery in Iwate has been higher than that in Miyagi, although the drop in Iwate's level is greater than in Miyagi in 2011. In 2014, the production level in Iwate was 95, while in Miyagiit was 64.

Besides, when observing further details of the seafood manufacturing industry's recovery, another contrast between the two prefectures can be found. Figure 7.3 shows the recovery of seafood manufacturing in further detail. Seafood manufacturing (920) comprises seven sectors according to the JSIC. In Iwate Prefecture, the degrees of recoveries were remarkable in the unprocessed and packaged frozen

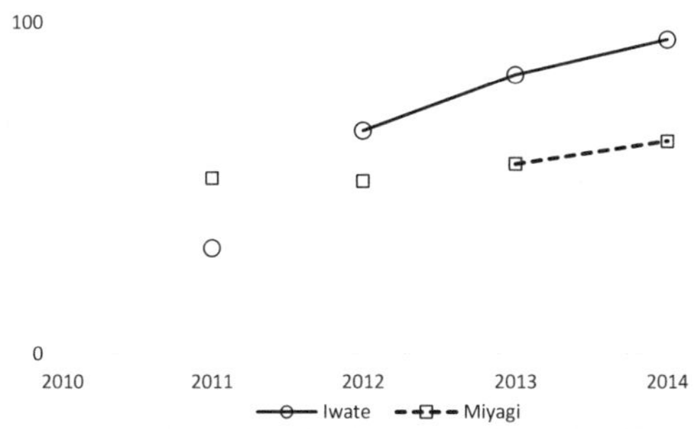

Fig. 7.2 Recovery of production in seafood manufacturing (JSIC: 920, 2010 = 100)
Note 1: The 2011 figures for Iwate spread three sectors over six sectors. It is an index against these three sectors in 2010
Note 2: The 2011 figures for Miyagi spread three sectors over seven sectors. The figure in 2012 spread six sectors over seven sectors. They are calculated as indices against the same sectors in 2010 (Source: Fisheries Agency "Annual Statistical Report of Fishing and Aquaculture")

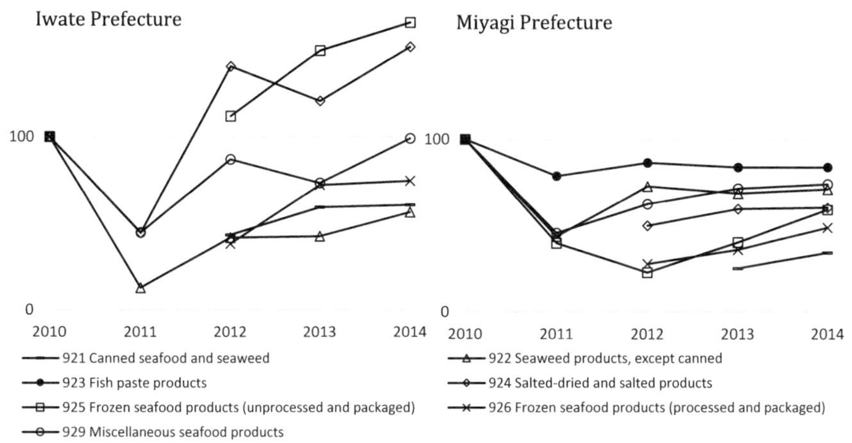

Fig. 7.3 Recovery of production in seafood manufacturing (SIC: 921–929, 2010 = 100) (Source: Ministry of Economy, Trade and Industry "Census of Manufacture")

seafood products sector (925) and the salted-dried and salted products sector (924). Their production level increased to 166 in the JSIC 925 and 152 in the JSIC 924 in 2014. The miscellaneous seafood products sector (929) returned to the same level as before the disaster. Recovery levels, conversely, were around 60 in the canned seafood and seaweed sector (921) and seaweed products sector (922) in 2014. Thus, the range of indices that indicate the disparity between different sectors' recovery rates consistently has expanded from 45 in 2011 to 110 in 2014. On the other hand, in Miyagi Prefecture, no sector had returned to the production level of 2014. The

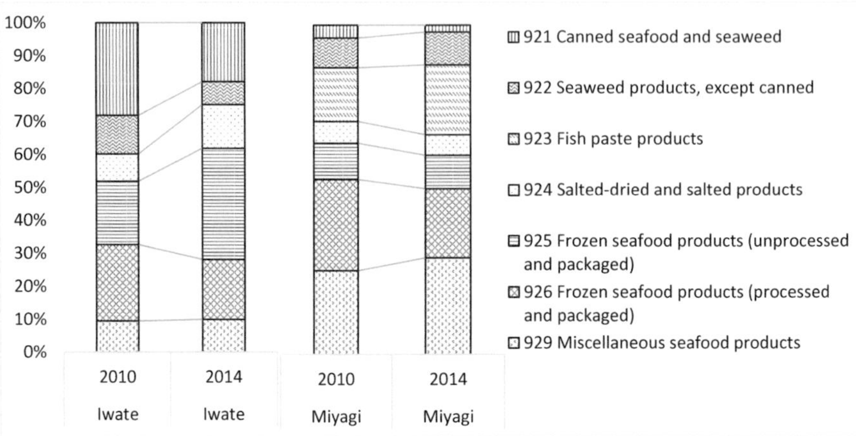

Fig. 7.4 Composition raito of production in seafood manufacturing (JSIC:920, 2010 = 100)

fish paste products sector (923), which has the highest recovery of all sectors in this prefecture, is at 84. In contrast with Iwate, the range between sector indices shows a gradual decline from 86 in 2012 to 59 in 2013, and then 50 in 2014. Therefore, we find that the disparity in the recovery of seafood manufacturing between the two prefectures is increasing in Iwate. In contrast, it is shrinking in Miyagi.

The recovery process caused sector formation to change the structure of before the disaster in Iwate Prefecture. Figure 7.4 shows changes in production structure from 2010 to 2014. It indicates that ratios of unprocessed and packaged frozen seafood products (925) and salted-dried and salted products (924) increased by 10 points and 5 points, respectively, while the ratio of canned seafood and seaweed (921) decreased by 10 points. Additionally, the rates of seaweed products (922) and unprocessed and packaged frozen seafood products (925) fell by 5 points. In Miyagi Prefecture, the sector that showed the greatest increase is seaweed products (922), which increased by 5 points, and the sector that showed the most significant decline is unprocessed and packaged frozen seafood products (925), which decreased by 7 points. Therefore, the production structure between sectors in seafood manufacturing has drastically changed in Iwate Prefecture.

A report by Tokyo University of Marine Science and Technology (TUMSAT) (2014) pointed out the difference of production scale and their decisions priority for recovery as reasons for these different recovery degrees between the two prefectures. It is necessary that many producers and associations related to a region's industry cooperate to achieve recovery since the availability of an individual producer is limited during natural disasters. In such a cooperative effort, the greater the number of participants, the longer it will take to communicate and build consensus. Table 7.1 shows production at fisheries and seafood product manufacturers in Iwate and Miyagi Prefectures in 2010. The production scale in Miyagi is larger than that of Iwate: 1.6 times in fishing, 2.4 times in fish farming, 4.3 times in oyster farming, and 4 times in seafood manufacturing. According to the

Table 7.1 Production of fishery and seafood manufacturing in 2010

		Aquaculture		
	Fishing (1000 t)	Fish (1000 t)	Oyster (1000 t)	Sales of seafood products (billion Yen)
Iwate	136	51	10	61
Miyagi	225	123	42	240

Source: Fisheries Agency "Annual Statistical Report of Fishing and Aquaculture," Ministry of Economy, Trade and Industry "Census of Manufacture"

Note: The data of seafood product manufacturers cover establishments with more than 4 employees.

TUMSAT's report, people engaged in fishery and seafood manufacturing in Iwate had concerned about the sustainability of their production due to their smaller production scale than Miyagi under depopulation. Besides, after the disaster, their concerns led their immediate decision that being able to show a smooth and rapid recovery was the top priority to maintain their presence in the domestic fish market.

The TUMSAT report also pointed out that prefectural measures differed regarding the priority they placed on promptness and consensus-building. For example, it relates to the application of Article 84 of the Building Standards Act (BSA), which is a rule on building restrictions in the affected areas. This article usually restricts the rebuilding of production facilities following a disaster. However, Iwate Prefectural Government did not apply this rule to the affected areas and handed over the reins of power to municipal governments under Article 39 of the BSA. In Ofunato City in Iwate, each producer was allowed to rebuild their facilities, as the local administration did not also apply rebuilding restrictions to areas at where seafood manufacturers located. On the other hand, Miyagi Prefectural Government applied rebuilding restrictions under Article 84 of the BSA. The TUMSAT report referred that this rule's application resulted in a long time to resume production by fishers and seafood manufacturers of Miyagi.

7.2.2 Labor and Productivity of Seafood Manufacturing

The distinct recovery patterns of seafood manufacturing in the two prefectures are reflected in the respective recovery of employment. Figure 7.5 shows the recovery of employment as of 2010 in the industry using standardized indices of 100. These levels dropped to 50 in Iwate, and 53 in Miyagi at the time of the GEJE. In Iwate, the employment index returned to around 80 in 2014, after being at 67 in 2012, and 73 in 2013. In contrast, in Miyagi, the index has been stable at around 60 since 2012.

We can find differences among sectors when observing further details of employment in each sector, as shown in Fig. 7.6. In Iwate, the unprocessed and packaged frozen seafood products sector (925) has been above the pre-disaster

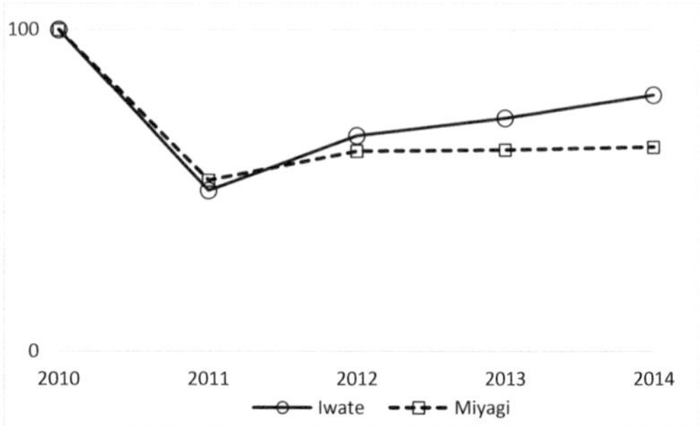

Fig. 7.5 Recovery of employment of in seafood manufacturing (JSIC:920, 2010 = 100) (Source: Ministry of Economy, Trade and Industry "Census of Manufacture")

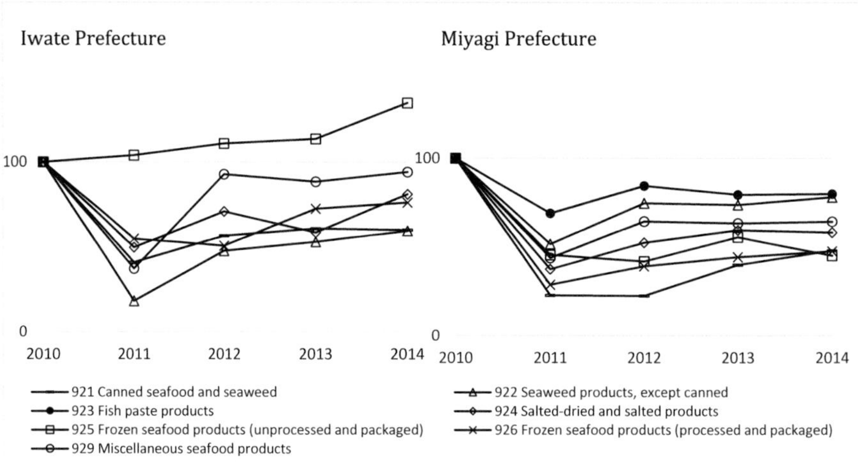

Fig. 7.6 Recovery of employment in seafood manufacturing (JSIC: 921–929, 2010 = 100) (Source: Ministry of Economy, Trade and Industry "Census of Manufacture")

level. The miscellaneous seafood products sector (927) exceeded 90 in 2012 and drew close to its pre-disaster level in 2014. On the other hand, the sector that showed the greatest increase is the fish paste products sector (923), which was at 80 in 2014 in Miyagi. Accordingly, in 2014, the range of the recovery of labor in each sector in Iwate was higher than that of Miyagi while the recovery rate of Iwate is faster than that of Miyagi.

Furthermore, there is no slackness in the cost of labor units in seafood manufacturing, which nearly means average wages, unlike the situations of production and employment. However, a difference also exists between the recovery degrees in both prefectures. Figure 7.7 indicates that the index of labor unit costs has been consistently increasing since the disaster in Iwate; it was at 120 in 2014. In contrast, the index in Miyagi has been lower than that of Iwate after being at 117 in 2011 and then remaining at 104 in 2014.

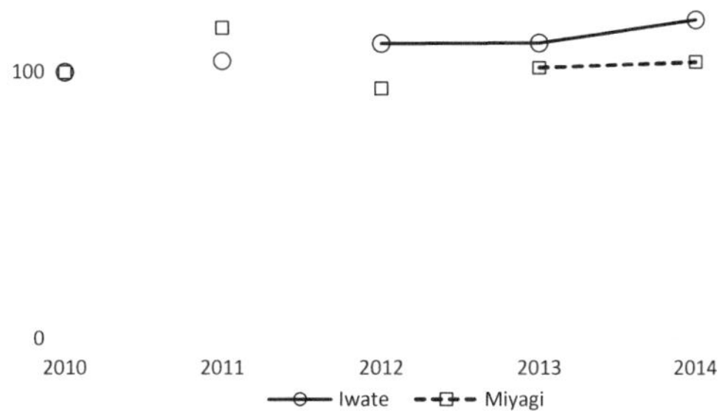

Fig. 7.7 Recovery of labor unit costs of seafood manufacturing (JSIC:920) (Source: Ministry of Economy, Trade and Industry "Census of Manufacture". Note: No-lines means that is lack sector in collecting data between markers. See notes at Fig. 7.2.)

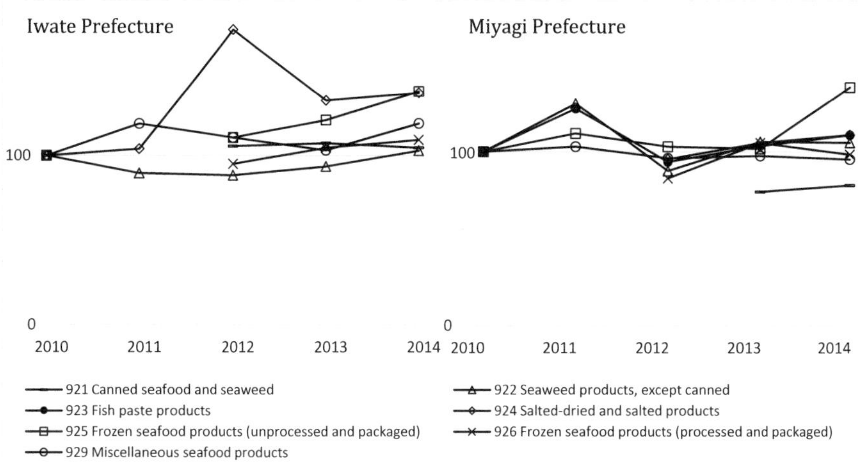

Fig. 7.8 Recovery of labor unit costs of seafood manufacturing (JSIC:921–929, 2010 = 100)

Figure 7.8 shows the labor unit costs in each sector in Iwate and Miyagi. Indices of all sectors in Iwate exceeded the level of before the disaster from 103 at JSIC 922 to 138 at JSIC 926. In contrast, in Miyagi, several sectors had indices below 100 in 2014.

Seafood manufacturers need to increase their productivity to enhance their sustainability. Figure 7.9, which are standardized indices of 100 in 2010, shows the recovery of labor productivity in Iwate and Miyagi. The contrast between the productivity recoveries of both prefectures is similar to that observed for other previously described data. Productivity fell by around 5 points at the time of the disaster in both prefectures in 2011. In Iwate, productivity has increased since the year following the catastrophe: it was at 100 in 2012, 116 in 2013, and 120 in 2014. On the other hand, productivity in Miyagi declined until 2012, then returned to the same level as before the disaster in 2014.

Figure 7.10 shows the labor productivity in seafood manufacturing sectors of Miyagi and Iwate using the same format as Fig. 7.9. The productivity of the salted-dried and salted products sector (924) in Iwate reached 200 in 2012.

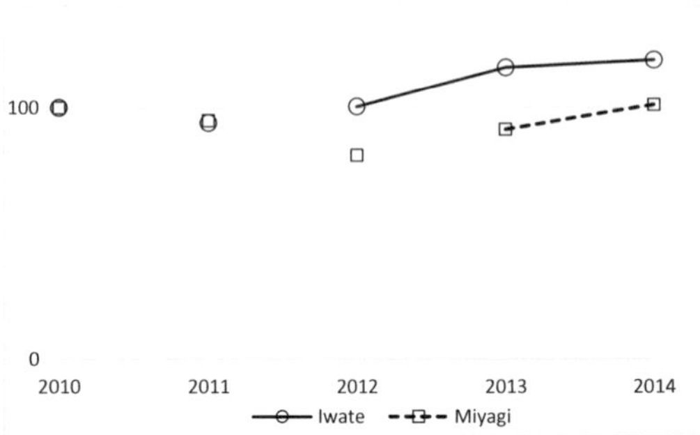

Fig. 7.9 Recovery of labor productivity of seafood manufacturing (SIC: 920, 2010 = 100) (Source: Ministry of Economy, Trade and Industry "Census of Manufacture". Note: No-lines means that is lack sector in collecting data between markers. See notes at Fig. 7.2.)

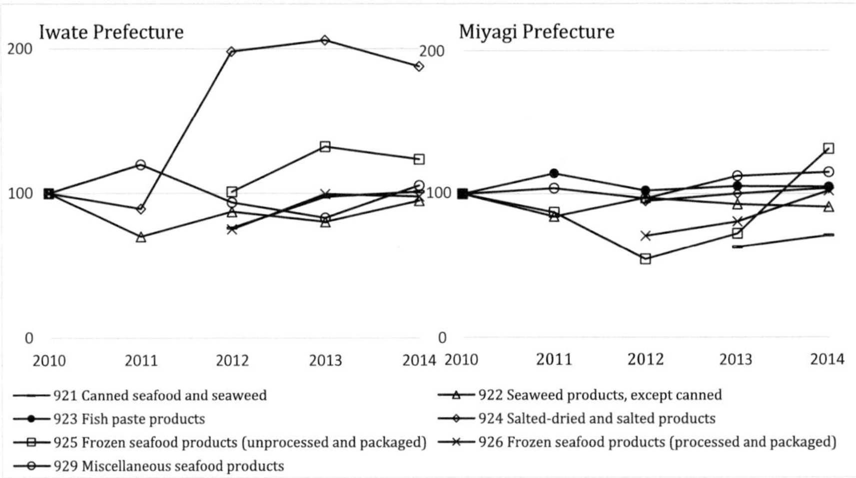

Fig. 7.10 Recovery of labor productivity in seafood manufacturing (SIC: 921–929, 2010 = 100) (Source: Ministry of Economy, Trade and Industry "Census of Manufacture")

This increase leads to an increase in the whole of seafood manufacturing, as shown in Fig. 7.9. On the other hand, no sector in Miyagi had a noticeable increase like that seen in Iwate. Thus, the range in productivity recovery among sectors in Iwate was 32, while in Miyagi it was 17 in 2014. The disparity in productivity across sectors in Iwate is larger than that in Miyagi, while the recovery degree of Iwate was faster than that of Miyagi.

7.2.3 *Geographic Proximity of Fishers and Seafood Manufacturers*

Several theories, such as the New Economic Geography (NEG) established by Fujita et al. (1999) and the cluster theory claimed by Porter (1998), emphasize that geographic proximity of activities serves to encourage various innovations by producers in a region. Not only individual producers but also regional producers will find it is necessary to form competitive advantages. Indices to indicate geographic proximity based on the NEG have been developed since the locational Gini coefficient by Krugman (1991), with various further indices proposed. Ellison and Glaeser (1997) proposed two kinds of indices for the geographic proximity of activities: geographic concentration within the same industry and co-agglomeration with different industries.

Spatial analysis, which has used a method of analyzing variables with spatial elements, provides another geographic proximity statistic. A statistic for geographic concentration within the same activity is covered by Moran's I (Moran 1950), which indicates a univariate spatial autocorrelation. Moreover, a bivariate version of Moran's I is employed for statistics of geographic concentration between different activities, and was developed as a measure of multivariate spatial autocorrelation by Wartenberg (1985), and then modified as a measure of bivariate spatial autocorrelation by Anselin et al. (2002). Univariate and bivariate Moran's I measures will be tested under a statistical hypothesis wherein the null hypothesis states that there is no spatial autocorrelation in the measured variables. Regarding judgment as to whether a spatial autocorrelation exists, the hypothesis testing for Moran's I is superior to indices based on the NEG. Hence, the geographic proximity of fishery and seafood manufacturing measurements indicates the use of the univariate Moran's I for geographic concentration in each sector of the industries, and the bivariate Moran's I for co-agglomeration of the industries.

Table 7.2 shows five univariate Moran's I results for fisheries and ten sectors of seafood manufacturing in Iwate and Miyagi Prefectures before and after the GEJE in 2011. There are four variables related to fisheries: number of management entities, workers, ships, and production. The former three variables are indicated as input in fishing activities and are sourced from the 2008 and 2013 Census of Fishers by Ministry of Agriculture, Forestry and Fisheries (MAFF). The remaining variable, fishing production, is indicated as output in fishing. Values for this variable before the disaster are taken from data in 2009. Meanwhile, there are ten sectors in seafood manufacturing. These sectors use data of facilities sourced from both year's censuses.

First, the comparison of geographic proximity within fisheries in Iwate and Miyagi shows a smooth recovery and the difference between the geographic concentrations in both prefectures. In Iwate, all of four Moran's I variables for fisheries are significantly positive, and these figures are at the same level both before and after the disaster. Hence, the geographic proximity of the fisheries recovered after the catastrophe. In Miyagi, similarly, three Moran's I values, indicating input variables in fisheries, are significantly positive and at the same level as in 2008 and 2013. The Moran's I of fish production is also significantly positive in 2013 but was not significant in 2008. Comparing the degree of geographic proximity between prefectures, the input variables of Miyagi are higher than those of Iwate. In contrast, fish production, indicating fisheries output, in Iwate is greater than in Miyagi.

Table 7.2 Moran's I of fishery and seafood manufacturing

		Iwate Prefecture				Miyagi Prefecture			
		2008		2013		2008		2013	
		I	z-value	I	z-value	I	z-value	I	z-value
Fishery	Management entities	0.291***	2.99	0.296***	2.99	0.410***	4.48	0.391***	4.26
	Workers	0.316***	3.33	0.309***	3.13	0.392***	4.82	0.437***	4.39
	Ships	0.344***	3.49	0.312***	3.24	0.479***	5.28	0.461***	4.92
	Production	0.307**	3.14	0.245***	2.86	0.072	1.37	0.083*	1.52
Seafood manufacturing	Freezing and refrigeration	0.272**	3.20	0.191**	2.37	0.104*	1.36	0.098*	1.34
	Frozen seafood products (unprocessed and packaged)	0.298***	3.40	0.181**	2.28	0.237***	3.70	0.155*	2.08
	Frozen seafood products (processed and packaged)	0.496***	4.98	0.316***	3.56	0.035	1.08	0.028	0.61
	Fish paste products	**−0.067**	**−0.47**	**−0.102****	**−0.82**	**−0.073**	**−0.57**	**−0.056**	**−0.49**
	Dried products	0.185**	2.71	0.333***	3.80	0.113*	1.58	0.158**	2.21
	Salted-dried products	0.219***	3.22	0.225***	3.82	0.272**	3.05	0.072	1.02
	Small dried sardines	0.227**	2.67	**−0.018**	0.25	0.474***	5.74	0.040	0.69
	Salted products	0.236***	3.17	0.162**	2.21	0.063	1.19	**−0.022**	0.06
	Smoke-dried products	0.340***	3.61	**−0.045**	**−0.25**	0.057	0.92	0.058	0.94
	Dried products	0.132**	2.34	**−0.045**	**−0.27**	**−0.122*****	**−1.38**	**−0.048**	**−0.36**

Note 1: *** is significant at 1%, ** is significant at 5%, and *is significant at 10%
Note 2: The spatial weight matrix is employed queen contiguity

Furthermore, concerning seafood manufacturing, different characteristics show on the geographic concentration of fisheries. In 2008, the number of significant and positive sectors in Iwate was nine, but in Miyagi, the number was five. The geographic concentration of seafood manufacturing in Iwate was higher than that in Miyagi before the disaster. After the disaster, the number decreased to six in Iwate, and three in Miyagi. Thus, the disaster caused the geographic proximity of seafood manufacturing in both prefectures to fall. This situation, namely the slower recovery of seafood manufacturing than that of the fishery, is a similar phenominon to the recovery of production of the industries described above.

Finally, the bivariate Moran's I values for fishery and seafood manufacturing, indicating co-agglomeration, are shown in Fig. 7.11. A supply linkage is found between fishery and each sector in seafood manufacturing in the coastal regions. In comparison with areas bounded by the bivariate Moran's I figures of Iwate and Miyagi Prefectures, Iwate's area is larger than Miyagi's. It indicates that the both industries of Iwate have higher co-agglomeration than that of Miyagi. In Iwate, the co-agglomeration of small dried sardines and salted product sectors returned to pre-disaster levels. Most geographic proximity levels of seafood manufacturing and fisheries, except for smoke-dried products, came close to pre-disaster levels. In contrast, in Miyagi, while co-agglomeration of activities related to frozen seafood returned or came close to pre-disaster levels, co-agglomeration levels were remarkably lower in several sectors, such as dried-products, small dried sardines, and smoke-dried products, despite these sectors having higher co-agglomeration levels than sectors related to frozen seafood. Thus, several sectors in seafood manufacturing, which showed earlier recovery of production, recovered regarding not only geographic concentration of their own sector, but also co-agglomeration with the fishery, which is an upstream industry in the fish food system.

7.3 Dynamic CGE Model

7.3.1 Activities Based on Supply Linkages of Fishery and Seafood Manufacturing

Table 7.3 shows ten activities from the Dynamic CGE model presented in this chapter. These activities fall into the following three categories: fisheries, seafood manufacturing, and others. The fisheries have two activities: fishing and aquaculture, which involves fish and oyster farming as described in 7.2. Seafood manufacturing has five activities: the manufacture of frozen seafood products, salted-dried and salted products, canned seafood products, fish paste products, and miscellaneous seafood products. The manufacture of frozen seafood products includes JSIC 925 and 926, and the manufacture of miscellaneous seafood products includes JSIC 922 in 929. The others include three activities: agriculture and

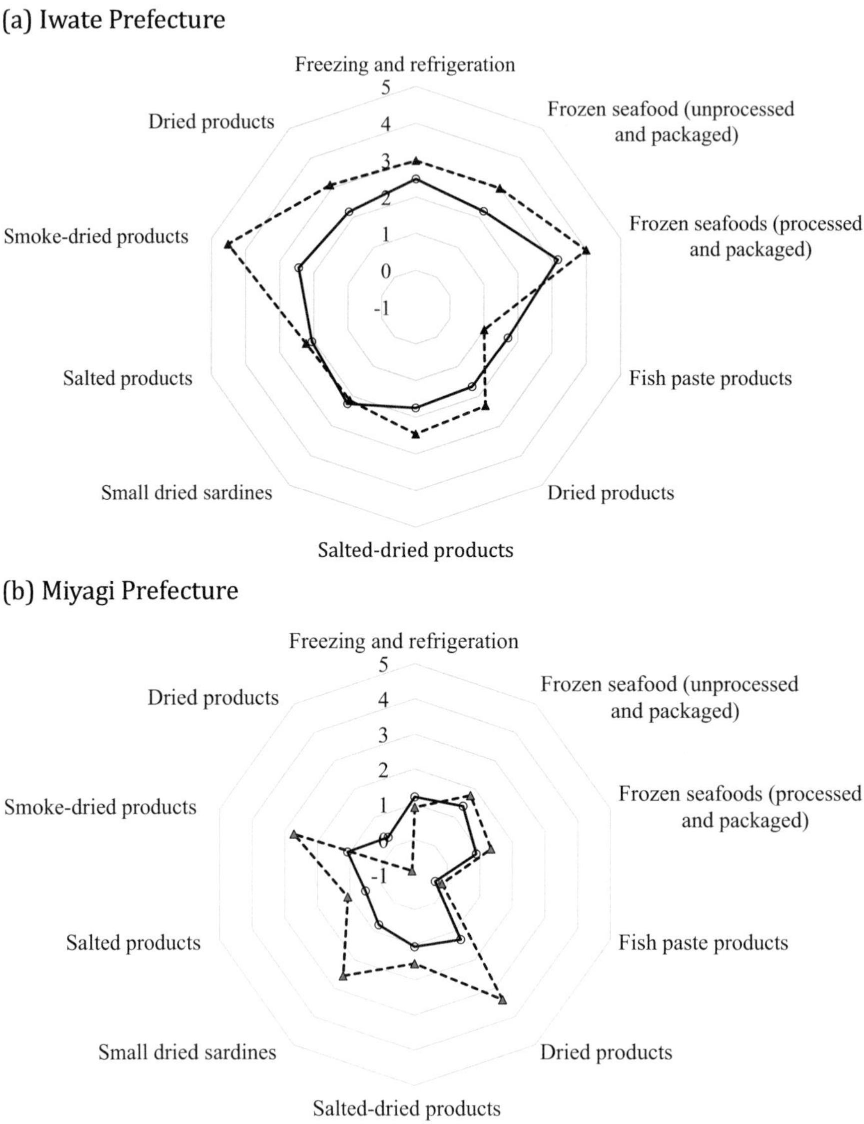

Fig. 7.11 Bivariate Moran's I for fishery and seafood manufacturing

forestry, food manufacture and restaurants excluding activities related to seafood manufacturing, and manufacturing and services.

One feature of this model is its explicit treatment of industrial linkages between fisheries and seafood manufacturing. These linkages are based on the demand destinations and input-output relationship. According to the Social Account Matrix in 2005, whose data the model uses, 54.5% of domestic output from fishery and seafood manufacturing is consumed as final products by households, and the

Table 7.3 Activities in the model

	Activities in the model	Activities in 7.2.
Fisheries	Fishing	Fishing
	Aquaculture	Fish and oyster farming
Seafood manufacturing	Manufacture of frozen seafood products.	JSIC 925 and 926
	Manufacture of salted-dried and salted products	JSIC 924
	Manufacture of canned seafood products	JSIC 921
	Manufacture of fish paste products	JSIC 923
	Manufacture of miscellaneous seafood products	JSIC 922 and 929
Others	Agriculture and forestry	–
	Food manufacturing and restraints	–
	Manufacturing and services	–

remaining 45.5% is used as intermediate products. In detail, around 90% of the produce of fishing and around 60% of the product of aquaculture is consumed as intermediates. Meanwhile, The ratio of final goods in seafood manufacturing is 73–80%, except for frozen seafood products, which has a final goods ratio of around 60%. Thus, the products of the fishery are largely consumed as intermediates, while the products of seafood manufacturing are principally consumed as final goods.

In the input-output between fishery and seafood manufacturing, 51% is intermediate products supplied between both industries. Of this, 60% is input from fishing, 27% is from frozen seafood manufacturing, and 11% is from aquaculture. In turn, 78% of the output from fishing is supplied to frozen seafood manufacturers, and in turn, 82% of produced frozen seafood is provided to other seafood manufacturers while 79% of the output of aquaculture is then input to other seafood producers. Hence, the fish food system is constructed of the following supply-linkages: fishing and aquaculture are at the upstream, the frozen seafood manufacturing is at the middle-stream, and other seafood manufacturers are at the downstream.

7.3.2 Model Structure

This model structure is based on Akune et al. (2015) and modified for the analysis purpose in this chapter. In the model, each producer is assumed to maximize profits subject to several production technologies, the model of which assumes perfect competition in these activities. Top-level technology is specified by a Leontief function of the quantities of aggregate intermediate inputs and value-added. At the bottom level, each activity evaluates the value added by composite factors under a constant elasticity of substitution (CES) function. Each activity produces one commodity.

In the trade sector, domestically-produced commodities are distributed for both export and domestic sales under a constant elasticity of transformation (CET) function. Meanwhile, imports comprise generated composite goods with domestic sales, as determined using the CES function under Armington's assumptions (Armington 1969) and are supplied to the domestic market.

Domestic demand comprises household and government consumption, investment, and intermediate inputs. A representative household's consumption is assumed to maximize the Cobb-Douglas utility function. Government consumption is considered as comprising a specific consumption share of each commodity. Private and government savings are determined by multiplying each average propensity to save. Investment demand is equal to total private, government, and foreign savings.

This model includes the following market equilibrium conditions: (1) domestic markets for goods and primary factors and (2) the external balance. Firstly, each composite good supplied equals the sum of demand from households, government, investment, and intermediate inputs. Secondly, there are two markets for primary factors: labor and capital. The aggregate demand for labor and capital activities equals the total supply from households. Finally, the current account is balanced under fixed foreign savings each year.

This model involves the use of a recursive dynamic mechanism. An activity's capital stock in a given year comprises the capital stock discounted by depreciation and investment in the activity during the previous year. Another important mechanism is the capital-growth function and the expected rate of return. In this function, if an activity's capital growth exceeds its historically normal capital-growth rate, its expected rate of return will exceed its historically normal rate of return, which implies that the activity is attracting sufficient investment.

7.4 Results of Simulation

7.4.1 Simulation Scenarios

The model deals with five simulation scenarios: business as usual (BAU), scenario 1, scenario 2, scenario 3, and scenario 4. These scenarios are developed in terms of the following five points: (i) depopulation, (ii) capital losses of fisheries and seafood manufacturing in a disaster, (iii) a productivity shock during the disaster, (iv) physical restoration after the disaster, and (v) production recovery of activities after the disaster. This section describes the combination of these five points for the scenarios as shown in Table 7.4 and then explores the five points in further detail.

(i) Depopulation is set to occur in all scenarios; BAU shows the effects of depopulation on fishery and seafood manufacturing without disaster. On the other hand, disaster is assumed to occur in other four scenarios except for BAU.

Table 7.4 Simulation scenarios

	(i) Depopulation	(ii) Physical loss (iii) Productivity shock at the disaster	(iv) Physical restoration	(v) Production recovery after the disaster
BAU	YES	NO	—	—
Scenario 1	YES	YES	NO	NO
Scenario 2	YES	YES	YES	YES: A criterion
Scenario 3	YES	YES	YES	YES: I criterion
Scenario 4	YES	YES	YES	YES: M criterion

Furthermore, the dimension of damage during the catastrophe and restoration and recovery after the disaster in the four scenarios are based on the scale of damage and recovery process of the GEJE. Scenarios 1–4 have points (ii) capital losses and (iii) a productivity shock due to the disaster. The result of scenario 1 shows disaster damage to fisheries and seafood manufacturing without any capital restoration policy and intentional production recovery. Scenarios 2–4 assume that point (iv) capital restoration policy will be conducted by the actual response to the GEJE. These scenarios have an additional three criteria type about (v) production recovery. Scenario 2 assumes that production recovery is conducted along the lines of the actual situation of the GEJE; we call this "A (actual) criterion." Scenario 3 has "I (innovative) criterion," which is based on evidence of production recovery in Iwate. Scenario 4 assumes moderate production recovery "M (moderate) criterion" based on the evidence from Miyagi.

Therefore, a comparison between scenarios 1 and 2 indicates the effects of the capital restoration policy and production recovery evidence based on the actual case of the GEJE. A comparison between scenarios 1 and 3 indicates the effects of innovative production recovery, which induces a change in production structure. A comparison between scenarios 1 and 4 shows the effect of moderate production recovery, which prioritizes to stabilize a production structure of before the disaster.

The five main points in all scenarios are as follows:

- (i) Depopulation: The Japanese population assumes to consistently decrease at a rate of 0.53% per annum in all scenarios. The rate is based on the Population Projection data for Japan (January 2012) forecasted by the National Institute of Population and Social Security Research. It is a version of medium-fertility (medium-mortality) projection. Migration is not assumed; the labor force declines along depopulation in all scenarios.
- (ii) Physical loss: The disaster-induced damage to fisheries and seafood manufacturing is based on evidence from the GEJE. Table 7.5 shows the damage ratio of the catastrophe. As physical damage to fisheries, it set at 10.4% for fishing, a level estimated from the damage to ships: the number of damaged ships was around 29,000 out of 276,000 in Japan. The damage to aquaculture facilities totaled 73.7 billion yen, and that to products reached 59.7 billion yen, which was 13.9% of total aquaculture output in 2010. Although it is better to use

Table 7.5 Ratio of physical losses in the scenarios

	Ratio of physical losses
Fishing	10.4%
Aquaculture	13.9%
Manufacture of frozen seafood products	0.6%
Manufacture of salted-dried and salted products	0.3%
Manufacture of canned seafood products	0.0%
Manufacture of fish paste products	0.1%
Manufacture of miscellaneous seafood products	1.0%

data on the damage to capital stock, it is hard to estimate the total values. Thus, it assumes that the physical damage to aquaculture is the same degree of damage as inflicted upon products. Moreover, the damage rate of the whole of seafood manufacturing industry was 5.7% in Japan. Around 1,725 official and cooperative facilities destroyed along with around 700 private facilities, out of the around 12,000 total facilities in Japan. The ratios of capital losses for each manufacturing as shown in Table 7.5 are calculated by the proportion of production before the GEJE in the affected regions.

- (iii) Productivity shock at the disaster: Disasters also cause human loss. It means not only the humans directly affected but also the motivation to retire from the labor market, especially for older adults. For example, although the recovery of ships as described above indicates that their restoration occurred faster than other physical equipment, this figure is based on ships belonging to those who are willing to restart and excludes retired fishers. Where retired fishers are excluded, the rate of restoration of ships against that of before the disaster is 64%. Table 7.6 shows productivity shock ratios for each activity, calculated as a decline in production in 2011 when the GEJE occurred.
- (iv) Physical restoration: It assumes that a physical restoration policy was intensively conducted for the 2 years following the disaster, since the restoration policy for facilities was intensively carried out between 2011 and 2012 in the case of the GEJE. The MAFF announced that 91% of the damaged ships restored as of December of 2015; 84% of seafood manufacturing facilities that desired to restart also completed. Table 7.7 shows these capital restoration rates against the losses given in Table 7.5 recalculated to ratios for each activity.
- (v) Production recovery after the disaster: The recovery of production assumes to bring about due to the efforts of producers and innovation among manufacturers as well as capital restoration policy. Their recovery is based on evidence of three criteria: actual (A), innovative (I), and moderate (M) as shown in Table 7.8. These figures relate to the productivity shock, as mentioned in (iii) productivity shock entries in Table 7.6. For example, the productivity of fishing at the second period after a disaster in the A criterion returned to −2.6% by adding 0.011 into −3.7% of the primary productivity.

Table 7.6 Productivity shock ratios

	Productivity shock ratios
Fishing	−3.7%
Aquaculture	−15.3%
Manufacture of frozen seafood products	−12.2%
Manufacture of salted-dried and salted products	−7.7%
Manufacture of canned seafood products	−22.5%
Manufacture of fish paste products	−2.0%
Manufacture of miscellaneous seafood products	−3.8%

Table 7.7 Rate of physical restoration against losses

	Ratio of restoration	
	First period	Second period
Fishing	34.7%	4.8%
Aquaculture	70.5%	5.0%
Manufacture of frozen seafood products	4.7%	0.3%
Manufacture of salted-dried and salted products	2.1%	0.1%
Manufacture of canned seafood products	0.4%	0.0%
Manufacture of fish paste products	1.1%	0.1%
Manufacture of miscellaneous seafood products	8.1%	0.6%

7.4.2 Results of Simulation

Figure 7.12 shows the trends in fishery and seafood manufacturing outputs. There are 25 periods of simulation in the scenarios. The first year, denoted as $t = 0$, is the year before the disaster. The disaster occurs at $t = 1$. The numbers are standardized indices of 100 in the year before the catastrophe.

We find the following five points in Fig. 7.12. First, all scenarios trends for both industries show falling tendencies. This reason is why the population, which supplies labor as a necessary production factor and provides demand, decreases, causing the whole economy to shrink. Besides, the second point is that the disaster reveals a structural problem, which is the depopulation that Japanese society is facing now. At $t = 25$ in BAU, the fishery is at 88, and seafood manufacturing is at 88, at the year of the disaster all lines of other scenarios are below the level at 25 years later of BAU. It shows that the catastrophe accelerates the problem caused by the social structural problem to be more severe.

Moreover, the third point is that in scenarios 1–4, the margins between these scenarios and BAU are decreasing toward the end of the simulation. It is due to cause an increase in demand for the capital for fishery and seafood manufacturing by rising ratios of capital returns depending on the degree of capital loss. Fourth, the pace of production recovery in the seafood manufacturing in scenario 3 with I criterion is faster and closer to BAU than that of scenarios 2 and 4. The degree of

Table 7.8 Production recovery points of A-I-M criteria

Productivity recovery points									
	A criterion			I criterion			M criterion		
Period after a disaster	2nd	3rd	4th	2nd	3rd	4th	2nd	3rd	4th
Fishing	0.011	0.010	−0.002	0.015	0.007	0.000	0.009	0.012	−0.003
Aquaculture	−0.003	0.032	0.032	0.000	0.032	0.031	−0.005	0.032	0.033
Manufacture of frozen seafood products	0.037	0.022	0.018	0.084	0.048	0.012	0.023	0.014	0.019
Manufacture of salted-dried and salted products	0.052	0.002	0.007	0.112	−0.017	0.027	0.034	0.008	0.001
Manufacture of canned seafood products	0.063	0.043	0.009	0.097	0.036	0.003	0.000	0.056	0.020
Manufacture of fish paste products	0.007	−0.003	0.000	0.007	−0.003	0.000	0.007	−0.003	0.000
Manufacture of miscellaneous seafood products	0.013	0.005	0.003	0.029	−0.009	0.018	0.012	0.006	0.002

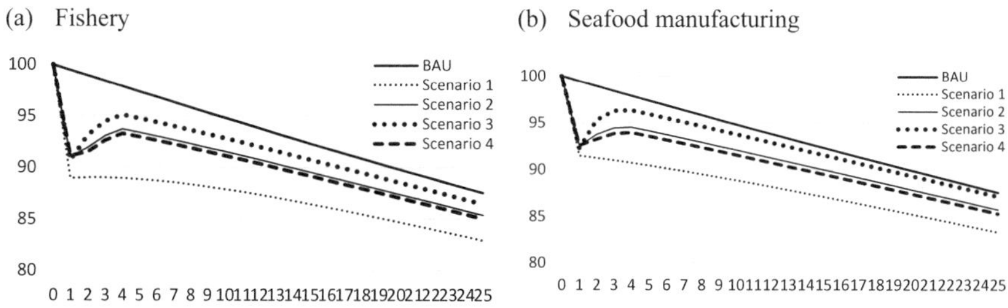

Fig. 7.12 Trends of output of fishery and seafood manufacturing (100 at t = 0)

output recovery in scenarios 2 and 4 is the same, while scenario 2 with the A criterion is a little higher than scenario 4 with M criterion. These results indicate the advantage of pursuing innovative production recovery.

Finally, the fifth point is different recoveries between fishery and seafood manufacturing. The lines at sea food manufacturing are higher than that of the fishery in Fig. 7.12. As mentioned in 7.3, the fishery is at upstream, and seafood manufacturing is a downstream industry. That is that the degree of production recovery of the downstream industry is faster than that of the upstream industry. The recovery of fish food system describes in detail later.

Figure 7.13 shows the production recovery of fishing and aquaculture in the fishery. Fishing productivity recovers faster than that of aquaculture in all scenarios because the capital losses and productivity shock of aquaculture are greater than that of fishing. Besides, the line of scenario 1 without any restoration policy is the lowest in the simulation results; the lines of scenario 2 to 4 rise sharply higher recoveries during t=1 to 4 by conducting the policy for recovery of capital losses. These results indicate that an intensive restoration policy is necessary for more effective production recovery.

Moreover, the comparison between fishing and aquaculture in scenario 3 shows another finding, namely that the difference in production recovery depends on the degree of supply-linkage. In Fig. 7.13, the recovery in production of seafood manufacturing is higher than that of fisheries in scenario 3. Other relationship among other activities also exist: 60% is input from fishing, while 11% is from aquaculture. Thus, an industry with a broader supply linkage is more likely to gain from a production recovery occurring in other sectors. Therefore, This result indicates that it is important to make a plan for production recovery regarding supply linkages.

Figure 7.14 shows the production recovery of five seafood manufacturers in each scenario. It shows the two following points. First, early production recovery after a disaster results in a shorter period of damage to the industry. For example, a few manufacturers exceed or come extremely close to BAU in scenario 3, which includes innovative production recovery based on the available evidence. One manufacturer exceeding BAU is a manufacturer of salted-dried and salted products, whereas a manufacturer of frozen seafood products is close to BAU. Moreover, the

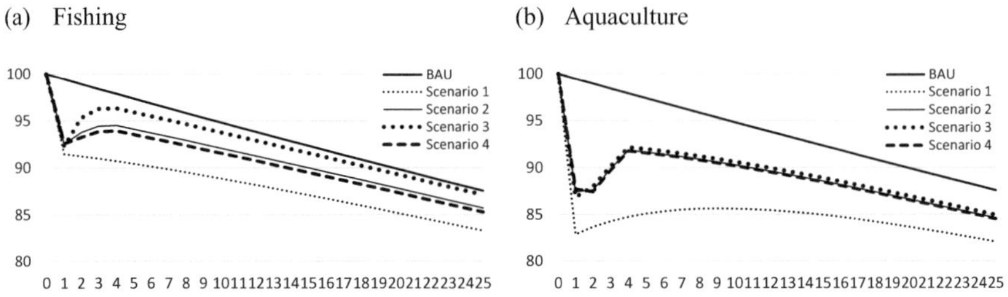

Fig. 7.13 Trends of output of fishery (100 at t = 0)

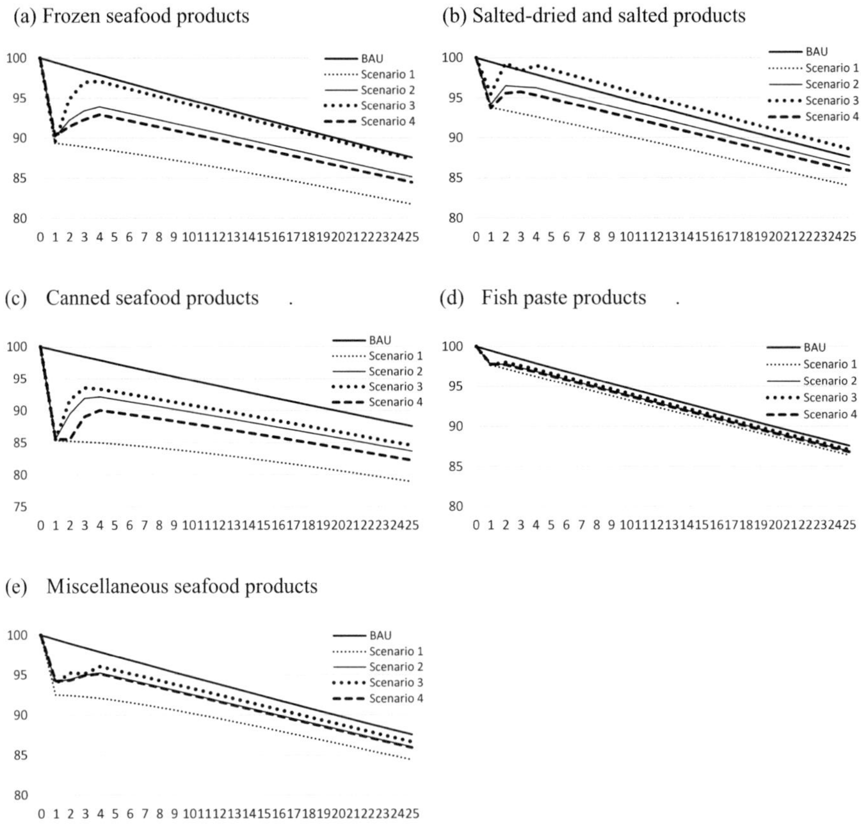

Fig. 7.14 Trends of output of seafood manufacturing (100 at t = 0)

second point is that an importance of production recovery for middle-stream industries in the fish food system demonstrates. The production recovery of the manufacture of frozen seafood products located in the middle among fisheries and seafood manufacturers contributes to accelerating production recovery for other

Table 7.9 Total economic welfare during simulation period (Equivalent variation)

	Scenario 1	Scenario 2	Scenario 3	Scenario 4
Total EV	−2.2	−1.1	−0.7	−1.3

Unit: trillion yen

activities. Particularly fishing situated above manufacturing of frozen seafood takes effects of this acceleration in scenario 3 of Fig. 7.13a.

Table 7.9 shows accumulated equivalent variation (EV) during the simulation period as economic welfare. All EV values are negative, even in scenario 3 that includes innovative production recovery. The interval between scenario 1 and 2 indicates the economic effects of a production restoration policy. In these simulations, the restoration policy and production recovery of fishery and seafood manufacturing based on actual evidence of the GEJE cut the economic loss in half. Moreover, the gap between scenario 3 and 4 is 0.6 trillion yen. It indicates that prompt and innovative production recovery actions after a disaster can reduce economic losses, though the recovery process transforms the production composition.

7.5 Seeking a Possibility for a Sustainability of Fishery and Seafood Manufacturing after a Disaster

As described, no scenario goes beyond the levels of production for BAU. However, the results indicate the necessity of longer-term supports for production recovery. Hence, I introduce an additional scenario, namely scenario 5, which utilizes a stepwise production recovery process. Scenario 5 has (i) depopulation, (ii) capital losses, and (iii) a productivity shock due to the disaster, as in scenario 1–4. It assumes that production recovery is conducted in two stages; first is the I criterion with the innovative recovery, and then the M criterion with the moderate recovery. In other words, production recovery occurs sequentially by combining the paths delineated in scenario 3 and 4.

Figure 7.15 shows the trends in fishery and seafood manufacturing output in scenario 5 with BAU and scenarios of 3 and 4. The production level of the fishery in scenario 5 reaches BAU level 7 years after the disaster. The output of seafood manufacturing reached BAU level 5 years later. Both production levels subsequently remain beyond the levels of BAU.

Figure 7.16 shows the production recovery of fishing and aquaculture in fisheries. The output level of the fishing in scenario 5 with a stepwise production recovery exceeds that of BAU, though the outputs of scenario 3 and 4 cannot reach BAU levels. The level of aquaculture in scenario 5 is not at BAU level unlike in the previous scenarios, though it is closer to BAU than others. These results indicate the production recovery of combined I and M criteria is sufficient for a recovery in fishing but insufficient for a recovery in aquaculture.

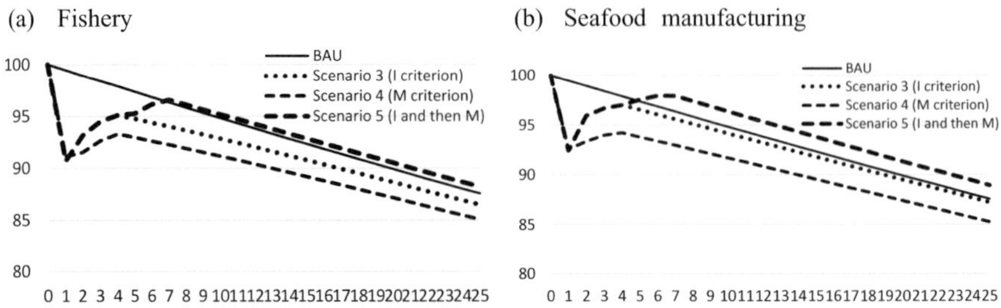

Fig. 7.15 Trends of fisheries and seafood manufacturing output (100 at t = 0)

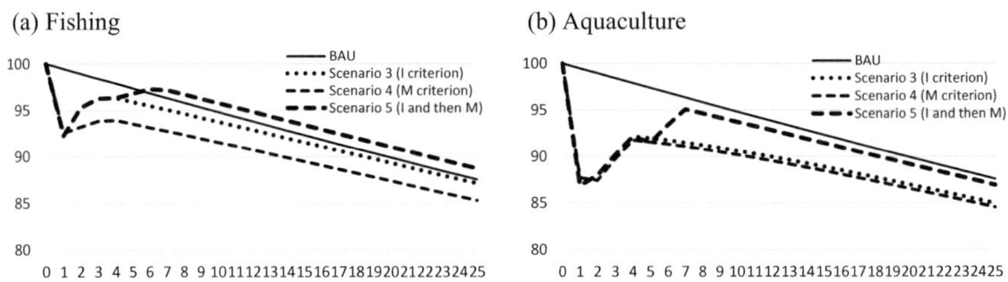

Fig. 7.16 Trends of fishery output (100 at t = 0)

Figure 7.17 shows the production recovery of five seafood manufacturers. This figure illustrates the following two points. First, two manufacturers, namely the manufacturer of frozen seafood products and the manufacturer of salted-dried and salted products, have recovery levels closer to BAU in scenario 3, and have exceeded BAU since the year after reaching BAU. The second point is the effect of this stepwise production recovery in scenario 5 with the manufacturers of canned seafood products and the manufacturers of miscellaneous seafood products. The sectors in this additional scenario arrive at BAU level, though production levels are lower than BAU in the previous scenario.

Finally, although the total EV in scenario 5 is negative at −0.22 trillion yen, a desirable trend did emerge during the simulation period. The EV has been positive since the seventh year after the disaster, after which it gradually increased as shown in Fig. 7.18. This result indicates the possibility that stepwise production recovery contributes to rises not only in specific industries but also to economic welfare in the long term.

7.6 Conclusion

This chapter evaluated the economic effects of production recovery based on evidence provided by fisheries and seafood manufacturers who experienced the GEJE under depopulation, by employing a dynamic CGE model. Japan is currently

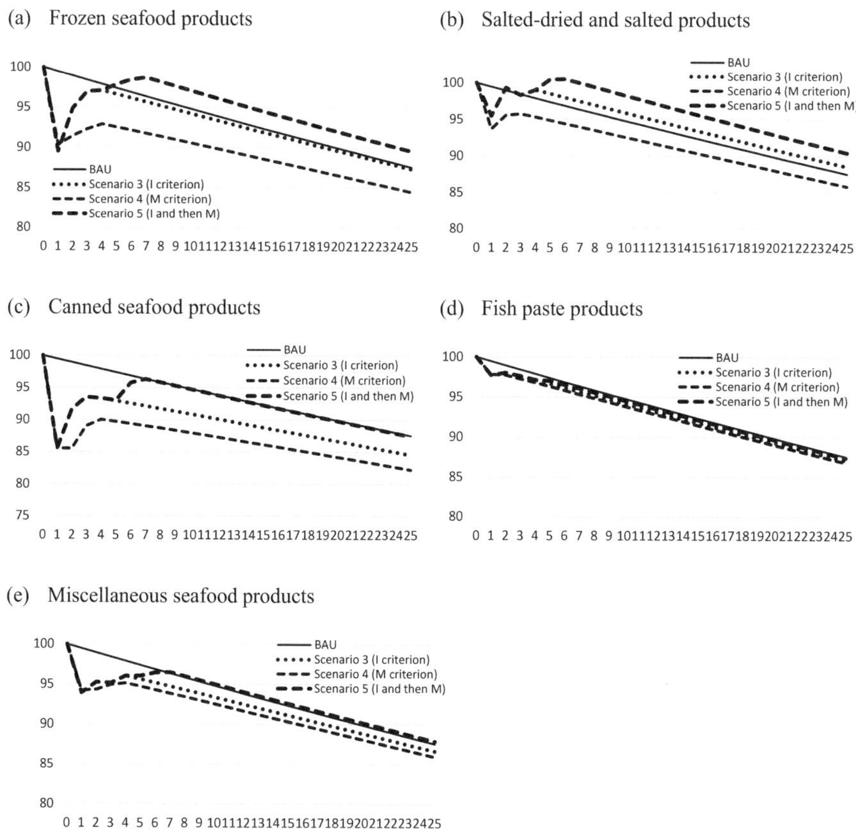

Fig. 7.17 Trends of seafood manufacturing output (100 at t = 0)

facing two serious problems; a decline in population and a risk of new natural disasters such as the Nankai earthquakes and Tokai earthquakes. The great catastrophe in 2011, which included a huge earthquake and tsunami, provided wide-ranging and highly significant lessons. They indicated that we cannot perfectly protect ourselves from damage caused by natural disasters and that the economic activity cannot evade taking damages from such catastrophes. In Japan, and the case of earthquakes and tsunamis, such damage is almost inevitable, particularly concerning fishery and seafood manufacturing, which locate in vulnerable coastal regions. Hence, it is necessary to consider how production by fishery and seafood manufacturing recover after disasters.

We confirmed a clear contrast between the production recovery by seafood manufacturers in Iwate and Miyagi Prefectures after the GEJE. The recovery speed in Iwate was faster than that in Miyagi. At the same time, the disparity between the recovery of each sector of seafood manufacturing in Iwate was greater than that in Miyagi. As a result, the composition of seafood product manufacturers in Iwate changed compared to before the disaster. Besides, the geographic

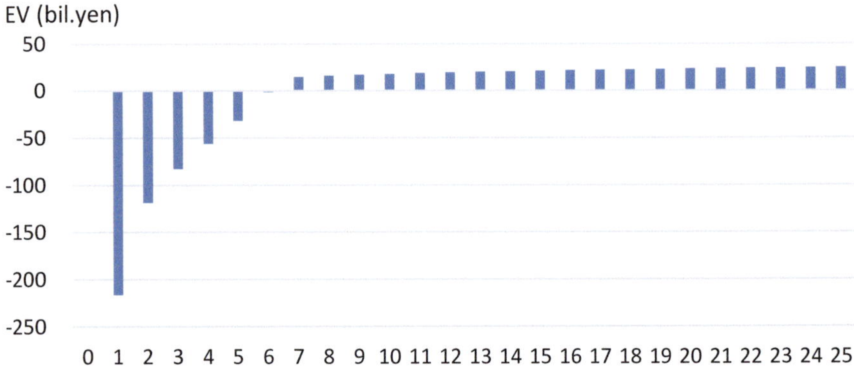

Fig. 7.18 Trends of economic welfare at in scenario 5

concentration of each industry and co-agglomeration between fisheries and each type of seafood manufacturing, which are considered to be one of the foundations of innovation, were similar regarding the degree of production recovery. The more production recovery activities that seafood manufacturers performed, the higher their area's own geographic concentration and co-agglomeration with fisheries they showed.

The results of four simulations based on the available evidence indicated the following four conclusions. First, the disaster accelerated the problem caused by depopulation: production declined to the same level as 25 years later in BAU assumed under depopulation with no catastrophe. Next, prompt capital restoration contributed to shrinking the period of output loss. For example, an intensive physical restoration policy in the short-term following the disaster reduced the decrease of the production and led to an increased pace of output recovery in scenario 2–4. Similarly, rapid production recovery also resulted in shrinking the damage period, as shown in scenario 3. Thirdly, the degree of output recovery of the downstream industry was faster than that of the upstream industry. The comparison between production recovery of fishery and seafood manufacturing in scenario 3 showed that the recovery progress of seafood manufacturing was greater than that of the fishery. Finally, although scenario 3 assumed more innovative production recovery than other scenarios, based on evidence from Iwate Prefecture, unfortunately, even this scenario was not sufficient to reach the level of BAU.

We considered an additional scenario that utilized a stepwise production recovery. Scenario 5 had (i) depopulation, (ii) capital losses, and (iii) a productivity shock due to the disaster, as with scenarios 1–4. This scenario assumed that production recovery was conducted in two stages (1) the I criterion with innovative recovery, and (2) M criterion with the moderate recovery. This result indicated the possibility that stepwise production recovery contributes to increases not only in specific industries but also to economic welfare in the long term. If damage inflicted by the next disaster is at the same level as that caused by the GEJE, stepwise recovery consisting of the recovery process used in Iwate and Miyagi Prefectures might be more effective than conducting them separately. Although it is hard to define a precise length of time that such supports will be needed, the simulation result showed that intensive recovery supports were necessary for at least approximately 7 years after a disaster.

Furthermore, I would like to refer to the possibility of the sustainability of activities in coastal regions regarding the following three issues: a decline in labor, shrinkage of the domestic market, and an increase in productivity. The depopulation causes the former two problems. The labor decline in regions where fishery and seafood manufacturers locate has started already and been more severe than that in other regions before the GEJE. At many seafood producers, foreigners under the Japanese training programs have worked; it help to alleviate the shortage of labor. However, the programs are impossible to provide the fundamental solution, as the programs have the aim of providing training rather than ongoing work, additionally, foreign trainees are allowed to stay for 3 years during the programs.

An increase in producers' profits could provide another solution. People might want to work for companies with higher profits if higher profits were translated into higher wages. An increase in exports and productivity could provide such a possibility. According to a report by the Tohoku Bureau of Economy, Trade and Industry (TBETI) (2016) producers engaged in fishery and seafood manufacturing have already noticed the importance of this and started these challenging, e.g., to explore opportunities to export or to collaborate with new business partners for producing goods with more value-added. An increase in exports might simultaneously provide a solution to the shrinkage of the domestic market.

The TBETI report also points out issues related to export promotion: increased costs due to small-scale operations, a lack of understanding regarding export processing, and a shortage of workers with both knowledge of products and English skills. The report emphasizes the necessity of extensive cooperation with producers at not only same sector but also other sectors, e.g., developing new products and sharing workers with knowledge/skills and mutual branding ideas. It is related to geographic proximity as mentioned in Sect. 7.2. Several sectors in seafood manufacturing, which showed earlier recovery of production, recovered not only regarding the geographic concentration of their own activities but also concerning co-agglomeration with fisheries. Thus, this promotion on a higher geographic concentration and co-agglomeration might provide a method to achieve increased productivity.

Finally, there are outstanding issues not resolved by the analyses conducted in this chapter. The first is to define boundaries of cooperation. The unit used for measurement in the data is the municipal organization. The higher indices of Iwate indicated that is prefecture's seafood manufacturers had greater similarity with contiguous cities and towns. Conversely, there were fewer similarities with neighbors in Miyagi. However, that might be caused by lower need to cooperate with nearby cities and towns, since Miyagi has a much higher production volume and linkages constructed within a city. Reconsiderations of geographic proximity's role in increasing productivity and conferring comparative advantages are one remaining issue. Moreover, this chapter showed the economic impact of the damage and production recovery in specific industries, while overall, disaster damage is complicated and related to financial and tax burden issues as well. A simultaneous analysis regarding more detailed policy for specific industries and comprehensive fiscal policy is necessary.

References

Akune Y, Okiyama M, Tokunaga S (2013) Evaluation of the restoration of fisheries and seafood manufacturers after the Great East Japan Earthquake: economic analysis utilizing a dynamic computable general equilibrium model, RIETI Political Discussion Paper, 13-P-022

Akune Y, Okiyama M, Tokunaga S (2015) Economic evaluation of dissemination of high temperature-tolerant rice in Japan using a dynamic computable general equilibrium model. Jpn Agric Res Q 49(2):127–133

Anselin L, Syabri I, Smiirnov O (2002) Visualizing multivariate spatial correlation with dynamically linked windows, REAL 02-T-8

Armington PS (1969) A theory of demand for products distinguished by place of production. Staff Pap Int Monet Fund 16(1):159–178

Ellison G, Glaeser EL (1997) Geographic concentration in U.S. manufacturing industries: a dartboard approach. J Polit Econ 105:889–927

Fujita M et al (1999) The spatial economy: cities, regions, and international trade. MIT Press, Cambridge, MA

Hallegtte S, Prsyluski V (2010) The economics of natural disaster: concepts and methods. Policy research working paper 5507. World Bank, doi:https://doi.org/10.1596/1813-9450-5507

Huang MC, Hosoe N (2016) Computable general equilibrium assessment of a compound disaster in northern Taiwan. Rev Urban Reg Dev Stud 28(2):89–106

Krugman P (1991) Geography and trade. MIT Press, Cambridge, MA

Moran PAP (1950) A test for the serial dependence of residuals. Biometrika 37:178–181

Porter ME (1998) On competition. Harvard Business School Press, Boston

Rose A, Liao SY (2005) Modeling regional economic resilience to disasters: a computable general equilibrium analysis of water service disruptions. J Reg Sci 45:75–112

Shibusawa H, Yamaguchi M, Miyata Y (2009) Evaluating the impacts of a disaster in the Tokai region of Japan: a dynamic spatial CGE model approach. Stud Reg Sci 39(3):539–551

Tohoku Bureau of Economy, Trade and Industry (2016) Report of society for regional revitalization by seafood processing in the affected regions (Suisan Kakougyou wo Syutai to shita Hisaiti ni okeru Tiiki Kasseika Kennkyuukai Torimatome Houkokusyo). http://www.tohoku.meti.go.jp/s_hukko_suisan/suisan_kyogikai/pdf/kenkyukai.pdf. (in Japanese)

Tokunaga S, Okiyama M, Akune Y (2013) Analyses of disrupted supply chains by the Great East Japan Earthquake and reconstruction of the disaster-affected region by the cluster of the automotive industry: utilizing the regional CGE model, RIETI Discussion Paper Series 13-J-068

Tokyo University of Marine Science and Technology (2014) Research on recovery of fisheries and seafood manufacturing from the Great East Japan Earthquake: evidence in third year since the disaster, (Higashi-Nihon-Daisinsai kara no Suisangyou oyobi Kanren Sangyou no Fukkou Taisaku ni kakaru Kenkyu: Daisinsai kara no 3 nen me no Genti no Fukkyu-Fukkou Jyoukyou). https://www.kaiyodai.ac.jp/project/r1/10/hokoku14/04.pdf (in Japanese)

Tsuchiya S, Tatano H, Okada N (2007) Economic loss assessment due to railroad and highway disruptions. Econ Syst Res 19(2):147–162

Wartenberg D (1985) Multivariate spatial correlation: a method for exploratory geographical analysis. Geogr Anal 17(4):263–283

Chapter 8
Economic Ripple Effects of a Biogas Electricity Power Plant as Part of Earthquake Disaster Restoration in the Coastal Area of Iwate Prefecture

Yoji Kunimitsu

Abstract The coastal area of Iwate Prefecture was seriously damaged by the Great East Japan Earthquake in 2011, and the revitalization of this region is an important issue in Japan. This chapter aims to estimate the input–output tables of this region before and after the earthquake and to use these tables to show the impacts of the earthquake and the impacts of a particular revitalization measure: the construction of a biogas electricity power plant. The results showed several distinct features. First, value-added production in 2011 was higher than the previous year because of recovery investment, but both intermediate inputs and the total production decreased. This occurred because the ratio of value-added production in the construction sector was higher than in other industries that were seriously damaged by the earthquake. Second, investment demand increased in 2011 and 2012, but the intermediate inputs and private and public consumption decreased because of the earthquake. The power-of-dispersion coefficients and production-inducement coefficients were slightly changed by the earthquake, but the increasing rate of induced production was low after the earthquake. Third, the multiplier value of induced production for the construction of a biogas electricity power plant would have been 0.64, if it had been constructed before the earthquake. However, that value after the earthquake was 0.17 because industrial linkage was damaged. Compared to the conventional electricity sector, the biogas electricity generation sector induced more production because the import rate of this sector was zero and most induced demand occurred within the region. Therefore, a biogas electricity power plant can contribute to the revitalization of a regional economy although the electricity generation capacity is small.

Y. Kunimitsu (✉)
Regional Resource Engineering Division, National Agriculture and food Research Organization, 2-1-6, Kannondai, Tsukuba, Ibaraki 305-8609, Japan
e-mail: ykuni@affrc.go.jp

S. Tokunaga, B.P. Resosudarmo (eds.), *Spatial Economic Modelling of Megathrust Earthquake in Japan*, New Frontiers in Regional Science: Asian Perspectives 11,
https://doi.org/10.1007/978-981-10-6493-7_8

Keywords Input–Output (I/O) analysis • Induced production • Induced labor • Biogas electricity power plant

8.1 Introduction

The coastal area of Iwate Prefecture suffered fatal damage due to the tsunami caused by the Great East Japan Earthquake in March 2011. After that, reconstruction has been conducted with government support and the cooperation of the local civic association. However, there are still many people living in temporary housing today—4 years after the earthquake disaster. Some industries, such as fishery, have not been able to reach the same level of production as before the earthquake disaster. After such a serious disaster, policy makers are determined to build a resilient country and community.

In addition to the need for recovery, the government is trying to revitalize the local economy, which faces an aging and declining population, by using local resources. The construction of a biogas electricity power plant is planned for such a purpose. This plant in the coastal city of Ohfunato would produce methane gas from food residues and generate electricity with it. This plant would also be carbon neutral and would not increase or decrease CO2 even after burning materials. However, this technology is expensive and low efficiency compared to traditional power plants. In order to operate a biogas electricity power plant, support from government is generally needed. For getting consensus of people for such support, an economic evaluation is important.

Many previous studies, as shown in the next section, evaluated the economic effects of biomass resources using an input–output (I/O) model. For example, Urbanchuk (2009) analyzed the macroeconomic effects of bioethanol production by using an I/O model and showed that bioethanol production induced $32.5 billion of added value and created 110 thousand jobs in the U.S. in 2006. Most studies have evaluated the macroeconomic effects of bioenergy production under a normal economic situation, but few studies have evaluated the economic impacts and ripple effects of biogas production on a local economy during the restoration period after a natural disaster. As shown in the Great East Japan Earthquake, the construction of new industries, especially those related to energy production, is critical for restoration after a disaster.

The present study aims to estimate the I/O tables of this region before and after the earthquake and use these tables to show the impact of the earthquake and the impact of a particular revitalization measure: the construction of a biogas electricity power plant. The features of this study are (i) that recent I/O tables after the earthquake are estimated by the RAS method based on statistics of net-residential income published by the prefectural government, (ii) that impacts of the earthquake on industrial linkages are shown by analyzing the inverse matrices of the I/O tables, and (iii) that the ripple effects of construction and operation of biogas electricity power plant are estimated by these I/O tables.

The following section shows the method of estimating recent I/O tables and the measurement of ripple effects. The third section shows the results, and the fourth section summarizes and concludes the analysis.

8.2 Method

8.2.1 Previous Studies and Scientific Questions

In addition to Urbanchuk (2009), introduced in the previous section, Polagye et al. (2007) showed the profit from a bioethanol production factory in the U.S. Hayashi (2010) analyzed the Japanese E10 policy using I/O analysis and pointed out that the economic benefit of bioethanol production can be positive in the near future if fossil oil prices increase along the recent trend. Kunimitsu and Ueda (2006) measured the economic effects of bioenergy production in Thailand using an I/O analysis. Kunimitsu et al. (2013) estimated economic and environmental effects of bioethanol production that used the edible parts of agricultural products as materials in ASEAN countries using an I/O analysis and showed that first-generation bioethanol production was economically beneficial but not environmentally effective. The authors noted that this technology also competed with food consumption and raised food prices.

Regarding bioenergy from woody biomass, Nakamura et al. (2012) used I/O analysis to show that regional biomass production and intraregional usage yielded sufficient circulation, substitution, and income-reservation effects from relatively lower prices. Nakamura et al. (2013) showed using an extended I/O model with an environmental aspect that the CO2 credit on woody biomass use could reduce income disparities among regions. Kunii et al. (2014) proposed an assessment tool for woody biomass projects and showed that some communities would have to import woody resources from other communities within the next 10 years even if they currently had enough resources available in their communities. Also, they found that transitioning from kerosene to firewood could increase money flow in a town and induce a positive economic impact on the local economy. Okiyama and Tokunaga (2009) used the SAMI-O linked model to show that development in the biofuel industry could reduce income disparities among people in different regions and different income classes in a developing country.

Based on these previous studies and our own I/O analysis, we measure the impacts of the 2011 earthquake and tsunami disaster on the coastal areas in Iwate Prefecture and evaluate the newly constructed biogas electricity power plant as part of the restoration of the economy in this area.

8.2.2 Changes in Industrial Linkage

The latest regional I/O table in the coastal region of Iwate Prefecture was published in 2005, and it consists of 35 industrial sectors and 9 towns, namely, Miyako, Ohfunato, Rikuzentakada, Kmaishi, Ohzuchi, Yamada, Iwaizumi, Sumida, and Tanohata. Industrial linkage was obviously changed by the 2011 earthquake, so we estimated the 2010, 2011, and 2012 I/O tables by assuming the following:

(i) Total regional demand (= intermediate production + final demand within region + export to other regions and abroad) changed in accordance with the net residential production of the town in 2010, 2011, and 2012 that was published by the prefectural government.
(ii) Industrial linkage changed subject to the total regional demand as the control total, with the least change from whole input coefficients. Under such assumption, new I/O data are estimated by the RAS method, which solves the matrix balance problem.
(iii) The contents of value-added production, i.e., "consumption expenditures outside of households," "labor income," "depreciation of capital stock," and "tax subsidies," are assumed to keep their share even after the total value-added production is changed. Therefore, we calculate these values according to their initial share rate in the 2005 I/O table.
(iv) The contents of final demand, i.e., "consumption expenditures outside of households," "household consumption," "public consumption," "public investment," "private investment," and "inventory investment," are also assumed to keep their share rate within the total final demand even after the total regional demand is changed. Therefore, we calculate these values in the same way as in (iii).
(v) The export to other regions and abroad was assumed to change in accordance with the annual change in the total regional demand of export industries. The import from other regions and abroad was then calculated by subtracting the export from the total regional production.
(vi) For initial settings, column-wise input coefficients and the total regional production of the biogas electricity generation sector are based on the plant design of the operation costs. Row-wise input coefficients of the biogas electricity generation sector are the same as the electricity generation sector, and the total regional demand of this sector corresponds to the total regional production. The total regional demand and total regional production of the biogas electricity generation sector are subtracted from the total of the electricity generation sector. After these settings, the RAS method is applied to balance the column-wise and row-wise totals.

8.2.3 Estimation Method for Induced Production

The production induced by an increase in construction investment and an increase in the final demand for biogas electricity products is calculated as follows (Miyazawa 2002):

$$\Delta \mathbf{X} = \left\{ \mathbf{I} - \left(\mathbf{I} - \bar{\mathbf{M}} \right) \mathbf{A} \right\}^{-1} \Delta \mathbf{F}_D \tag{8.1}$$

where $\triangle\mathbf{X}$ is the vector of total production change (size is 35 × 1), $\mathbf{I}$ is the unit matrix, $\bar{\mathbf{M}}$ is the import coefficient matrix from both other regions and abroad that is defined as $\mathbf{M}/(\mathbf{AX}+\mathbf{F}-\mathbf{E})$, (35 × 1), $\mathbf{A}$ is the input coefficient matrix (35 × 35), $\mathbf{F}=(\mathbf{F}_D + \mathbf{E})$ is the final demand vector (35 × 1), $\boldsymbol{F}_D$ is the final demand vector within the region, $\triangle\mathbf{F}_D$ is its change, and $\mathbf{E}$ is the export matrix (1 × 35).

8.3 Results

8.3.1 Estimation Results on an I/O Table

Figure 8.1 shows the net residential production of each town published by the Iwate prefectural government. As shown in this table, net income increased even in 2011, when the earthquake happened. This increase was caused by an increase in construction for recovery. However, other industries, such as fishery, manufacturing, and the service sector, showed a decrease in income due to the demolition of physical capital or capital stock.

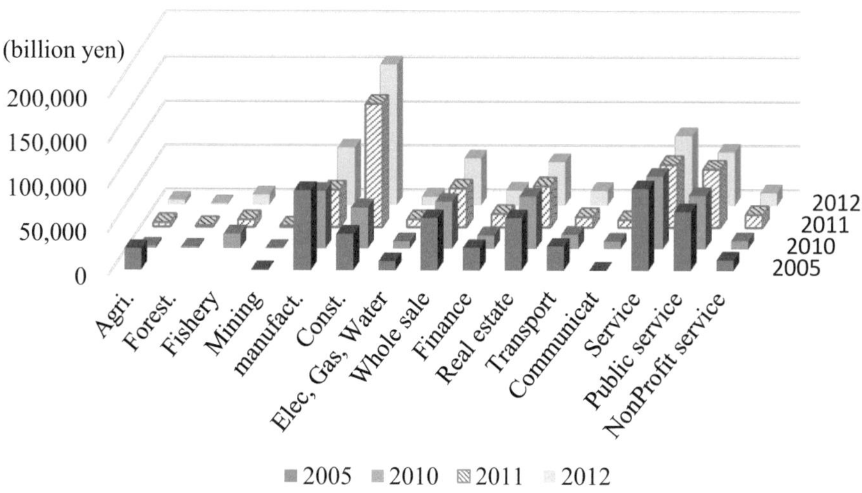

Fig. 8.1 Net residential income

Figure 8.2 shows the estimation results of the intermediate inputs and value-added production in each year. These values were estimated by applying the RAS method.

Even though value-added production in 2011 was higher than the previous year due to the recovery efforts after the earthquake, both the intermediate inputs and the total production were lower than in the previous year. This happened because the ratio of value-added production in the construction sector was higher than in the other industries that were seriously damaged by the earthquake.

Figure 8.3 shows the chronological changes in final demand by contents. Final demand in 2011, consisting of consumption and investment, was larger than the previous year because of an increase in value-added production. Both private and public consumption decreased, but public and private investments increased due to the recovery investment for damaged capital stock caused by the earthquake. Investment in 2012 was also much higher than before the earthquake. Imports and exports decreased in 2011, and these components recovered afterward. This is owing to the recovery construction work that continued after 2011.

8.3.2 Changes in Industrial Linkage Structure

Figure 8.4 shows the chronological tendencies in the power-of-dispersion coefficients in the top five and bottom five industrial sectors most significantly increased or decreased by the earthquake. Although no values were shown in this figure, the highest coefficient value in 2005 was 1.12 in the food sector among 34 sectors

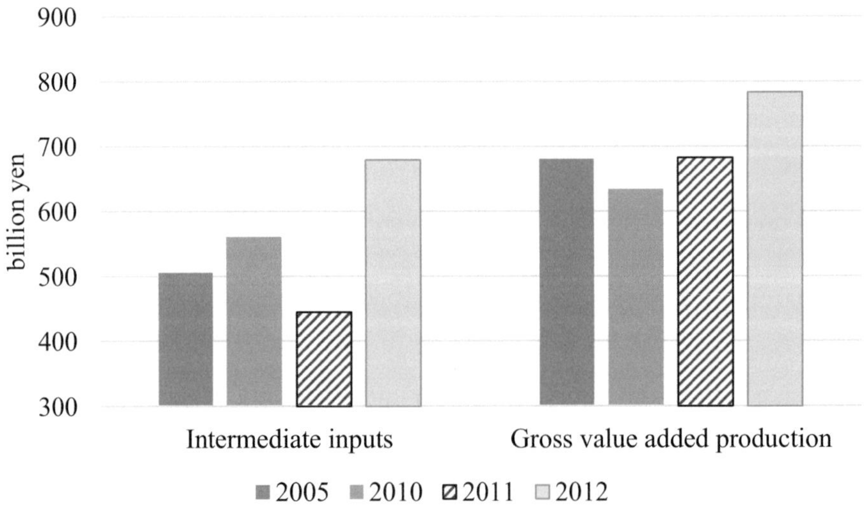

Fig. 8.2 Chronological changes in cost components

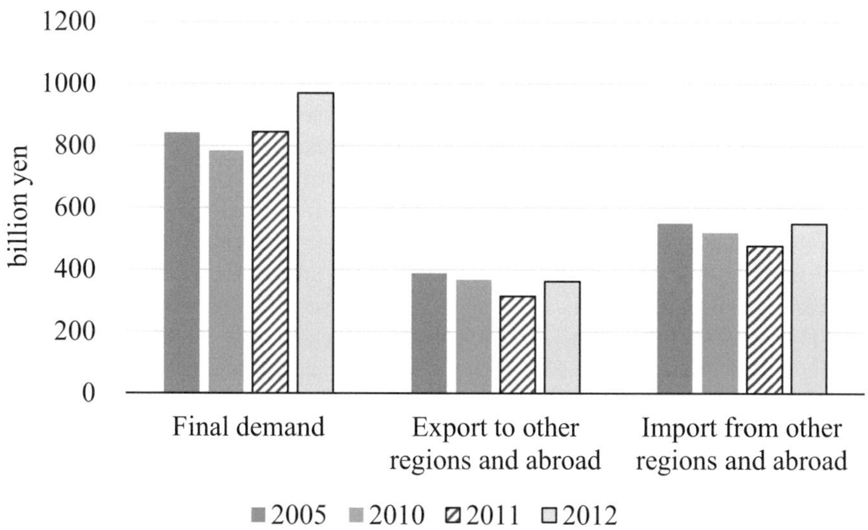

Fig. 8.3 Chronological changes in final demand

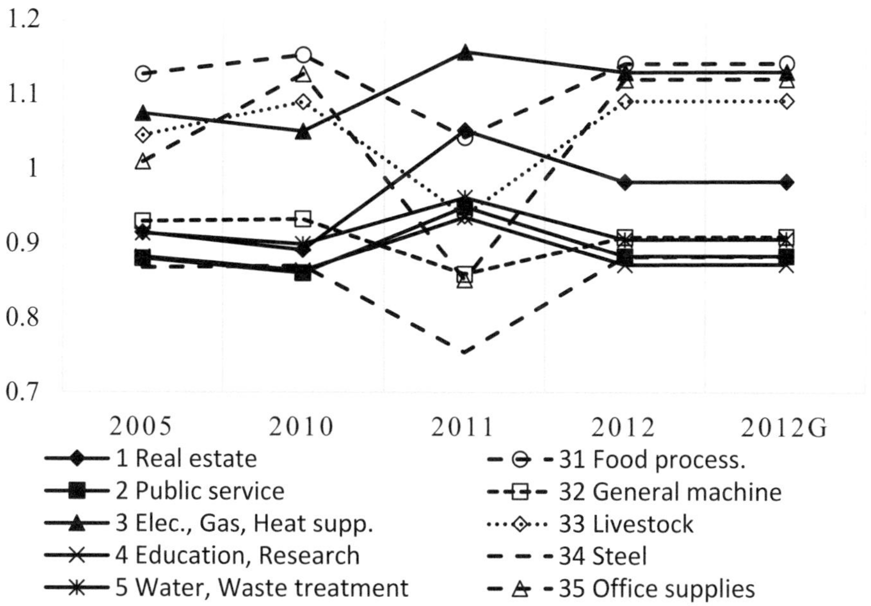

Fig. 8.4 Chronological changes in the power-of-dispersion coefficients
Note: The highest and lowest five sectors with regard to changes from 2010 to 2011 were selected. "2012G" is the year 2011 with installation of the biogas sector

except for the “others” sector. The lowest coefficient value was 0.867 in the steel sector.

The changes in the power-of-dispersion coefficient before and after the earthquake were large in the manufacturing sector whose level of coefficient was relatively high. This is because industrial linkage was weakened by the earthquake disaster. However, such linkage was recovered in 2012.

Contrarily, the power-of-dispersion coefficient in the service sector increased after the earthquake disaster. This is probably because the service sector rose in this coefficient against the manufacturing sector where this coefficient decreased after the earthquake. Furthermore, the change in the power-of-dispersion coefficient by the introduction of the biogas electricity generation sector was low.

Figure 8.5 indicates the chronological changes in the production-inducement coefficients in the top (or bottom) five industrial sectors whose coefficient most significantly increased (or decreased) because of the earthquake as similar to Fig. 8.4. The level of this coefficient in 2005 was the highest, 2.13, in the finance and insurance sector, and it was the lowest, 0.81, in the micro machine sector.

The coefficient of the construction sector was greatly raised owing to the amount of reconstruction investment after the earthquake. Moreover, this coefficient stayed at a high level for 2012. Water and waste treatment was the second highest sector where those coefficients rose after the earthquake. Such an influence occurred because the transactions between the construction sector and the water and waste treatment sector were comparatively big.

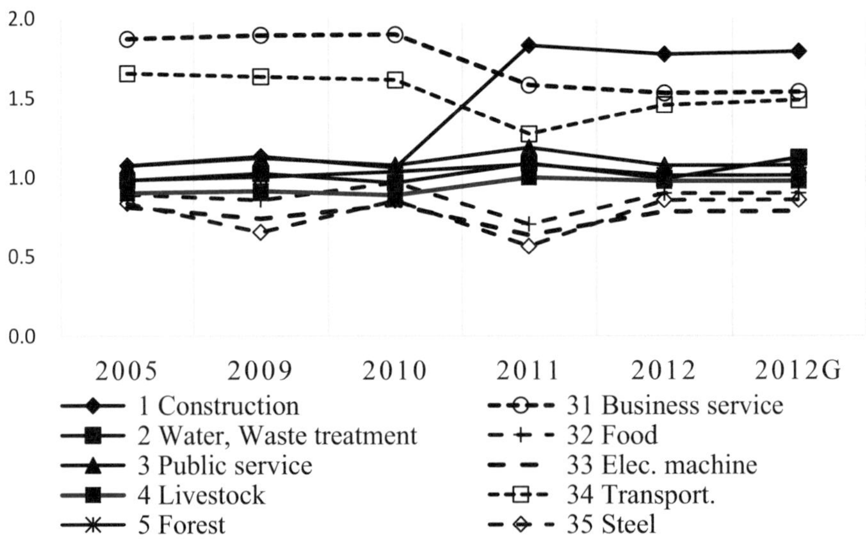

Fig. 8.5 Chronological changes in production-inducement coefficients
Note: The highest and lowest five sectors with regard to changes from 2010 to 2011 were selected. “2012G” is the year 2011 with the installation of the biogas-electricity-generation sector

On the other hand, in the business service sector, the production-inducement coefficient declined after the earthquake. This happened because the share of intermediate inputs with other industries was large; moreover, transactions with other sectors decreased greatly due to the earthquake. The decline in the production-inducement coefficient of the manufacturing sector, including food, electric machinery, and steel, was serious. This was probably because the inter-industrial linkage became weak in the manufacturing sector after the earthquake.

Figure 8.6 shows the summation of elements in the inverse matrix, with the total effects of induced production caused by an increase in final demand. Those numbers decreased in 2011 when the earthquake happened, but after that year, the total effects of induced production increased and reached a higher level than before the earthquake. Especially, when a biogas electricity power plant was installed in 2012, that coefficient became the highest. In this sense, new investment related to the electricity sector can stimulate the domestic economy by switching demand from imports to domestic production.

8.3.3 *Induced Production of the Biogas Electricity Power Plant*

Table 8.1 shows the design specification of the newly constructed biogas electricity power plant in Ohfunato city in the coastal region of Iwate Prefecture. This plant was planned to use food residues in and around Ohfunato as fuel material.

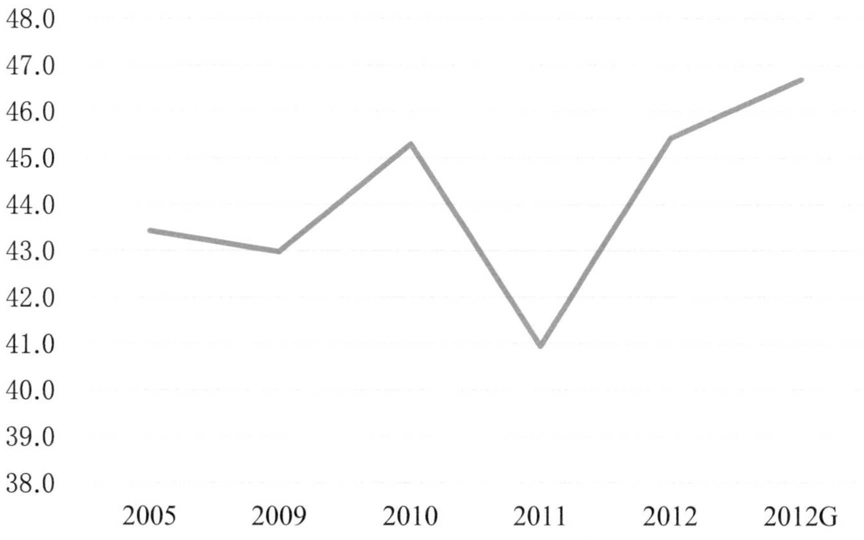

Fig. 8.6 The sum of elements in the inverse matrix
Note: "2012G" is the year 2011 with the installation of the biogas-electricity-generation sector

Table 8.1 Blueprint of the biogas electricity power plant

Items	Explanation
Collecting area	Ohfunato city and neighoboring cities
Amount of biomas materials	Total 13 t/day
	Sewage sludge (3 t/day)
	Food residues (10 t/day) 3 t/day are from Ohfunato, rest of them is from neighboring cities
Energy use	Electricity (sell), Remaining heat (used in neighboring area)
	39 yen/kWh
Waste treatment	Sewage disposal

Table 8.2 Economic feasibility of the plant

Items	Contents
Initial cost	600 million yen
Payback period of investment	18 years

Compared with a typical biogas electricity power plant constructed in Hokkaido, the scale of this plant was small, but it was built to maximum capacity to accommodate material collection. If the plant had been larger, the cost efficiency would have decreased due to the large collection areas that would need to be assigned in order to provide materials for the plant.

Table 8.2 shows the profitability of the biogas electricity power plant in Ohfunato. It will take 18 years to recover the initial costs of this plant. This duration is almost the same as the expected lifetime of the facility. Hence, the efficiency of this plant is not high although the total revenue during its lifetime is higher than the costs. The reason why the efficiency is not high is that this plant was planned to be installed with sewage treatment to accommodate the digested liquid after methane fermentation. If the digested liquid can be disposed of in the fields as fertilizer, this part of the costs should become small and would make the cost–benefit ratio higher.

Table 8.3 shows the contents of the input cost for the construction of the plant. Table 8.4 concerns the running costs during the operation of this plant. These values were used for the simulation shown in Tables 8.5 and 8.6. Since the materials for biogas electricity generation are collected without any payments, the initial structure of running costs in other sectors was assumed to be the same as the original I/O table.

Table 8.5 shows the induced production and labor, which increased because of the construction of the biogas electricity power plant. To estimate the induced production, the investment for the plant was assumed to be conducted in two ways, i.e., construction in 2010 before the earthquake and construction in 2011 after the earthquake. Each induced production was calculated using each I/O table and the inverse matrix. Since the import rate from other regions and abroad was high, leakage of effects was large in this region; the induced multiplier (=induced production/investment) was 0.64 and lower than 1. If the induced multipliers in 2010 and 2011 were compared, the latter (0.17) was much lower than the former

Table 8.3 Cost components for the construction of the biogas electricity power plant

(thousands yen)	
Input sectors	Inputs amount for construction of the biogas plant
Ceramic	20,000
Metal	11,000
General machine	225,000
Electronic machine	70,000
Construction	230,000
Transportation	10,000
Business srvice	40,000
Subtotal	606,000
Value added	0
Total production	606,000

Table 8.4 Operational cost for the biogas electricity power plant

(thousands yen)	
Input sectors	Operation stage
Gneral machine	13,500
Electronic machine	7,000
Construction	500
Electricity, Gas, Heat supply	2,500
Water, Recicle	16,000
Finance	360
Transportation	5,100
Communication	120
Business service	120
Office supplies	120
Subtoal	45,320
Wage	16,960
Capital depreciation	35,650
Value added	52,610
Total production	97,930

Table 8.5 Production and labor induced by the construction of the biogas electricity power plant

(million yen, person)					
	Investment	Induced production		Induced labors	
Constraction year	a	b	b/a	c	c/a
2010	606	388	0.64	32	0.053
2011	606	106	0.17	13	0.022

Table 8.6 Production and labor induced by the annual operation of the biogas electricity power plant

(million yen, person)			
Sectors	Operation cost	Induced production	Net increase in labors
Electricity, Gas, Heat supply	−98	−20	−0.53
Biogas Electricity	98	98	2.59
Others	0	16	0.57
Total	0	94	2.63

(0.64). This happened because of the damage in industrial linkage in this economy caused by the earthquake. In this sense, the effects of recovery investment are lower than investment in a normal time even if the investment amount is the same.

Table 8.6 shows the induced production and labor during operation after the construction of the plant. In this estimation, we set 98 million yen as the increase in the final demand of the biogas electricity generation sector, but at the same time, −98 million yen for the conventional electricity generation sector. The total increase in the final demand in the whole economy was zero, but the induced process was different for both sectors. As compared to the conventional electricity sector, the biogas electricity generation sector induced more production because the import rate of this sector was zero and most induced demand occurred within the region. Induced labor was at 2.6 persons per year, and this is a pure increase. Hence, a biogas electricity power plant is effective for the revitalization of a regional economy.

8.4 Summary and Conclusion

This chapter estimates the input–output (I/O) tables of this region before and after the 2011 earthquake and to use these tables to show the impact of the earthquake and the impact of a particular revitalization measure, i.e., the construction of a biogas electricity power plant. The results showed distinct features, as described below.

First, value-added production in 2011 was higher than in the previous year because of the recovery construction investment, but both intermediate inputs and total production decreased. This happened because the ratio of value-added production in the construction sector was higher than in the other industries that were seriously damaged by the earthquake.

Second, investment demand increased in 2011 and 2012, but intermediate inputs and private and public consumption both decreased because of the earthquake. The power-of-dispersion coefficients and production-inducement coefficients were slightly changed by the earthquake, but the increasing rate of induced production became low after the earthquake.

Third, if the biogas electricity power plant had been constructed before the earthquake, the multiplier value of the induced production for the construction would have been 0.64. However, that value after the earthquake was 0.17 because industrial linkage was damaged. Compared with the conventional electricity sector, the biogas electricity generation sector induced more production because the import rate of this sector was zero and most induced demand occurred within the region. Therefore, it was shown that a biogas electricity power plant can contribute to the revitalization of a regional economy although the electricity generation capacity is small.

There are many remaining issues in this study. The estimated I/O table should be secured by comparing an actual data survey or official table that will be published in the future. We need to consider other possible investments rather than just a biogas electricity power plant in order to compare other options to revitalize a regional economy after a serious natural disaster.

Acknowledgements The data of this study has been provided by Toshihiko Nakata (Professor, Tohoku University) and Masaharu Yoshimi (Rinergen Co. Ltd.). The authors greatly appreciate their support.

References

Hayashi T (2010) Evaluating the economic-environment co-benefit of bio-fuel production: an application of eco-efficiency. J Agric Policy Res 18:41–57

Kunii D, Sawauchi D, Hayashi T (2014) Multi-dimensional assessment, supply and demand matching, woody biomass, geographical information system (GIS), Nishiwaga. Stud Reg Sci 44(4):481–493

Kunimitsu Y, Ueda T (2006) Economic evaluation on agricultural biomass resources use in Thailand-case of Rice husk electricity power plants. Stud Reg Sci 36(3):561–573

Kunimitsu Y, Takahashi K, Furubayashi T, Nakata T (2013) Economic ripple effects of bio-ethanol production in ASEAN countries: application of inter-regional input-output analysis. Jpn Agric Res Q (JARQ) 47(3):307–317

Miyazawa K (2002) Introduction of input–output analysis. Bunko, Nikkei

Nakamura R, Nakazawa J, Matumoto A (2012) Regional economic effects of CO_2 reduction activities with wood biomass: application and extension of a regional IO model. Stud Reg Sci 42(4):799–817

Nakamura R, Ishikawa Y, Matsumoto A (2013) Endogenous correction of regional economic disparities by making use of regional environmental resources (wooden biomass): extension of inter-regional input-output model. Input-Output Anal 20(3):228–242

Okiyama M, Tokunaga S (2009) Analysis of disparity reduction among people in different regions and different income classes by exporting products of biofuel industry: utilizing a SAM/I-O linked model for Thailand. Pap Reg Sci 39(4):893–909

Polagye B, Hodgson KT, Malte PC (2007) An economic analysis of bio-energy options thinning from overstocked forests. Biomass Bioenergy 31(2, 3):105–125

Urbanchuk JM (2009) Contributions of the ethanol industry to the economy of the United States. RFA Reports and Studies, Renewable Fuels Association, Wayne

Part III
Evaluating Regional Impacts of Megathrust Earthquakes and Tsunamis

Chapter 9
Economic Impacts of Population Decline Due to the Great East Japan Earthquake: An Inter-regional Input–Output Approach

Yoshifumi Ishikawa

Abstract The Great Earthquake of March 11, 2011, significantly damaged the Tohoku region in Japan. Particularly, Fukushima Prefecture is still suffering from the impacts of the Great Earthquake and the nuclear power plant accident and the consequent evacuation of numerous people. It is considered that the population decline due to deaths, those declared missing, and evacuation has a negative impact on the regional economy through a decline in consumer demand. Therefore, using a 47-region inter-regional input–output table at prefecture level, we analyzed the economic impacts of the population decline for the past 5 years after the Great Earthquake. Consequently, it was observed that the impact was spread across the country. In particular, while the amount of production in the Tohoku region, including Fukushima, decreased, that of the Kanto region, including Tokyo, increased because many evacuees moved from the three affected prefectures and then became consumers in their new region. In addition, it is considered that the population decline will continue in the affected area in the long term. Therefore, based on population projections that consider the impact of the earthquake, we analyzed its long-term economic impacts. Consequently, it is shown that in the worst case, production in Fukushima in 2030 will decline by 15.4% compared with 2010. Thus, there is concern that the population decline in affected regions, such as Fukushima, will continue in the long term, and the regional economy will decline.

Keywords Input–output analysis • Earthquake • Population decline • Consumer demand • Economic impact

---

Y. Ishikawa (✉)
Faculty of Policy Studies, Nanzan University, 18 Yamazato-cho, Showa-ku, Nagoya, Aichi 466-8673, Japan
e-mail: yishi@ps.nanzan-u.ac.jp

S. Tokunaga, B.P. Resosudarmo (eds.), *Spatial Economic Modelling of Megathrust Earthquake in Japan*, New Frontiers in Regional Science: Asian Perspectives 11,
https://doi.org/10.1007/978-981-10-6493-7_9

9.1 Introduction

The Great East Japan Earthquake (The Great Earthquake) of 2011 affected many people's lives as well as Tohoku's regional economy, including prefectures such as Iwate, Miyagi, and Fukushima. The number of deaths exceeded 20,000, including earthquake-related deaths and individuals reported missing, and approximately 174,000 people were still unable to return to their homes as of 2015. Evacuation destinations were spread across the country, and many people were evacuated to the Kanto region, including Tokyo. Thus, a large number of evacuees are living as refugees outside their prefectures. As of 2015, 43,497 people have been relocated from Fukushima to other regions, along with 6,533 and 1,496 people from Miyagi and Iwate, respectively. Thus, the population of the affected areas has decreased owing to deaths, those declared missing, and evacuation of people from their hometowns.

The population of the Tohoku region was already in decline before the Great Earthquake, and the population of Fukushima Prefecture has been declining since the late 1990s, decreasing by about 10,000 each year in the late 2000s. However, just after the Great Earthquake in 2011, the population declined by about 40,000 followed by a further decline in 2012. Factors contributing to this decline include not only deaths due to the Great Earthquake but also those due to the subsequent relocation of individuals. Following the Great Hanshin–Awaji Earthquake in 1995, the population of Kobe City declined from 1.52 million to 1.42 million, and it took 10 years to recover to the pre-earthquake level. Thus, there is concern that the population decline will continue in the long term, particularly in Fukushima Prefecture, if reconstruction and recovery from the Great Earthquake is delayed. Deguchi (2012) predicted the population by prefecture for Japan as a whole after considering the impact of the earthquake disaster. According to those projections, the population of Fukushima Prefecture in 2030 will be lower by 340,000 in comparison with the hypothetical case where the disaster did not occur. Such a population decrease impacts the regional economy through a reduction in consumer demand.

In this chapter, we present the impact of the population decline due to the Great Earthquake on the regional economy at the prefectural level. Because the population decline is considerably advanced in areas affected by the earthquake, it has had a significant negative impact on the regional economy through a rapid decline in consumer demand. Therefore, we analyze the economic impacts of changing consumer demand attributable to the population decline in each prefecture for the past 5 years after the Great Earthquake. In addition, as mentioned by Deguchi (2012), it is considered that the population decline will continue in the affected areas in the long term. Therefore, based on population projections that consider the impact of the earthquake, we analyze the economic impacts in the case where population decline continues in the future.

9.2 Population Change in the Disaster Areas

9.2.1 Population Change in Japan and Affected Prefectures

As shown in Table 9.1, the decline in the Japanese population began in 2009 although the populations of the affected areas, such as Iwate, Miyagi, and Fukushima, had already begun to decline prior to that (see Fig. 9.1). In 2010, prior to the earthquake, the population of Iwate was 1.33 million and those of Miyagi and Fukushima were 2.35 million and 2.03 million, respectively (Table 9.1). The populations of Iwate and Fukushima had been decreasing by 10,000 each year; however, in 2011 they dropped significantly by 17,391 and 40,069, respectively (Table 9.2). The population of Miyagi had been decreasing by about 5000 each year prior to the Great Earthquake although it increased by 8000 in 2010. However, after the earthquake, the population decreased by 24,941 (Table 9.2). Thus, the Great Earthquake is considered to have had a negative impact on the population in the affected areas, particularly in Fukushima.

Factors contributing to the population changes resulting from the earthquake were natural decreases, such as deaths and those declared missing, and social changes due to the evacuations. The natural decrease and social change can be estimated using data on deaths and evacuees outside the prefecture provided by the National Police Agency and the Reconstruction Agency. Deaths include direct deaths from the earthquake (15,893) and earthquake-related deaths (3,407). In all, 2,565 people were declared missing, bringing the total of the natural decrease to 21,865 (Table 9.3).

Many people died and many were declared missing in three prefectures: Iwate, Miyagi, and Fukushima. However, the damage from the earthquake also includes long-term evacuations. The number of evacuees had reached 520,000 on March 16, 2011, just after the earthquake, with 180,000 still not able to return home by November 2015 (Fig. 9.2). Many people were forced to evacuate to areas outside

Table 9.1 Population of Japan and affected prefectures before and after the Great Earthquake

	Japan (Thousand persons)	Iwate (persons)	Miyagi (persons)	Fukushima (persons)
2006	127,901	1,374,699	2,354,992	2,080,186
2007	128,033	1,363,702	2,348,999	2,068,352
2008	128,084	1,352,388	2,343,767	2,055,496
2009	128,032	1,340,852	2,340,029	2,042,816
2010	128,057	1,330,147	2,348,165	2,029,064
2011	127,799	1,312,756	2,323,224	1,988,995
2012	127,515	1,303,351	2,325,407	1,962,333
2013	127,298	1,294,453	2,328,143	1,947,580
2014	127,083	1,284,384	2,327,993	1,936,630
2015	126,890	1,272,891	2,324,683	1,925,605

Source: Statistics Bureau (2015)

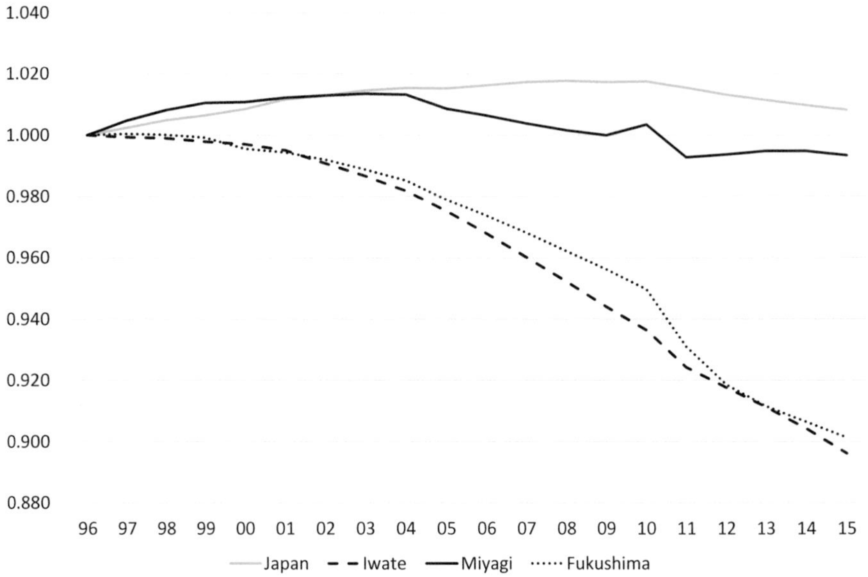

Fig. 9.1 Changes in population of affected prefectures (Source: Statistics Bureau 2015)

Table 9.2 Increase/decrease in population of affected prefectures

	Japan	Iwate	Miyagi	Fukushima
	(Thousand persons)	(persons)	(persons)	(persons)
2007	132	−10,997	−5,993	−11,834
2008	51	−11,314	−5,232	−12,856
2009	−52	−11,536	−3,738	−12,680
2010	25	−10,705	8,136	−13,752
2011	−258	−17,391	−24,941	−40,069
2012	−284	−9,405	2,183	−26,662
2013	−217	−8,898	2,736	−14,753
2014	−215	−10,069	−150	−10,950
2015	−193	−11,493	−3,310	−11,025

Source: Statistics Bureau (2015)

Table 9.3 Human toll due to the Great Earthquake

	Deaths		Earthquake-related deaths		Missing		Injured	
Prefectures	(Persons)	Ratio	(Persons)	Ratio	(Persons)	Ratio	(Persons)	Ratio
Iwate	4,673	29.4%	455	13.4%	1,125	43.9%	213	3.5%
Miyagi	9,541	60.0%	918	26.9%	1,237	48.2%	4,145	67.4%
Fukushima	1,612	10.1%	1,979	58.1%	199	7.8%	183	3.0%
Other	67	0.4%	55	1.6%	4	0.2%	1,611	26.2%
Overall	15,893	100.0%	3,407	100.0%	2,565	100.0%	6,152	100.0%

Source: National Police Agency (2015)

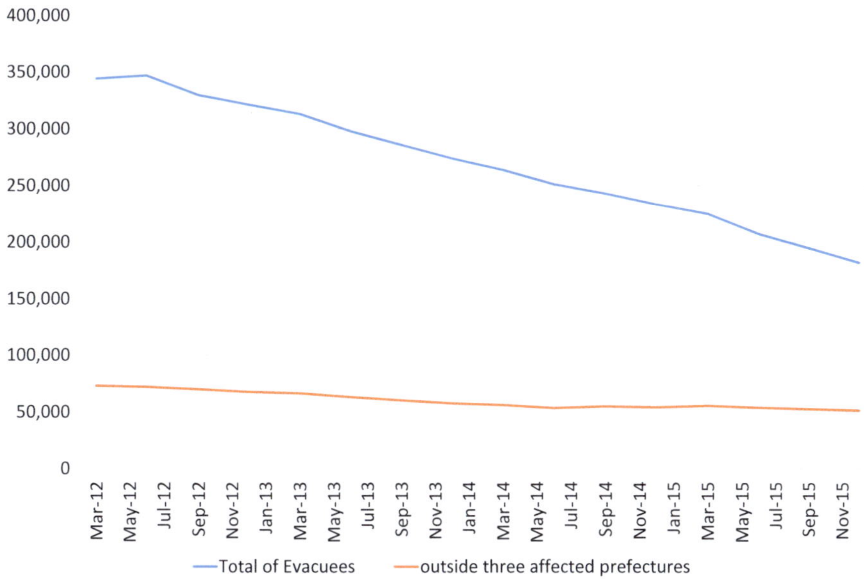

Fig. 9.2 Change in the number of evacuees (Source: Reconstruction Agency 2015)

the prefecture where they lived. As shown in Table 9.4, in September 2011, the number of evacuees living outside Fukushima stood at 55,608 while the figures for Miyagi and Iwate were 8,449 and 1,437, respectively. Despite 4 years having passed, as of 2015, there were still 43,497 evacuees from Fukushima, 6,533 from Miyagi, and 1,496 from Iwate although the total number of evacuees has been decreasing each year. Evacuation destinations were spread across the country. In 2015, 7,118 evacuees were still living in Tokyo, and 5,138 and 3,755 were living in Saitama and Ibaraki, respectively (Fig. 9.3). Thus, significant depopulation had occurred in the Tohoku region.

9.2.2 *Population Decline in Prefectures Affected by the Great Earthquake*

In this study, we analyze the economic impacts of the population decline in areas affected by the Great East Japan Earthquake, namely, Iwate, Miyagi, and Fukushima. Therefore, we need to estimate the population decreases in these areas.

The population figures before and after the earthquake are publically available (Table 9.5), which means that it is relatively straightforward to estimate the differences. However, as mentioned above, because the populations of the three prefectures in the Tohoku region were already decreasing prior to 2011, the depopulation calculated using this method includes a natural decrease and a social decrease not related to the earthquake (Table 9.6).

Table 9.4 The number of evacuees in the Tohoku region

Prefectures	2011/10/6		2012/10/4		2013/10/10		2014/9/11		2015/8/13	
	Acceptance	Extra-	Acceptance	Extra-	Acceptance	Extra-	Acceptance	Extra-	Acceptance	Extra-
Iwate		1,437	41,969	1,702	36,825	1,531	31,714	1,451	25,761	1,496
Miyagi		8,449	114,787	8,177	95,163	7,373	77,836	6,925	57,565	6,533
Fukushima		55,608	99,229	59,031	89,924	50,633	78,577	46,645	62,773	43,497
Other			70,888		60,199		54,913		27,954	
Overall		65,494	326,873	68,910	282,111	59,537	243,040	55,021	174,053	51,526

Source: Reconstruction Agency (2015)

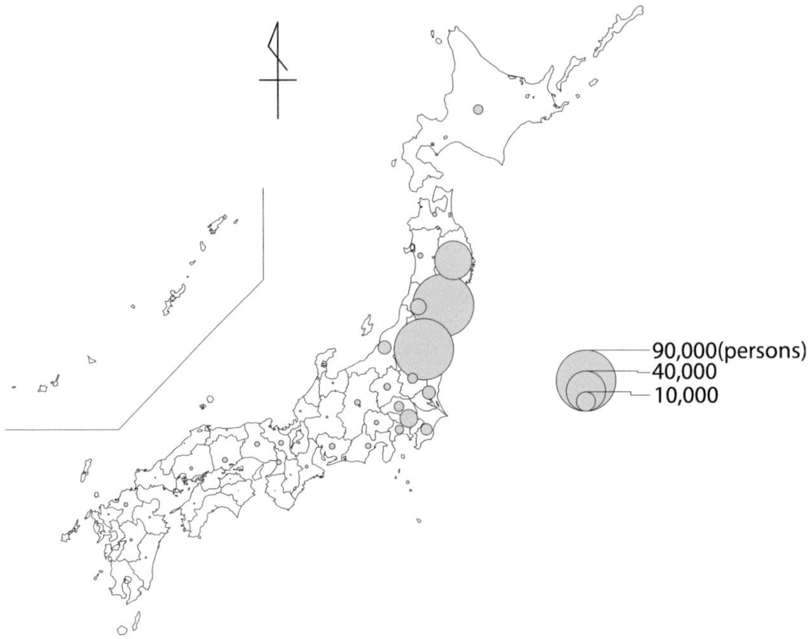

Fig. 9.3 The number of evacuees by prefecture (Source: Reconstruction Agency 2015)

Table 9.5 Population changes before and after the Great Earthquake in Iwate, Miyagi, and Fukushima

	10/1/2010	10/1/2011	10/1/2012	10/1/2013	10/1/2014	10/1/2015
Iwate	1,330,147	1,314,076	1,303,154	1,294,535	1,284,426	1,272,891
Miyagi	2,348,165	2,326,735	2,325,247	2,327,811	2,327,700	2,324,683
Fukushima	2,029,064	1,989,834	1,961,705	1,946,202	1,935,456	1,925,605
	2010–2011	2011–2012	2012–2013	2013–2014	2014–2015	total
Iwate	−16,071	−10,922	−8,619	−10,109	−11,535	−57,256
Miyagi	−21,430	−1,488	2,564	−111	−3,017	−23,482
Fukushima	−39,230	−28,129	−15,503	−10,746	−9,851	−103,459

Source: Statistics Bureau (2015)

Table 9.6 Social decrease due to evacuation

	2011	2012	2013	2014	2015
Iwate	1,437	1,702	1,531	1,451	1,494
Miyagi	8,449	8,177	7,373	6,925	6,650
Fukushima	55,608	59,031	50,633	46,645	44,094
Total	65,494	68,910	59,537	55,021	52,238

Source: Reconstruction Agency (2015)

Table 9.7 Natural decrease due to deaths, related deaths, and missing persons

	2011-
Iwate	6,253
Miyagi	11,696
Fukushima	3,790
Total	21,739

Source: National Police Agency (2015)

Therefore, we estimated the population decline due to the Great Earthquake using data on deaths and evacuees from the prefectures provided by from the National Police Agency and the Reconstruction Agency to analyze the economic impacts. As shown in Table 9.7, the population of the three affected prefectures declined by 21,739 due to death and those declared missing, with a subsequent continued decline. In addition, 65,494 people who lived in the affected prefectures had to move out of the region. As of 2015, this figure was 52,238 (Table 9.6). Consequently, the population of Fukushima has decreased more due to evacuations than due to natural decreases, such as death and people going missing in Fig. 9.4.

9.3 Analysis Method

9.3.1 Input–Output Table

Local governments and the Ministry of Economy, Trade and Industry (METI), among others, make regional input–output tables available for all prefectures in Japan and for several regions. However, local governments mainly provide intra-regional input–output tables. Thus, we are unable to analyze inter-regional economic impacts using these data. In addition, the inter-regional input–output table provided by METI is not suitable for analyzing the economic impacts of Iwate, Miyagi, and Fukushima as it divides the whole country into nine regions, such as the Tohoku region, and the three affected prefectures are included in the Tohoku region.

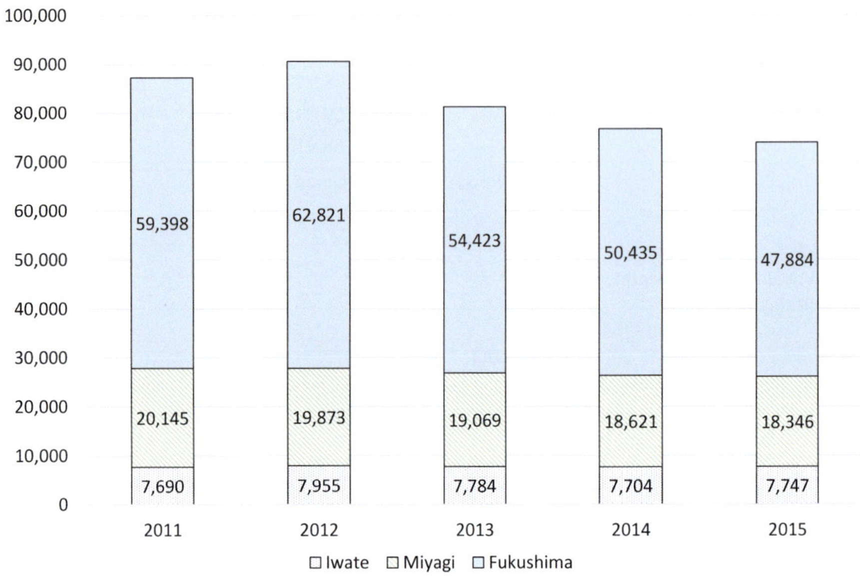

Fig. 9.4 Total population decreases by affected prefectures

Therefore, we used an inter-regional input–output table for 47 prefectures that we had constructed in a previous study for the year 2005.[1] Several researchers have based their studies on these data. For instance, Ishikawa and Miyagi (2004) analyzed the regional relations among all prefectures by measuring the inter-regional technical coefficients and the Leontief inverse matrix. Then, they estimated the effects of an increase in demand for automobiles in Aichi Prefecture on each prefectural economy in Japan. In addition, Sasayama (2011) analyzed the economic effects of port investment on all prefectures. An example of a study related to the Great Earthquake is that of Ishikura and Ishikawa (2011), who analyzed the impacts of constraints on the supply of electricity in the metropolitan area using the inter-regional input–output table for 47 prefectures.

Thus, our inter-regional input–output table has been widely used and is used here to enable us to analyze the economic impacts of the population decrease in each prefecture.

9.3.2 *Model*

To analyze the economic impacts of population decline at the prefectural level considering inter-regional feedback effects, we use the inter-regional input–output

[1] First, the interregional input–output table for 47 prefectures was constructed for the year 1995 in Ishikawa and Miyagi (2004); after that, the input–output tables for 2000 and 2005 are constructed.

table for 47 prefectures in 2005. While there are a number of models for handling imports, here, the analytical model of the inter-regional non-competitive inflow and competitive import-type tables previously created are shown as follows:

$$X = \left[I - \left(A - \hat{M}A^*\right)\right]^{-1}\left(Y - \hat{M}Y^* + E\right) \tag{9.1}$$

where
X: regional production column vector
A: input coefficient matrix
$\hat{M}$: diagonal matrix of import coefficients
A^*: diagonal matrix of input coefficients
Y: final demand column vector
Y^*: self-sufficient in final demand column vector
E: export column vector

The number of sectors is 45 and the country is divided into 47 prefectures. In this study, we also analyze the impacts on the value added by multiplying production value by value-added coefficients:

$$V = \bar{V}X \tag{9.2}$$

where V: value-added vector

$\bar{V}$: diagonal matrix of value-added coefficients

9.4 Economic Impacts of Population Decline Over the Past 5 Years After the Great Earthquake

9.4.1 Change in Consumption Demand as Population Declines in Affected Areas

The purpose of this chapter is to present the economic impacts of the past 5 years' decline in population due to the Great Earthquake. It can be considered that consumer demand decreases consequent to depopulation. Therefore, we first need to estimate the demand drop attributable to population decrease. Consumer demand can be estimated by multiplying the population by per capita consumption. The decrease in population reduces consumer demand in the affected prefectures provided that per capita consumption does not increase. In this study, we assume that per capita consumption by prefectures is constant before and after the Great Earthquake. Additionally, because of evacuations, the population of the affected prefectures decreases; however, that of the other prefectures increases. Therefore, with regard to social change, we estimate consumer demand for all prefectures on the basis of the number of evacuees by prefecture.

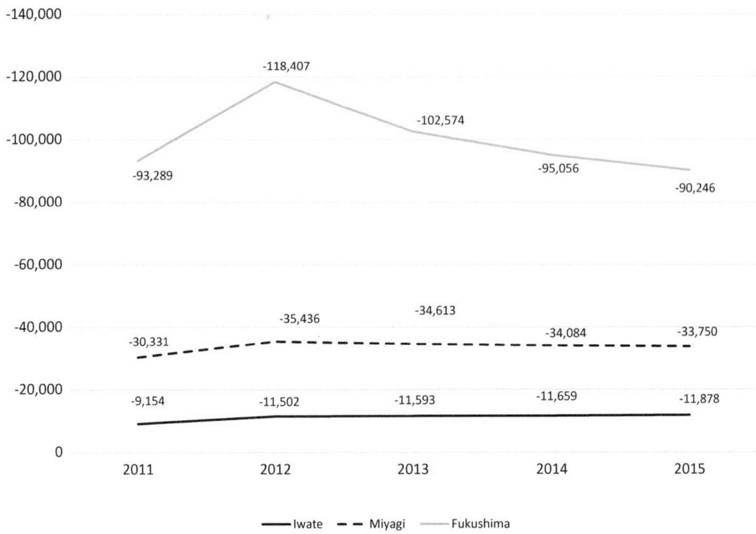

Fig. 9.5 Changes in consumption demand after the Great Earthquake (over a 5-year period) (million yen)

Changes in consumption demand after the Great Earthquake are shown in Fig. 9.5. The consumption demand of Fukushima decreased by 93 billion yen in 2011, immediately after the Great Earthquake, and further declined by 118 billion yen in 2012 due to evacuation from Fukushima. As of 2015, this figure had reached 90 billion yen as people had still not returned to their hometowns. The figure for Miyagi decreased by 30 billion yen in 2011, with a further decrease in 2012; after that, the negative impact in Miyagi gradually reduced. However, as of 2015, consumer demand decreased by 34 billion yen compared with levels prior to the earthquake. In Iwate, consumer demand decreased by 9 billion yen, and then the gradual decline was rendered to 12 billion yen in 2015.

As shown in Table 9.8, the total consumer demand in Fukushima decreased by around 500 billion yen over the past 5 years and that of Miyagi and Iwate by 168 billion yen and 56 billion yen, respectively. Fukushima accounts for 69% of the negative impacts in the three affected prefectures. Conversely, the Kanto region, including Tokyo, shows positive impacts of the earthquake, and consumer demand increased by 324 billion yen in the same period. Other than the three affected prefectures, consumer demand in Tohoku as well increased by 78 billion yen.

Table 9.9 shows changes in consumer demand by prefecture in terms of the type of population change. The population of Miyagi and Iwate declined due to natural reasons rather than social reasons and that of Fukushima shrank due to social reasons rather than natural reasons. Prefectures with increasing populations and Kanto prefectures, such as Tokyo, Ibaraki, Chiba, and Saitama, ranked in the top 10, and so did Tohoku prefectures, including Yamagata and Niigata (other than the three affected prefectures).

Table 9.8 Total changes in consumer demand due to the population decline in affected prefectures over the past 5 years after the earthquake (million yen)

	Iwate	Miyagi	Fukushima	Total	
Iwate	−71,866	10,902	5,179	−55,786	7.7%
Miyagi	1,588	−194,002	24,200	−168,214	23.2%
Fukushima	55	484	−500,109	−499,571	69.0%
Hokkaido	679	4,791	16,056	21,526	4.2%
Tohoku*	1,092	9,832	66,987	77,911	15.2%
Kanto	8,095	28,269	287,901	324,265	63.2%
Chubu	799	4,632	16,316	21,747	4.2%
Kinki	707	7,888	22,188	30,783	6.0%
Chugoku	285	2,158	8,317	10,759	2.1%
Shikoku	138	837	2,195	3,170	0.6%
Kyusyu	523	4,502	9,761	14,786	2.9%
Okinawa	138	1395	6,449	7,982	1.6%
Outside of afected area	12,455	64,305	436,169	512,928	–
Whole of Japan	−57,769	−118,311	−34,562	−210,642	–

Note: Tohoku in this table excludes the three affected prefectures

9.4.1.1 Economic Impacts of Change in Consumer Demand

Changes in production and value added are shown in Tables 9.10 and 9.11. With regard to production, the results are as follows (Fig. 9.6).

Immediately after the Great Earthquake, in 2011, production in Fukushima declined by 82 billion yen and that of Miyagi and Iwate declined by 39 billion yen and 11 billion yen, respectively. Conversely, Kanto, including Tokyo, increased production by 44 billion yen, with production increasing by 75 billion yen outside the affected prefectures. In 2012, the production amount of the three affected prefectures had further declined but increased in other regions. In Fukushima, the impacts of population decline have been gradually diminishing since 2013; however, the amount of production had declined by 80 billion yen by 2015 compared with the pre-earthquake level. After all, the total amount of production in Fukushima declined by 442 billion yen over the past 5 years, which accounts for 61% of the production decrease in the three affected prefectures. Conversely, changes in the composition of the population as well as absolute population due to the Great Earthquake brought positive impacts on regions such as Kanto, Tohoku (other than the affected prefectures), and Kinki.

The same tendency can be seen in the value–added data shown in Table 9.11. Thus, the value added of Fukushima declined by 52 billion yen in 2011 in comparison with the pre-earthquake level and by 281 billion yen over the following 5 years. However, the value added of the Kanto region increased in comparison with other regions of Japan.

Table 9.12 shows changes in production in the three affected prefectures by sector and year. In 2011, immediately after the earthquake, the amount of

Table 9.9 Changes in consumer demand by prefecture in terms of the type of population change (million yen)

	Natural change				Social change				Total			
	Iwate	Miyagi	Fukushima	Total	Iwate	Miyagi	Fukushima	Total	Iwate	Miyagi	Fukushima	Total
Hokkaido	0	0	0	0	679	4,791	16,056	21,526	679	4,791	16,056	21,526
Aomori	0	0	0	0	532	2,019	4,543	7,094	532	2,019	4,543	7,094
Iwate	−57,769	0	0	−57,769	−14,098	10,902	5,179	1,983	−71,866	10,902	5,179	−55,786
Miyagi	0	−118,311	0	−118,311	1,588	−75,690	24,200	−49,903	1,588	−194,002	24,200	−168,214
Akita	0	0	0	0	294	3,181	7,499	10,974	294	3,181	7,499	10,974
Yamagata	0	0	0	0	266	4,632	54,945	59,843	266	4,632	54,945	59,843
Fukushima	0	0	−34,562	−34,562	55	484	−465,547	−465,008	55	484	−500,109	−499,571
Ibaraki	0	0	0	0	193	577	35,982	36,751	193	577	35,982	36,751
Tochigi	0	0	0	0	257	679	28,127	29,063	257	679	28,127	29,063
Gumma	0	0	0	0	110	716	14,044	14,871	110	716	14,044	14,871
Saitama	0	0	0	0	1,661	5,293	26,924	33,878	1,661	5,293	26,924	33,878
Chiba	0	0	0	0	1,138	2,856	32,083	36,077	1,138	2,856	32,083	36,077
Tokyo	0	0	0	0	2,570	8,651	63,367	74,588	2,570	8,651	63,367	74,588
Kanagawa	0	0	0	0	1,450	5,256	20,378	27,084	1,450	5,256	20,378	27,084
Niigata	0	0	0	0	9	1,033	44,664	45,706	9	1,033	44,664	45,706
Toyama	0	0	0	0	28	409	1,993	2,429	28	409	1,993	2,429
Ishikawa	0	0	0	0	46	595	2,907	3,548	46	595	2,907	3,548
Fukui	0	0	0	0	0	428	1,887	2,315	0	428	1,887	2,315
Yamanashi	0	0	0	0	101	521	6,247	6,869	101	521	6,247	6,869
Nagano	0	0	0	0	248	633	8,865	9,746	248	633	8,865	9,746
Gifu	0	0	0	0	28	605	2,108	2,740	28	605	2,108	2,740
Shizuoka	0	0	0	0	358	2,056	7,219	9,633	358	2,056	7,219	9,633
Aichi	0	0	0	0	587	2,595	7,191	10,373	587	2,595	7,191	10,373
Mie	0	0	0	0	110	428	2,118	2,656	110	428	2,118	2,656
Shiga	0	0	0	0	46	623	1,896	2,565	46	623	1,896	2,565
Kyoto	0	0	0	0	55	1,265	6,016	7,336	55	1,265	6,016	7,336

Osaka	0	0	0	0	349	2,809	5,901	9,059	349	2,809	5,901	9,059
Hyogo	0	0	0	0	229	2,158	5,179	7,566	229	2,158	5,179	7,566
Nara	0	0	0	0	18	316	934	1,268	18	316	934	1,268
Wakayama	0	0	0	0	9	288	375	673	9	288	375	673
Tottori	0	0	0	0	28	167	1,145	1,340	28	167	1,145	1,340
Shimane	0	0	0	0	0	121	953	1,074	0	121	953	1,074
Okayama	0	0	0	0	46	558	3,071	3,675	46	558	3,071	3,675
Hiroshima	0	0	0	0	156	1,042	2,320	3,518	156	1,042	2,320	3,518
Yamaguchi	0	0	0	0	55	270	828	1,153	55	270	828	1,153
Tokushima	0	0	0	0	0	112	385	497	0	112	385	497
Kagawa	0	0	0	0	18	47	481	546	18	47	481	546
Ehime	0	0	0	0	110	502	857	1,469	110	502	857	1,469
Kochi	0	0	0	0	9	177	472	658	9	177	472	658
Fukuoka	0	0	0	0	193	2,205	3,167	5,564	193	2,205	3,167	5,564
Saga	0	0	0	0	0	242	1,001	1,243	0	242	1,001	1,243
Nagasaki	0	0	0	0	55	391	818	1,264	55	391	818	1,264
Kumamoto	0	0	0	0	64	428	1,040	1,532	64	428	1,040	1,532
Oita	0	0	0	0	147	419	1,222	1,788	147	419	1,222	1,788
Miyazaki	0	0	0	0	64	446	1,309	1,820	64	446	1,309	1,820
Kagoshima	0	0	0	0	0	372	1,203	1,575	0	372	1,203	1,575
Okinawa	0	0	0	0	138	1,395	6,449	7,982	138	1,395	6,449	7,982

Table 9.10 Changes in production by region (million yen)

	2011	2012	2013	2014	2015	Total	
Iwate	−10,606	−13,209	−13,223	−13,247	−13,433	−63,717	8.8%
Miyagi	−39,968	−47,911	−45,374	−44,048	−43,227	−220,527	30.3%
Fukushima	−82,483	−104,525	−90,849	−84,351	−80,198	−442,407	60.9%
Hokkaido	4,420	5,582	4,741	4,332	4,083	23,157	6.0%
Tohoku*	9,091	11,587	9,625	8,679	8,077	47,060	12.1%
Kanto	43,838	56,595	46,521	41,697	38,741	227,392	58.5%
Chubu	2,997	3,890	3,005	2,575	2,312	14,778	3.8%
Kinki	6,326	8,036	6,659	5,989	5,580	32,590	8.4%
Chugoku	2,402	3,043	2,552	2,312	2,167	12,477	3.2%
Shikoku	267	359	249	195	165	1,235	0.3%
Kyusyu	3,956	4,983	4,227	3,858	3,635	20,659	5.3%
Okinawa	1,704	2,140	1,850	1,710	1,622	9,026	2.3%
Outside of afected area	75,001	96,215	79,428	71,347	66,382	388,374	–
Whole of Japan	−58,056	−69,429	−70,018	−70,298	−70,475	−338,276	–

Table 9.11 Changes in value-added by region (million yen)

	2011	2012	2013	2014	2015	Total	
Iwate	−6,495	−8,097	−8,112	−8,130	−8,249	−39,084	8.7%
Miyagi	−23,008	−27,465	−26,136	−25,429	−24,990	−127,028	28.4%
Fukushima	−52,370	−66,386	−57,661	−53,516	−50,866	−280,799	62.8%
Hokkaido	2,490	3,140	2,674	2,448	2,310	13,062	5.3%
Tohoku*	5,961	7,575	6,366	5,784	5,415	31,101	12.5%
Kanto	28,460	36,540	30,468	27,560	25,774	148,803	59.9%
Chubu	1,810	2,322	1,865	1,643	1,508	9,148	3.7%
Kinki	3,931	4,973	4,172	3,782	3,544	20,403	8.2%
Chugoku	1,493	1,886	1,600	1,461	1,377	7,817	3.1%
Shikoku	212	277	213	181	163	1,046	0.4%
Kyusyu	2,286	2,874	2,450	2,242	2,117	11,969	4.8%
Okinawa	960	1,206	1,042	963	914	5,085	2.0%
Outside of afected area	47,603	60,792	50,851	46,064	43,122	248,432	–
Whole of Japan	−34,269	−41,157	−41,059	−41,011	−40,983	−198,479	–

production in the three affected prefectures declined by 133 billion yen and by 727 billion yen over the past 5 years. Moreover, decreases are apparent in a wide range of sectors. The largest decrease (157 billion yen) is in *real estate* followed by *personal services* (85 billion yen), *food and tobacco* and *commerce* (65 billion yen), and *finance and insurance* (50 billion yen).

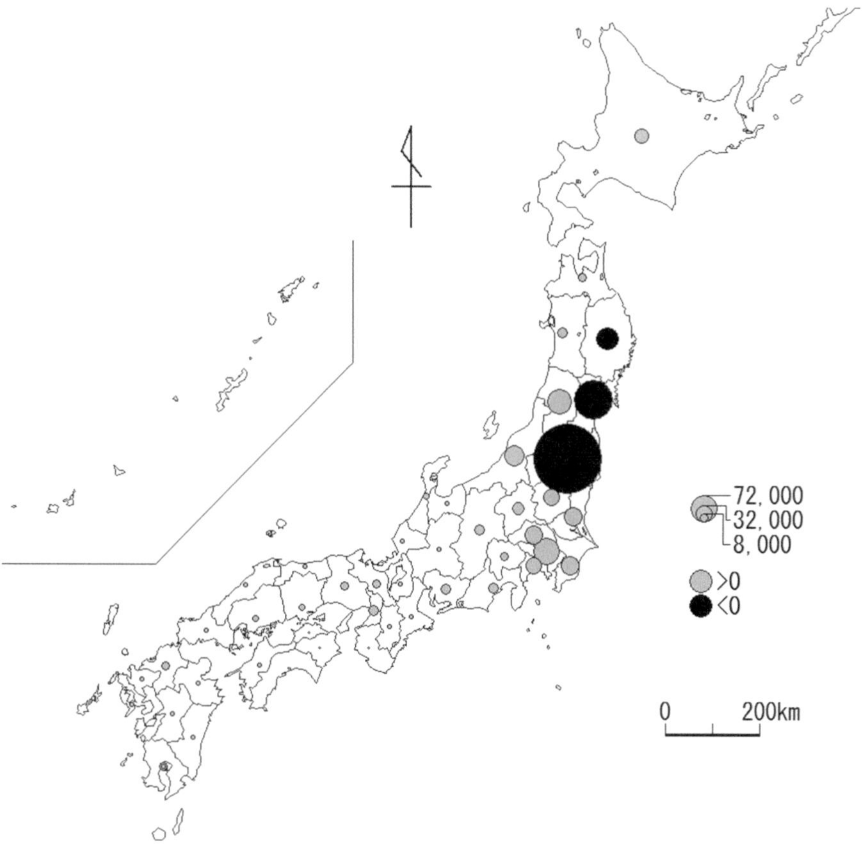

Fig. 9.6 Production changes by prefecture (million yen)

9.5 Population Projections and Economic Impacts of the Great Earthquake

9.5.1 Population Projections Considering the Impacts of the Great Earthquake

As mentioned in Sect. 9.2, Japan's population by region has been altered since the Great Earthquake, and it is considered that the population decline will continue in affected areas in the future. Therefore, Deguchi (2012) predicted the future population in 2030 by prefecture for Japan after considering the trend in population movements after the earthquake. This research predicted the population in the case without the disaster (case A) and that with the disaster (case B) in Table 9.13. In addition, with regard to case B, the population was predicted on the basis of two scenarios: (1) where net migration rate is assumed to continue at 70% of that

Table 9.12 Production changes by sector in the three affected prefectures (million yen)

	2011	2012	2013	2014	2015	Total
Agriculture	−4,741	−5,953	−5,332	−5,036	−4,856	−25,918
forestry	−177	−218	−200	−191	−185	−971
fishery	−425	−526	−485	−466	−454	−2,357
Mining	−330	−410	−377	−361	−352	−1,831
Foods, Tabacco	−11,892	−14,883	−13,413	−12,710	−12,288	−65,185
Textile products	−958	−1,190	−1,089	−1,040	−1,011	−5,288
Lumber and wood products	−98	−120	−112	−108	−106	−544
Furniture and fixtures	−120	−149	−134	−127	−122	−652
Pulp, paper, Paper products	−677	−833	−765	−731	−711	−3,716
Printing, plate making and book binding	−402	−502	−453	−430	−415	−2,203
Chemical products	−999	−1,253	−1,122	−1,059	−1,020	−5,453
Petroleum and coal products	−1,071	−1,335	−1,196	−1,128	−1,086	−5,816
Plastic products	−325	−406	−372	−356	−347	−1,805
Rubber products	−126	−158	−141	−134	−129	−687
Miscellaneous manufacturing products	−301	−376	−337	−318	−306	−1,638
Ceramic, stone and clay products	−158	−194	−179	−172	−168	−871
Iron and steel	−47	−58	−54	−52	−50	−260
Non-ferrous metals	−132	−165	−150	−143	−139	−728
Metal products	−335	−420	−378	−359	−347	−1,839
General-purpose machinery	−61	−75	−70	−67	−65	−338
Business oriented machinery	−34	−42	−38	−36	−35	−185
Electronic components	−139	−173	−161	−156	−153	−781
Electrical machinery	−1,863	−2,341	−2,075	−1,949	−1,869	−10,096
Miscellaneous electrical machinery	−95	−118	−107	−102	−99	−522
Moter Vehicle	−274	−336	−321	−314	−311	−1,556
Miscellaneous vehicle products	−40	−49	−45	−43	−42	−218

(continued)

Table 9.12 (continued)

	2011	2012	2013	2014	2015	Total
Precision machine	−440	−554	−491	−461	−442	−2,389
Miscellaneous manufacturing products	−361	−450	−405	−384	−370	−1,969
Building construction	−993	−1,220	−1,133	−1,090	−1,066	−5,501
Public construction	0	0	0	0	0	0
Electricity	−7,014	−8,844	−7,798	−7,300	−6,988	−37,945
Gas and heat supply	−307	−374	−347	−334	−326	−1,687
Water supply	−2,783	−3,482	−3,103	−2,921	−2,808	−15,097
Commerce	−11,906	−14,760	−13,397	−12,738	−12,333	−65,133
Finance and insurance	−9,147	−11,380	−10,274	−9,741	−9,414	−49,956
Real estate	−28,824	−35,830	−32,354	−30,676	−29,645	−157,328
Transport and postal services	−7,698	−9,541	−8,646	−8,212	−7,944	−42,041
Information and communications	−4,288	−5,324	−4,813	−4,567	−4,414	−23,406
Public administration	−491	−613	−551	−521	−503	−2,678
Education and research	−3,223	−3,991	−3,622	−3,443	−3,333	−17,611
Medical, health care and welfare	−6,242	−7,789	−6,996	−6,614	−6,380	−34,020
Miscellaneous non-profit services	−1,794	−2,239	−2,012	−1,903	−1,837	−9,785
Business services	−5,477	−6,766	−6,188	−5,907	−5,737	−30,074
Personal services	−15,667	−19,479	−17,558	−16,631	−16,057	−85,391
Office supplies	−583	−726	−654	−619	−597	−3,179
Total	−133,057	−165,645	−149,446	−141,645	−136,857	−726,650

observed just after the earthquake (case B1) and (2) where the net migration rate continues as the same as in 2012, just after the earthquake (case B2).

According to the projection, the future populations in Iwate, Miyagi, and Fukushima will decline rapidly due to the decreasing birthrate and aging population in any scenario. In particular, the population of Fukushima will decline considerably in the case considering the earthquake due to a population outflow of people under their forties and a high population aging rate. In the Tohoku region (other than Fukushima and Miyagi), the population in case B (with the disaster) will increase in comparison with case A (without the disaster). However, in the South Kanto region, including Tokyo, Kanagawa, and Chiba the population will decline compared with case A (without the disaster).

Table 9.13 Results of population projections in Deguchi (2012): Population growth rate 2010–2030

	Case A	Case B1	Case B2
	2030年	2030年	2030年
the whole of Japan	−11%	−11%	−11%
Hokkaido	−20%	−15%	−15%
Aomori	−29%	−21%	−23%
Iwate	−27%	−21%	−22%
Miyagi	−14%	−14%	−16%
Akita	−31%	−25%	−27%
Yamagata	−26%	−18%	−19%
Fukushima	−23%	−34%	−40%
Ibaraki	−14%	−15%	−16%
Tochigi	−15%	−13%	−14%
Gunma	−16%	−14%	−14%
Saitama	−6%	−7%	−6%
Chiba	−4%	−11%	−12%
Tokyo	8%	−5%	−3%
Kanagawa	−2%	−6%	−5%

Source: Deguchi (2012)

9.5.2 *Economic Impacts of Changing Consumer Demand Attributable to Future Population Decline in Each Prefecture*

It is considered that population changes impact consumer demand provided that per capita consumption does not increase. As shown in the previous section, the population changes resulting from the earthquake differ in each region. Consumer demand in Fukushima will reduce due to depopulation, whereas it will increase in regions with population increases. We analyzed the economic impacts of changing consumer demand attributable to future population change.[2] The results by the scenario of population change are as follows.

As shown in Table 9.14, assuming that the disaster did not occur (case A), final demand in Iwate, Miyagi, and Fukushima will decline by 12.6%, 7.7%, and 11.5%, respectively, when compared with the level in 2010. In any case, it is predicted that the population in the affected prefectures will decrease sharply over 20 years (2010–2030) and that production in Fukushima and Iwate will decline by around 10%. Thus, even if the disaster had not occurred, it is predicted that regional economies will decline in the affected areas due to depopulation. In the scenario that the net migration rate will continue at 70% of that observed just after the earthquake (case B1), final demand in Fukushima will further decline and production will decline by 17.4% in comparison with 2010. In case B2, assuming that the net migration rate just after the earthquake will continue, final demand in 2030 will

[2]In this study, it is assumed that per capita consumption remains constant.

Table 9.14 Economic impacts of the Great Earthquake on affected prefectures

		Case A(2010) Case A(2030)	Case A(2030) Case B1(2030)	Case A(2030) Case B2(2030)
Iwate	Final demand	−12.6%	3.2%	2.4%
	Production	−11.0%	2.5%	1.8%
Miyagi	Final demand	−7.7%	0.0%	−0.7%
	Production	−7.6%	0.0%	−0.7%
Fukushima	Final demand	−11.5%	−5.9%	−9.5%
	Production	−9.0%	−4.1%	−6.4%

decline by 21% compared with 2010 and by 9.5% compared with 2030 in case A. Production will also decline, which is attributable to the decrease in final demand. However, the change of final demand in each prefecture does not necessarily move in tandem with production change by prefecture because inter-regional feedback effects are estimated. Therefore, it is shown that production in Fukushima in the case with the disaster (case B1 in 2030) will decline by 4.1% compared with case A in 2030 and that in the case B2 will decline by 6.4% compared with case A in Table 9.15. In Iwate Prefecture, it is predicted that production in 2030 in case B1 will increase by 2.5% compared with case A in 2030. The factor influencing this result is population growth attributable to an increase in the population in their productive age. That is, it is considered that people will move out of Fukushima. In Miyagi, it is estimated that the disaster will have a negligible impact on production in 2030.

9.6 Conclusion

The Great Earthquake of March 11, 2011, significantly damaged the Tohoku region, especially the prefectures of Iwate, Miyagi, and Fukushima. In particular, as of 2015, Fukushima Prefecture is still suffering from the effects of the Great Earthquake and the nuclear power plant accident and the consequent evacuation of numerous people. Population decline due to deaths, those declared missing, and evacuations have had a significant negative impact on the regional economy through a decline in consumer demand.

In this chapter, we analyzed the economic impacts of the population decrease for the past 5 years after the Great Earthquake using an inter-regional input–output model. Utilizing a 47-region input–output table at the prefectural level developed in our previous study, economic impacts were estimated by prefecture, considering the inter-regional feedback effects. It was observed that the impact was spread across the country. Specifically, although the amount of production in the Tohoku region, including Fukushima, decreased, that in the Kanto region, including Tokyo, increased due to the many evacuees arriving from the three affected prefectures and consuming in the region. Further, it is considered that the population decline will

Table 9.15 Changes to rates of production by prefecture based on their future populations in each scenario

	A(2030)-A(2010)	B1(2030)-A(2030)	B2(2030)-A2(2030)
	2010–2030	2030	2030
Hokkaido	−10.1%	2.6%	2.4%
Aomori	−12.3%	3.2%	2.6%
Iwate	−11.0%	2.5%	1.8%
Miyagi	−7.6%	0.0%	−0.7%
Akita	−13.6%	2.9%	2.1%
Yamagata	−9.8%	2.9%	2.6%
Fukushima	−9.0%	−4.1%	−6.4%
Ibaraki	−5.6%	−0.6%	−1.0%
Tochigi	−5.5%	0.2%	−0.1%
Gumma	−5.8%	0.6%	0.4%
Saitama	−3.1%	−0.5%	−0.1%
Chiba	−2.3%	−3.0%	−3.1%
Tokyo	2.5%	−5.5%	−4.6%
Kanagawa	−1.6%	−2.0%	−1.6%
Niigata	−8.6%	1.6%	1.3%
Toyama	−6.5%	0.9%	0.7%
Ishikawa	−5.5%	0.6%	0.4%
Fukui	−7.1%	1.7%	1.3%
Yamanashi	−7.0%	1.4%	0.9%
Nagano	−7.6%	2.0%	1.8%
Gifu	−6.2%	1.1%	0.7%
Shizuoka	−6.0%	0.9%	0.7%
Aichi	−2.1%	−0.3%	−0.1%
Mie	−5.2%	0.7%	0.6%
Shiga	−2.5%	0.2%	0.5%
Kyoto	−6.5%	1.1%	1.1%
Osaka	−5.5%	0.7%	0.9%
Hyogo	−5.6%	0.9%	0.9%
Nara	−8.3%	2.1%	1.8%
Wakayama	−8.6%	1.4%	0.9%
Tottori	−9.7%	2.2%	1.6%
Shimane	−8.8%	1.7%	1.3%
Okayama	−5.4%	1.0%	1.1%
Hiroshima	−5.6%	1.1%	0.9%
Yamaguchi	−7.6%	1.2%	0.9%
Tokushima	−9.2%	1.5%	1.0%
Kagawa	−7.9%	1.8%	1.7%
Ehime	−8.5%	1.4%	1.0%
Kochi	−11.2%	2.4%	1.7%
Fukuoka	−5.2%	1.6%	2.1%
Saga	−8.2%	2.5%	2.1%

(continued)

Table 9.15 (continued)

	A(2030)-A(2010)	B1(2030)-A(2030)	B2(2030)-A2(2030)
	2010–2030	2030	2030
Nagasaki	−10.6%	2.9%	2.0%
Kumamoto	−7.4%	1.9%	1.7%
Oita	−6.4%	0.6%	0.4%
Miyazaki	−8.1%	1.7%	1.2%
Kagoshima	−10.7%	3.0%	2.5%
Okinawa	−1.5%	1.8%	3.8%

continue in the affected areas in the long term. Therefore, based on population projections that consider the impacts of the earthquake, we analyzed the long-term economic impacts of the Great Earthquake. We show that in the worst case, production in Fukushima in 2030 will decline by 15.4% compared with 2010. Thus, there is concern that the population decline in affected regions such as Fukushima will continue in the long term and that the regional economy will decline. Therefore, the government needs to make an effort to increase the population in the affected regions in order to provide economic recovery from the earthquake damage.

References

Deguchi K (2012) Estimation of future population by prefecture considering impact of the disaster. JCER Discussion paper 133 Japan Center for Economic Research (in Japanese)

Ishikawa Y, Miyagi T (2004) An interregional industrial linkage analysis in Japan, using a 47-region interregional input-output table. Studies in Regional Science 34(1):139–152 (in Japanese)

Ishikura T, Ishikawa Y (2011) Spatial economic impacts of power supply shortage by the Great East Japan Earthquake. Input-Output Analysis 19(3):51–59 (in Japanese)

Sasayama H (2011). Estimating the economic effects of port investment with interregional input-output table at the prefecture level technical note of national institute for land and infrastructure management 630 (in Japanese)

Chapter 10
An NEG Analysis of Megathrust Earthquakes in Japan

Ryusuke Ihara

Abstract This chapter investigates the impact of megathrust earthquakes on Japanese regional structure. Even after the Great East Japan Earthquake in 2011, there are serious concerns about predicted earthquakes in the near future; for instance, Tokyo Metropolitan Area (M. A.) earthquakes (expected to occur just in Tokyo M. A.), Tokai earthquakes (expected to occur in the area facing the Pacific Ocean on the west side of Tokyo M.A.), and Nankai megathrust earthquake (expected to occur in the area facing the Pacific Ocean on the west side of Kansai M.A.). In this chapter, we focus on the Great East Japan Earthquake and the Nankai megathrust earthquakes and the associated tsunami and consider how they changed the regional economies of Japan. Constructing an NEG model composed of the 47 prefectures of Japan, simulation results show that the predicted labor distribution approaches the actual distribution and as transportation costs decrease, labor distribution changes from dispersion to agglomeration in the metropolitan areas and then to re-dispersion in rural areas. In addition, adapting the damage data of megathrust earthquakes to this model, we predict the impact of the Great East Japan Earthquake and a Nankai megathrust earthquake on the regional potential and labor distribution among the prefectures. The simulation result of a Nankai megathrust earthquake predicts that the indirect utility levels in the affected prefectures fall by 1%, and the resulting outflow of refugees reduces the prefectural population by 3% in the worst case.

Keywords New economic geography • Megathrust earthquake • Regional potential • Labor distribution

---

This work was supported by JSPS KAKENHI Grant Number 24243036.

R. Ihara (✉)
Faculty of Economics, Asia University, 5-24-10 Sakai, Musashino-shi, Tokyo 180-8629, Japan
e-mail: ihara@asia-u.ac.jp

S. Tokunaga, B.P. Resosudarmo (eds.), *Spatial Economic Modelling of Megathrust Earthquake in Japan*, New Frontiers in Regional Science: Asian Perspectives 11,
https://doi.org/10.1007/978-981-10-6493-7_10

10.1 Introduction

To investigate the impact of natural disasters on regions, New Economic Geography (NEG) presents a helpful framework. The NEG framework explains labor distribution in terms of agglomeration economies and transportation costs of goods. Agglomeration economies are the benefit of being located in close proximity with customers, consumers, and workers, which stimulates the formation of cities. However, transportation costs have a negative effect on agglomeration. If a company concentrates its production in a specific region, it would incur a cost for shipping to other areas. Considering the relationship between the agglomeration economies and transportation costs, NEG reveals a key feature, location; when transportation costs are sufficiently high, economic activities disperse among regions so as to reduce transportation costs. However, when transportation costs become sufficiently low, economic activities change their location choice and concentrate in cities to enjoy the agglomeration economies. If we introduce land use into this framework, then we can explain a re-dispersion of economic activities when transportation costs decrease further.

Recently, a considerable number of empirical studies have been conducted to verify the theory of NEG in an actual geography. The pioneering studies are those of Hanson (2005) and Redding and Venables (2004). Hanson (2005) explained the wage structure in the NEG framework in a single-step estimation using US county-level data. Redding and Venables (2004) presented a different approach consisting of two steps: (i) trade estimation with regional dummies and (ii) wage estimation with regional potential derived from the trade estimation. Most recent empirical studies follow one of these two approaches. For instance, Brakman et al. (2006) estimated the wage equation with data from the NUTS II EU regions, following Hanson's single-step approach. The two-step approach is followed by Head and Mayer (2006) and Breinlich (2006) for the EU regions, Knaap (2006) for the US, Ottaviano and Pinelli (2006) for Finland, and Nakamura et al. (2010) for Japan. Several articles focus on the multiple equilibria in NEG, and investigate whether a large disaster, such as World War II, changes regional structure; as explored by, for instance, Davis and Weinstein (2008) and Bosker et al. (2007). In addition to these empirical studies, an alternative approach has been developed with the recent progress of computer technology: simulation of a multi-regional NEG model in an actual geographical space. Stelder (2005) constructed an NEG model on a map of Western Europe and predicted the location of cities, and Brakman et al. (2006) simulated the NEG model of the EU region. Ihara (2011) also derived the location of economic activities in Asian countries.

Following the above stream of NEG literature, this chapter investigates labor distribution and the impact of megathrust earthquakes in Japan.[1] The remainder of

[1]The Central Disaster Management Council in the Cabinet Office of Japan expects various cases for an NME. Extracting a part of the cases expected, Ihara (2014) studied the impact of the earthquakes on regional structure in Japan. This chapter expands Ihara (2014) and investigates all the cases that have applicable estimated data for an NME as well as the Great East Japan Earthquake.

this chapter is organized as follows. In Section 10.2, we construct an NEG model with multiple regions. Section 10.3 shows the predicted labor distribution in 47 prefectures by applying Japanese prefectural data to this model. Sections 10.4 and 10.5 consider the Great East Japan Earthquake (GEJE) and a Nankai megathrust earthquake (NME), respectively, and reveal their impacts on regional potential and labor distribution. Section 10.6 concludes this chapter.

10.2 Model

10.2.1 N-Region Model

In this section, we construct an NEG model with land for housing consumption and factor input. The structure of this model is based on the works of Fujita et al. (1999) and Helpman (1998). The utility function is expressed by

$$u = \left(C^M\right)^{\alpha^M}\left(C^A\right)^{\alpha^A}\left(C^H\right)^{\alpha^H}, \alpha^M + \alpha^A + \alpha^H = 1, \tag{10.1}$$

where $C^M$is a composite index of the consumption of manufactured goods and services (MS goods), C^A means the consumption of agricultural goods, and C^H means the land consumption for housing. The parameters α^M, α^A, and α^H, respectively represent the expenditure shares. The composite index C^M is defined as a CES function:

$$C^M = \left(\int_0^n m(i)^{(\sigma-1)/\sigma} di\right)^{\sigma/(\sigma-1)}, \tag{10.2}$$

where $m(i)$ is a differentiated good produced by firm i, n is the number of firms or varieties, and $\sigma(>1)$is the elasticity of substitution between varieties.

Production follows the standard framework of NEG. In the manufacturing and services (MS) sector, we assume that individual firms produce their own differentiated goods in a monopolistic competition of Dixit–Stiglitz type. The production function of each firm is defined by the following equation:

$$\left(l^M\right)^{\beta_L^M}\left(k^M\right)^{\beta_K^M} = F + c^M q, \beta_L^M + \beta_K^M = 1. \tag{10.3}$$

That is, producing a quantity q of any variety requires labor input, l^M, and land input, k^M. The expenditure shares for labor and land are given by β_L^M and β_K^M. F and c^M are the fixed cost of starting up a factory and the marginal input requirement, respectively. On the other hand, agricultural goods are homogenous and produced with a constant-returns technology under the condition of perfect competition. The production function is

$$\left(L^A\right)^{\beta_L^A}\left(K^A\right)^{\beta_K^A} = c^A A, \beta_L^A + \beta_K^A = 1, \tag{10.4}$$

where L^A and K^A are the labor input and land input, respectively for the agricultural sector, β_L^A and β_K^A are their respective expenditure shares, and c^A is the marginal input requirement.

Now we consider the regional structure. There are N regions and MS goods and agricultural goods can be shipped among the regions, and the shipment of MS goods is assumed to require the so-called "iceberg" transportation costs. That is, to deliver one unit of MS goods to region r from the producing region s, $1+T_{sr}$ units must be dispatched. Finally, we assume that land is equally owned by the residents in each region; thus, the revenue from land usage is distributed to the residents. As a result, the total income of each region is expressed as

$$Y_r = w_r^M L_r^M + w_r^A L_r^A + R_r^M K_r^M + R_r^A K_r^A + R_r^H K_r^H, \tag{10.5}$$

where w_r^M and w_r^A are labor wages, respectively in the MS and agricultural sectors; R_r^M, R_r^A, and R_r^H are revenues from land usage for the MS sector, agricultural sector, and housing, respectively; and L_r^M, L_r^A, K_r^M, K_r^A, and K_r^H are the regional total amounts of labor and land for the sectors.

10.2.2 Equilibrium

Before we consider labor migration, we confirm the short-run equilibrium in which all the workers are not able to move among the regions. In this case, consumer's utility maximization yields total sales of each firm as

$$q_r = \alpha^M \sum_{s=1}^{N} \left[p_r^{-\sigma} P_s^{\sigma-1} Y_s (1+T_{rs})^{1-\sigma}\right], \tag{10.6}$$

where

$$P_s = \left(\sum_{r=1}^{N} n_r \left(p_r^M (1+T_{rs})\right)^{1-\sigma}\right)^{1/(1-\sigma)} \tag{10.7}$$

is the price index of MS goods. Profit maximization of each firm obtains the price of MS goods produced in region r as

$$p_r = \frac{\sigma c^M}{\sigma - 1} \frac{w_r^M}{\beta_L^M} \left(\frac{L_r^M}{K_r^M} \right)^{\beta_K^M}, \tag{10.8}$$

where we have used the relation between land rent and wage, $R_r^M = w_r^M (\beta_K^M L_r^M) / (\beta_L^M K_r^M)$. By assuming that with free entry and exit, which implies the zero-profit condition, the output of each firm becomes $q^* = F(\sigma - 1)/c^M$ and the number of firms becomes $n_r = (L_r^M)^{\beta_L^M} (K_r^M)^{\beta_K^M} / F\sigma$ in equilibrium. As a result, the wage of the workers in the MS sector in region r is expressed as

$$w_r^M = \beta_L^M \frac{\sigma - 1}{\sigma c^M} \left(\frac{K_r^M}{L_r^M} \right)^{\beta_K^M} \left(\frac{\alpha^M}{q^*} \sum_{s=1}^{N} \left[P_s^{\sigma-1} Y_s (1 + T_{rs})^{1-\sigma} \right] \right)^{1/\sigma}. \tag{10.9}$$

In the long-run equilibrium, the MS workers are able to move to any region that offers a higher indirect utility. As the income of each worker is given by the sum of labor wage and revenue from land usage, the indirect utility is expressed as

$$v_r^M = y_r^M (P_r)^{-\alpha^M} (R_r^H)^{-\alpha^H}, \tag{10.10}$$

where

$$y_r^M = w_r^M + (R_r^M K_r^M + R_r^A K_r^A + R_r^H K_r^H) / (L_r^M + L_r^A) \tag{10.11}$$

and

$$R_r^H = \alpha^H Y_r / K_r^H. \tag{10.12}$$

Consequently, the dynamics of MS workers are defined as

$$\dot{L}_r^M = \kappa (v_r^M - \bar{v}^M) L_r^M / L^M, \tag{10.13}$$

where $\bar{v}^M$ is the weighted average of the indirect utility, and L^M is the total number of MS workers. In the long-run equilibrium, labor distribution in the MS sector is obtained so as to equalize the levels of indirect utility of all the regions.

10.2.3 The Core–Periphery Model

Let us now look at a two-region model to understand the feature of migration behavior. As our model structure is based on that of Fujita et al. (1999) and Helpman (1998), the result is a composite of their studies. As transportation costs decrease, the distribution of workers changes from dispersion to agglomeration and

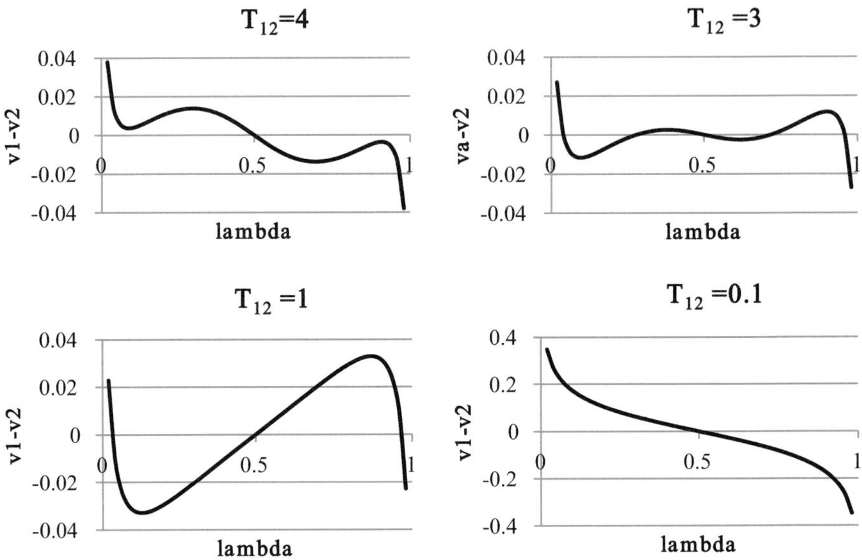

Fig. 10.1 Indirect utility differentials in the short-run

then to re-dispersion. Figure 10.1 plots the short-run indirect utility differentials $v_1^M - v_2^M$ with four different levels of transportation costs.[2] The horizontal axis is $\lambda = L_1^M / (L_1^M + L_2^M)$, which is the share of labor amount of region 1. If $v_1^M - v_2^M > 0$, then workers will move to region 1, and if $v_1^M - v_2^M < 0$, they will move to region 2. Therefore, the stable equilibrium of labor distribution is given by $v_1^M - v_2^M = 0$, with a negative slope. If the slope of the indirect utility differentials is positive, then the equilibrium is unstable. An unstable equilibrium means that if labor distribution deviates slightly from the equilibrium distribution by an accidental migration of workers, then the consequent increase in the difference of indirect utility causes a cumulative labor migration, and the equilibrium will break. Consequently, we find that workers equally disperse between the two regions in the cases of $T_{12} = 4$ and $T_{12} = 0.1$, whereas most workers concentrate in region 1 or 2 in the case of $T_{12} = 1$. These results are summarized in the bifurcation diagram depicted in Fig. 10.2. The solid and broken lines represent a stable equilibrium and an unstable equilibrium, respectively. We can observe that when transportation costs are sufficiently high, workers are evenly divided between the regions. When transportation costs reduce beyond $TB1$, the symmetric equilibrium breaks, and the regions get divided into core and periphery. Finally, when transportation costs become sufficiently low, such as $T_{12} < TB2$, workers disperse again so as to avoid the high

[2]Figures 10.1 and 10.2 are constructed with $\sigma = 4$, $\alpha^M = 0.7$, $\alpha^A = 0.2$, $\beta_K^M = \beta_K^A = 0.1$, $L_r^M = L_r^A = 1$, and $K_r^M = K_r^A = K_r^H = 1$. In addition, we set $F = 1$, $c^M = c^A = 1$ in this chapter for analytical simplicity.

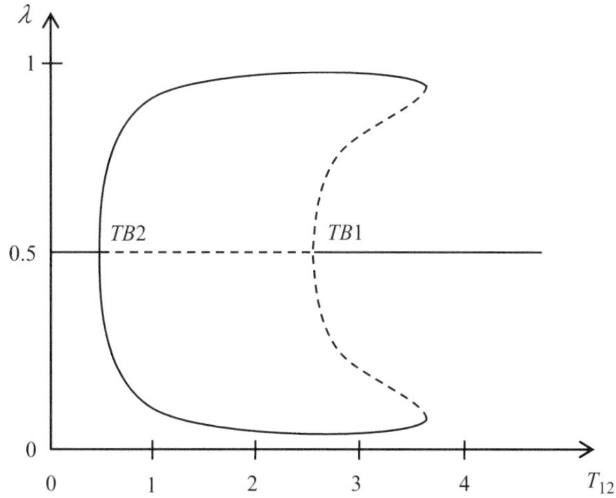

Fig. 10.2 Bifurcation diagram in the long-run

land rent in the core region, and a symmetric structure will emerge again as a stable equilibrium. The re-dispersion can be interpreted as a suburbanization.

10.3 Prediction of Labor Distribution in Japan

10.3.1 Regional Data and Calibration

In this section, we begin with a description of the regional data from the 47 regions (i.e., prefectures). The land areas are referenced from the Private Land Area by Land Category in the "Summary Report on the Fixed Asset Price," reported by the Minister's Secretariat of Japan in 2010. Agricultural land area is given as the sum of paddy fields, fields, forests, and pastures; the land area for MS industry use is given by the area for non-dwelling buildings; and the housing land area is given by the area for dwelling buildings. The expenditure shares of consumers are referenced from the "Annual Report on the Family Income and Expenditure Survey" reported by the Minister's Secretariat of Japan in 2010. Thus, the expenditure share for agricultural goods is given as the food ratio of the total disbursement, the same for housing is given by the housing ratio, and the remainder is for MS goods. The input shares of MS production are referenced from the Input–Output Table of Japan in 2010. That is, land input share is given by the operating surplus, and the remaining share is for labor input. Table 10.1 presents the results of the expenditure share. The actual distribution of workers among the 47 prefectures and industries is referenced from the National Census of Japan (2000). Finally, we define the transport costs as

Table 10.1 Expenditure share

$\alpha^M = 0.687$	$\beta_L^M = 0.892$
$\alpha^A = 0.236$	$\beta_K^M = 0.108$
$\alpha^H = 0.077$	$\beta_L^A = 0.913$
	$\beta_K^A = 0.087$

$$T_{rs} = (dist_{rs})^t, \tag{10.14}$$

where $dist_{rs}$ ($r \neq s$) is the distance in a straight line between government offices of prefectures r and s (unit: 100 km). The intra-regional distance of prefecture r is obtained by $dist_{rr} = \sqrt{S_r/\pi}$, where S_r is the total area of the private properties obtained in our prefectural data. The parameter t means the resistance effect due to distance.

10.3.2 Calibration and the Prediction of Labor Distribution

Using the regional data obtained above, we derive the labor distribution in the long-run equilibrium. First, we specify the remaining parameters, σ and t, in the following manner. Suppose the actual labor distribution is the result of labor migration in the long-run, which minimizes the indirect utility difference among the regions. Thus, taking the actual labor distribution as given, the parameters are obtained by the calibration expressed as

$$(\sigma, t) = \operatorname{argmin} \sum_{r=1}^{47} \frac{\left| v_r^M - \bar{v}^M \right|}{\bar{v}^M}, \tag{10.15}$$

where $\bar{v}^M$ is the weighted average of the indirect utility levels of the regions. As a result, we have $\sigma = 8.3$ and $t = 0.7$.

We are now able to predict the labor distribution in the long-run equilibrium in Japan. The simulation of labor migration is as follows. First, taking the actual labor distribution as initial distribution, we derive the indirect utility in the short-run equilibrium. Then, we consider the labor migration between regions. If some regions offer a higher level of indirect utility while other regions offer a lower level, then the MS workers in the latter regions will move to the former regions. Such a migration reduces the wage difference among regions. Continuing this process, the labor distribution in the long-run equilibrium is obtained by equalizing the indirect utility levels of all the regions.[3]

[3]For the simulation in this chapter, the long-run equilibrium is given by the condition where the difference of the indirect utility level from the national average becomes less than 0.1% for all regions.

Table 10.2 shows the results of equilibrium labor distribution with different levels of transportation costs (i.e., distance resistance): $t = 2$, 1.5, 1, 0.7, 0.5, and 0.1. First, let us examine the benchmark case of $t = 0.7$, which is the result of our calibration. In this case, the prediction for the whole region basically approaches the actual distribution.[4] That is, the correlation coefficient of the predicted and actual distributions is approximately 0.945 (although the predicted labor population of Tokyo is less than the actual distribution). Next, comparing this result with other cases of transportation costs, we observed that the feature of the transition of labor distribution is the same as the two-region core–periphery model in the previous section. That is, looking at the transition of the coefficient of the variation of labor distribution, we find that (i) in the range of t from 2 to 0.7, the decline of transportation costs raises the agglomeration tendency of workers, and (ii) in the range of t from 0.7 to 0.1, the decline of transportation costs lowers the agglomeration tendency. The representative examples of the transition are Aichi and Fukuoka, which are the economic and political centers of the Tokai area and Southern area of Japan, respectively. As transportation costs decrease, the population in these prefectures initially increases and then decreases. Conversely, a contrastive transition is observed in local areas such as Aomori, Shimane, and Okinawa. When transportation costs are high, the decline of transportation costs decreases the rural population, whereas when transportation costs become sufficiently low, the decrease in the costs raises rural population. The former case corresponds to the ongoing depopulation in rural areas in Japan. Finally, looking at Tokyo, the lasting decline of the predicted population does not seem to correspond to an actual concentration in Tokyo. However, focusing on the Tokyo Metropolitan Area (composed of Saitama, Chiba, Tokyo, and Kanagawa), we can confirm that the total population increases with the decrease of transportation costs from $t = 2$ to $t = 0.7$. A similar pattern is seen in Kansai Metropolitan Area (composed of Kyoto, Osaka, Hyogo, and Wakayama). These metropolitan areas imply the coexistence of concentration and suburbanization in a megalopolis.

10.4 The Impact of the Great East Japan Earthquake

Sections 10.4 and 10.5 consider how megathrust earthquakes affect regional structure. Specifically, we focus on (i) the indirect utility levels (following the empirical studies of NEG, we name it *regional potential*) in the short-run equilibrium and (ii) the labor distribution in the long-run equilibrium.

[4]The highest correlation coefficient is given in the case of $t = 1$, not in the case of $t = 0.7$. This is because the calibration takes labor distribution as given, whereas the simulation considers interregional labor migration.

Table 10.2 Prediction of labor distribution by prefecture

		t						
	Prefecture	2	1.5	1	0.7	0.5	0.1	Actual labor distribution
01	Hokkaido	832.5	1308.1	1758.1	1821.5	1756.9	2493.7	2190.8
02	Aomori	1370.9	1289.9	1006.5	723.9	600.3	809.4	541.3
03	Iwate	534.9	574.9	580.2	576.4	575.3	884.3	546.6
04	Miyagi	1383.6	1371.5	1243.9	1106.5	1019.8	1193.5	981.0
05	Akita	884.7	815.2	613.1	513.1	475.6	710.7	445.9
06	Yamagata	947.3	957.0	900.4	822.2	767.0	843.0	500.6
07	Fukushima	791.3	948.3	1099.3	1141.7	1175.8	1460.0	832.9
08	Ibaraki	1261.0	1386.1	1657.1	1947.5	2219.9	2288.4	1264.3
09	Tochigi	894.6	957.8	1138.2	1337.7	1535.1	1530.1	883.0
10	Gunma	1221.9	1130.9	1092.0	1210.5	1419.0	1511.3	883.3
11	Saitama	3284.2	3399.7	3588.7	3638.1	3541.5	2237.0	3169.2
12	Chiba	1480.4	1855.5	2536.9	3023.9	3246.3	2517.8	2631.5
13	Tokyo	4685.5	4453.8	4184.6	3822.7	3406.3	1838.4	5168.4
14	Kanagawa	2537.7	2650.2	2866.7	2939.2	2880.8	1941.3	3908.1
15	Niigata	1112.4	1074.3	962.2	944.0	1016.3	1571.1	1056.4
16	Toyama	763.3	626.1	462.6	428.0	488.7	803.8	516.5
17	Ishikawa	433.9	377.1	307.0	297.7	351.2	635.6	536.4
18	Fukui	339.3	272.8	229.1	251.0	327.6	565.6	379.6
19	Yamanashi	522.8	405.1	340.4	381.2	484.2	617.9	376.2
20	Nagano	1150.6	1100.9	1050.1	1108.0	1285.9	1756.1	950.8
21	Gifu	958.5	1038.7	1172.6	1299.7	1386.4	1295.9	957.1
22	Shizuoka	1290.8	1200.7	1098.0	1167.4	1411.3	2001.0	1770.2
23	Aichi	2984.1	3041.1	3200.5	3366.5	3449.2	3035.9	3359.9
24	Mie	529.1	548.8	697.0	925.6	1129.0	1236.4	815.1
25	Shiga	1292.8	1118.6	987.3	984.8	984.3	811.9	620.8
26	Kyoto	1271.4	1108.0	972.6	952.6	937.9	765.1	1086.3

27	Osaka	3502.3	3351.9	3089.8	2848.2	2552.7	1744.2	3488.9
28	Hyogo	1243.9	1603.4	1998.0	2217.4	2266.3	1975.4	2296.0
29	Nara	365.2	426.9	511.6	570.4	586.7	486.9	556.0
30	Wakayama	364.6	373.7	403.5	452.2	495.4	554.2	395.4
31	Tottori	426.0	307.4	216.6	198.1	219.1	377.6	244.9
32	Shimane	299.3	241.1	190.8	182.9	203.7	406.0	309.1
33	Okayama	709.5	792.2	818.2	847.8	894.1	1142.1	812.5
34	Hiroshima	847.2	852.4	774.6	758.4	768.8	1043.2	1234.8
35	Yamaguchi	625.1	601.5	540.3	536.2	550.9	764.7	615.5
36	Tokushima	334.7	306.4	288.8	302.7	332.7	441.8	304.5
37	Kagawa	974.4	765.2	583.2	518.6	497.1	561.8	424.8
38	Ehime	704.9	633.7	543.8	531.5	542.5	723.4	580.2
39	Kochi	243.4	214.3	195.3	201.0	222.6	355.9	286.1
40	Fukuoka	2581.2	2870.6	3008.0	2561.4	1989.7	1616.9	2071.8
41	Saga	718.3	729.7	754.6	678.2	554.1	453.5	359.0
42	Nagasaki	773.1	651.3	551.9	494.8	452.1	519.3	577.9
43	Kumamoto	964.2	956.2	987.3	968.0	900.9	907.7	727.1
44	Oita	669.9	590.5	498.3	479.0	483.1	614.9	492.6
45	Miyazaki	1054.9	977.2	754.0	592.2	512.8	645.7	452.2
46	Kagoshima	1000.1	999.6	900.8	771.4	681.3	876.1	668.7
47	Okinawa	607.9	513.5	414.9	328.0	191.5	202.9	499.5
Tokyo M.A.		11,988	12,359	13,177	13,424	13,075	8534	14,877
Kansai M.A.		6747	6864	6975	7041	6839	5526	7823
Coefficient of variation		0.794	0.807	0.845	0.845	0.818	0.594	0.956
Correlation coefficient		0.906	0.935234	0.949	0.945	0.921	0.770	

Unit: thousand
Tokyo M.A.: Saitama, Chiba, Tokyo, and Kanagawa
Kansai M.A.: Kyoto, Osaka, Hyogo, and Wakayama

10.4.1 A Survey of the Damage Situation and Reconstruction Stages

Using the NEG model presented above, this section investigates the impact of the Great East Japan Earthquake (GEJE) and the resultant tsunami in 2011. Expanding on Ihara et al. (2013), who study the effect of GEJE in the Miyagi prefecture, we consider the effects of the loss of land area and population on the indirect utility levels and labor distribution in 47 prefectures.

First, we describe the damage resulting from the GEJE. Due to the great tsunami that struck the Pacific coast of Eastern Japan, the coastal areas suffered serious damage.[5] In our model, the damages are captured by the loss of land and labor population.[6] The area lost due to the great tsunami is calculated from the "Land Category of Private Land Area Flooded by the Tsunami" reported by the Geospatial Information Authority of Japan. Specifically, damage to agricultural land area is calculated as the ratio of flooded area to the total agricultural area (paddy fields, fields, and forests), and that for the MS land area is calculated as the ratio of flooded area to the total land area for building. Besides, considering Fukushima Daiichi nuclear disaster in the Fukushima prefecture, we add the "areas under evacuation orders" to the flooded areas in Fukushima. The decline of labor population is calculated from the share of deaths from the GEJE.

Next, we classify the restoration from the GEJE into three stages. For the purpose of agricultural restoration, we refer to the "re-usable area for farming" in the "Current State and Issues of the Reconstruction" reported by the Reconstruction Agency of Japan. The Stage 1 restoration will be done by 2012, and both Stage 2 and Stage 3 restoration will be done[7] by 2013. For the restoration of MS production and housing land, we refer to the "Survey Result of Damage Situation of the Great East Japan Earthquake" (the first report) by the Ministry of Land, Infrastructure, Transport and Tourism of Japan. In Stage 1, the flooded areas with slight damage are restored, but the areas where buildings are "partially or

[5]The hardest hit prefectures were Aomori, Iwate, Miyagi, Fukushima, Ibaraki, and Chiba, which comprise 10.9% of the GDP of Japan (by industry, 19.6% of agricultural sector, 11.8% of manufacturing sector, and 10.5% of services sector), and 12.7% of the population of Japan in 2010. Miyagi prefecture is the economic and political center of Tohoku area while Chiba constitutes the Tokyo metropolitan area with Tokyo, Kanagawa, and Saitama.

[6]The impact of the tsunami differs among prefectures, depending on the land use of coastal areas. That is, some prefectures with large cities in coastal area, such as Miyagi, may suffer larger damage than others. However, our 47-prefecture model cannot directly handle such a difference based on coastal land use because there is no geographical information inside the prefectures in our model. Instead, we can refer to the detailed information about the damages to land use and population reported by the government. On the other hand, Ihara et al. (2013) also presented a model using 35 counties of the Miyagi Prefecture as a case study of the impact of the GEJE inside the prefectures.

[7]The Reconstruction Agency reports that all agricultural area will be recovered by 2015, except the areas sunk under the water and the warning zone in Fukushima.

completely destroyed" remain damaged. In Stages 2 and 3, the flooded areas where buildings are "partially destroyed" are restored. In addition, we assume that in Fukushima, the areas affected by the Fukushima Daiichi nuclear disaster are not recovered in Stages 1 and 2, and only "the zone in preparation for the lifting of the evacuation order" starts restoration in Stage 3. Table 10.3 shows the percentage of damaged land and population in the total amount in each stage of restoration. Miyagi and Iwate suffer large damage in both population and land area, whereas Fukushima mainly suffers a loss of land area. Other prefectures will almost recover through the process of restoration. On the other hand, we assume that the population of workers does not recover in the restoration stages.

10.4.2 Impact of the GEJE on Regional Potential and Labor Distribution

We investigate the impact of the GEJE in the following two steps.

1. Short-run analysis of regional potential: given the labor distribution, we investigate the change of indirect utility after the GEJE.
2. Long-run analysis of labor migration: we investigate the changes in the equilibrium labor distribution after the GEJE and in the restoration stages.

First, considering the short-run equilibrium, Table 10.4 shows the change in indirect utility level caused by the GEJE. The first point to note is the large decline of indirect utility in Miyagi (1.49%), Iwate (1.19%), and Fukushima (1.83%) in the MS sector. These prefectures were directly affected by the earthquake and tsunami. The second point is that there is a negative impact on inland prefectures that do not directly suffer from the tsunami. For instance, Yamagata faces a 0.06% drop in the indirect utility due to the influence of major damage in its adjoining prefectures, Miyagi and Fukushima. The whole country indirect utility declines by 0.07%. Similarly, looking at the agricultural sector, we observe large declines in Miyagi (0.90%), Iwate (0.57%), and Fukushima (1.56%).

Next, we consider the change of labor distribution in the following four cases: (i) the damage situation immediately after the GEJE, (ii) restoration Stage 1, (iii) restoration Stage 2, and (iv) restoration Stage 3. Table 10.5 shows the results of the percentage change in the long-run labor distribution after the GEJE. First, column (i) explains labor distribution changes from the initial distribution if there is no restoration. In this case, the number of residents declines by 10.10% in Fukushima, 9.45% in Miyagi, and 6.52% in Iwate. On the other hand, labor population in other prefectures increase, except in Yamagata, due to the acceptance of refugees. Next, we consider the restoration stages. In this case, it is reasonable to suppose refugees do not start migration immediately after the disaster and that they start to move to other prefectures when some of the restoration begins. Columns (ii)–(iv) show the change of labor distribution from the initial distribution during the restoration

Table 10.3 Percentage of damage from the Great East Japan earthquake by prefecture

Affected prefectures	Non-restoration		Restoration						Percentage of victims
			Stage 1		Stage 2		Stage 3		
	Agri. area	Building area	Agri. area	Building area	Agri. area	Building area	Agri. area	Building area	
02 Aomori	0.05	0.20	0.00	0.07	0.00	0.01	0.00	0.01	0.00
03 Iwate	0.12	6.96	0.08	3.95	0.06	3.35	0.06	3.35	0.44
04 Miyagi	3.20	8.28	1.70	2.77	0.77	1.73	0.77	1.73	0.46
07 Fukushima	8.92	9.76	8.83	8.73	8.57	8.66	4.88	4.97	0.09
08 Ibaraki	0.10	0.42	0.00	0.04	0.00	0.00	0.00	0.00	0.00
12 Chiba	0.19	0.30	0.00	0.01	0.00	0.00	0.00	0.00	0.00

Table 10.4 Percentage change of indirect utility following the Great East Japan earthquake by prefecture

Affected prefectures		MS workers	Agricultural workers
01	Hokkaido	0.00	−0.02
02	Aomori	−0.03	−0.06
03	Iwate	−1.19	−0.57
04	Miyagi	−1.49	−0.90
05	Akita	0.00	−0.05
06	Yamagata	−0.06	−0.06
07	Fukushima	−1.83	−1.56
08	Ibaraki	−0.08	−0.04
09	Tochigi	0.00	0.00
10	Gumma	0.00	0.00
11	Saitama	0.00	0.00
12	Chiba	−0.05	−0.04
13	Tokyo	0.00	0.00
14	Kanagawa	0.00	0.00
	Whole country	−0.07	−0.10

Table 10.5 Percentage change of labor population following the Great East Japan earthquake by prefecture

Affected prefectures		(i) Non-restoration	(ii) Restoration stage 1	(iii) Restoration stage 2	(iv) Restoration stage 3
01	Hokkaido	0.67	0.39	0.35	0.24
02	Aomori	0.57	0.40	0.40	0.26
03	Iwate	−6.52	−3.76	−3.20	−3.29
04	Miyagi	−9.45	−3.38	−2.21	−2.15
05	Akita	0.70	0.39	0.34	0.21
06	Yamagata	−1.35	−0.81	−0.71	−0.48
07	Fukushima	−10.10	−9.11	−9.03	−5.21
08	Ibaraki	0.32	0.42	0.39	0.26
09	Tochigi	0.82	0.47	0.40	0.28
10	Gunma	0.79	0.45	0.39	0.28
11	Saitama	0.85	0.47	0.43	0.31
12	Chiba	0.46	0.43	0.41	0.30
13	Tokyo	0.85	0.47	0.43	0.31
14	Kanagawa	0.82	0.46	0.42	0.30
	Other area	0.39	0.24	0.20	0.13

stages. In these cases, the population decline in affected prefectures decreases with the progress of restoration. For instance, the decline of labor population in Miyagi reduces from 9.45 to 2.15% in the final stage of restoration. However, we must draw attention that Fukushima maintains a large population decline during Stages 1 and 2 because of the "area under evacuation orders" (i.e., unrestored).

10.5 Impact Prediction for a Nankai Megathrust Earthquake

10.5.1 Damage Estimate

Following the methods described above, this section predicts the impact of a Nankai megathrust earthquake (NME) of magnitude 9 that is expected to hit Western Japan in the near future. Concerning the possibility of a NME, the Central Disaster Management Council in the Cabinet Office established the Committee for Policy Planning on Disaster Management, which compiled reports on the ideas and concrete measures for future disaster management. Under the committee, the working group estimates 11 cases of tsunami disaster according to the expected focal regions of the earthquake. In this chapter, we focus on Cases 1, 3, 4, and 5 that provide estimated data on human damage. Case 1 expects that large and super-large fault slips will occur in a region between Suruga Bay and offshore of Kii Peninsula. In this case, a great tsunami will mainly strike the Tokai area (i.e., Shizuoka, Aichi, and Mie). Similarly, Case 3 expects the occurrence of large and super-large fault slips between the Shikoku area and the offshore areas of Kii Peninsula, which will mainly affect the Tokai and Kansai areas (i.e., Mie and Wakayama). Case 4 expects large and super-large fault slips offshore of Shikoku, mainly affecting the Shikoku area (i.e., Wakayama, Kochi, and Miyazaki). Case 5 expects large and super-large fault slips between the offshore areas of Shikoku and Kyushu, mainly impacting the Kyushu area (i.e., Kochi, Miyazaki, and Kagoshima).[8]

Land damage from a tsunami disaster is calculated based on the estimated area and depth of the flood, which are reported in the "Estimated Damage of Nankai Megathrust Earthquake" (the second report). To specify the loss of agriculture, we

[8]To understand the impact of an NME, it is helpful to see some economic features of the expected stricken areas. The following table shows the regional share of the GDP (total and by industry) and the population of Japan in 2010. Tokai area specializes in the manufacturing sector and the Kansai area is the economic and political center of Western Japan.

		GDP Share (%)				Population
Area	Prefecture	Total	Agriculture	Manufacture	Services	(%)
Tokai	Shizuoka, Aichi, Mie	12.4	9.0	18.5	10.5	11.8
Kansai	Kyoto, Osaka, Hyogo, Nara, Wakayama	15.7	6.0	15.8	15.7	16.3
Shikoku	Tokushima, Kagawa, Ehime, Kochi	2.7	5.7	2.8	2.7	3.1
Kyushu	Fukuoka, Saga, Nagasaki, Kumamoto, Oita, Kagoshima	8.8	17.4	8.1	8.9	10.3

refer to areas flooded to a depth of 1 cm due to salt damage from the tsunami and ground subsidence. Damage in MS production and housing is calculated based on areas flooded to a depth more than 30 cm, which means water inundation above the floor level of buildings. The population damage refers to casualty estimates based on the earthquake occurring during the daytime and in summer.

The restoration stage is described as follows. First, referring to the restoration progress in agricultural fields following the GEJE, we assume that 50% of the flooded fields are recovered during restoration. With regard to buildings, we assume that the flooded area of a depth less than 1 m is recovered in the restoration stage. This is because (as the Cabinet Office of Japan reported in the "Damage Estimation of Buildings") wooden buildings are "partially or completely destroyed" by a flood depth over 1 m. The obtained data of the damages is shown in the tables listed in the Appendix.[9] The maximum loss of agricultural fields and building areas is approximately 5%, and the amount of damage is basically reduced by half during the restoration stage.

10.5.2 Change of Regional Potential and Labor Distribution

As in the previous section, we begin with an investigation of the impact of an NME on the short-run equilibrium. Table 10.6 shows the percentage changes in the indirect utility immediately after an NME (i.e., non-restoration). In Case 1, serious damage is observed in the Tokai area; the declines in indirect utility level of MS workers are 1.04% in Mie, 0.61% in Shizuoka, 0.58% in Aichi, and 0.81% in Miyazaki. The decline in the national average is 0.14%, which is the highest among our four cases. This is because the Tokai and Kansai areas that have large populations are hit by the tsunami. Case 3 shows that indirect utility in the Kansai area worsens, and Tokushima records the worst level in the four cases. The decline in the national average is −0.12%. In Case 4, the indirect utility levels in the Shikoku area become worse, but the decline in the national average is the least of the four cases. This is because the damage in Case 4 is the least in the four cases. Finally, Case 5 shows serious damage in Tokushima, Miyazaki, and Kochi, and the decline in the national average is about −0.12%.

Finally, we consider the change in labor distribution after an NME. In the same way as the GEJE, we suppose that affected workers start their migration to other regions after some time has passed and restoration has begun. Therefore, we consider the migration behavior under the condition of partial restoration, with results shown in Table 10.7 and Fig. 10.3. First, we observe that the scale of labor migration following an NME is similar to that in the restoration stages of GEJE (except Fukushima). Then, when examining the four cases, Case 1 shows a large decline of population in Tokai area: 3.18% (Mie), 2.09% (Shizuoka), and 0.62% (Aichi). In addition, the populations of Miyazaki (Kyushu area) also decreased by

[9]We do not consider the damage of the island area of the Tokyo prefecture, which is small and far from Honshu Island.

Table 10.6 Percentage change of indirect utility following a nankai megathrust earthquake by prefecture

		Case 1		Case 3		Case 4		Case 5	
Affected prefectures		MS workers	Agricultural workers	MS workers	Agricultural workers	MS workers	Agricultural workers	MS workers	Agricultural workers
08	Ibaraki	−0.02	−0.03	−0.02	−0.02	−0.02	0.00	−0.02	−0.03
12	Chiba	−0.19	−0.20	−0.10	−0.08	−0.04	−0.04	−0.07	−0.08
13	Tokyo	−0.02	−0.02	−0.02	−0.01	−0.01	0.01	−0.18	−0.09
14	Kanagawa	−0.20	−0.22	−0.12	−0.10	−0.03	−0.03	−0.06	−0.07
22	Shizuoka	−0.61	−0.54	−0.15	−0.13	−0.08	−0.06	−0.10	−0.10
23	Aichi	−0.58	−0.59	−0.40	−0.32	−0.08	−0.14	−0.20	−0.22
24	Mie	−1.04	−0.95	−0.95	−0.59	−0.24	−0.25	−0.40	−0.37
27	Osaka	−0.15	−0.17	−0.30	−0.41	−0.18	−0.35	−0.42	−0.45
28	Hyogo	−0.04	−0.03	−0.08	−0.05	−0.05	−0.01	−0.07	−0.06
30	Wakayama	−0.34	−0.30	−0.51	−0.48	−0.43	−0.40	−0.54	−0.49
33	Okayama	−0.03	−0.02	−0.04	−0.02	−0.05	0.00	−0.06	−0.04
34	Hiroshima	−0.04	−0.04	−0.04	−0.02	−0.03	0.01	−0.04	−0.03
35	Yamaguchi	−0.07	−0.06	−0.07	−0.05	−0.04	−0.03	−0.07	−0.07
36	Tokushima	−0.54	−0.52	−0.87	−0.90	−0.63	−0.71	−0.81	−0.75
37	Kagawa	−0.24	−0.26	−0.28	−0.29	−0.16	−0.23	−0.33	−0.34
38	Ehime	−0.28	−0.20	−0.17	−0.13	−0.13	−0.10	−0.20	−0.17
39	Kochi	−0.40	−0.35	−0.41	−0.43	−0.79	−0.68	−0.84	−0.73
40	Fukuoka	−0.02	−0.01	−0.02	0.00	−0.02	0.02	−0.02	−0.01
41	Saga	0.00	0.00	0.00	0.01	−0.01	0.03	0.00	0.01

42	Nagasaki	−0.13	−0.12	−0.13	−0.11	−0.10	−0.08	−0.16	−0.14
43	Kumamoto	−0.02	−0.01	−0.02	0.00	−0.02	0.02	−0.02	−0.01
44	Oita	−0.30	−0.28	−0.31	−0.27	−0.27	−0.25	−0.41	−0.36
45	Miyazaki	−0.81	−0.73	−0.80	−0.69	−0.54	−0.52	−0.89	−0.78
46	Kagoshima	−0.18	−0.16	−0.18	−0.16	−0.16	−0.14	−0.25	−0.22
47	Okinawa	−0.44	−0.44	−0.40	−0.40	−0.25	−0.25	−0.48	−0.48
Whole country		−0.14	−0.13	−0.12	−0.10	−0.06	−0.06	−0.12	−0.11

Table 10.7 Percentage change of labor distribution following a nankai megathrust earthquake by prefecture

Affected prefectures		Case 1	Case 3	Case 4	Case 5
08	Ibaraki	0.24	0.15	0.14	0.14
12	Chiba	−0.18	0.05	0.06	0.06
13	Tokyo	0.22	0.13	0.10	0.13
14	Kanagawa	−0.13	0.07	0.07	0.08
22	Shizuoka	−2.09	−0.20	−0.16	−0.14
23	Aichi	−0.62	−0.11	−0.05	−0.05
24	Mie	−3.18	−0.80	−0.59	−0.70
27	Osaka	0.30	−0.18	−0.17	−0.11
28	Hyogo	0.29	−0.09	−0.02	0.04
30	Wakayama	−0.97	−3.06	−1.47	−1.48
33	Okayama	0.25	0.08	0.04	0.07
34	Hiroshima	0.29	0.14	0.11	0.13
35	Yamaguchi	0.18	0.07	0.03	0.05
36	Tokushima	−0.51	−2.55	−1.83	−1.36
37	Kagawa	0.13	−0.09	−0.13	−0.10
38	Ehime	−0.27	−0.37	−0.48	−0.59
39	Kochi	−1.20	−1.94	−3.22	−3.02
40	Fukuoka	0.35	0.36	0.34	0.40
41	Saga	0.54	0.48	0.42	0.46
42	Nagasaki	0.19	0.09	0.08	0.10
43	Kumamoto	0.39	0.33	0.24	0.27
44	Oita	−0.49	−0.58	−0.75	−1.18
45	Miyazaki	−2.75	−2.46	−2.24	−3.26
46	Kagoshima	−0.07	−0.14	−0.36	−0.56
47	Okinawa	0.14	−0.10	0.03	0.10

2.75%. In Case 3, a large decline is observed in Wakayama (3.06%), Tokushima (2.55%), Kochi (1.94%), and Miyazaki (2.46%). Case 4 shows a result similar to Case 3, where the decline of Kochi's population worsens. Finally, Case 5 shows a large decline in the prefectures on the Pacific coast: Wakayama (1.48%), Tokushima (1.36%), Kochi (3.02%), and Miyazaki (3.26%). These results suggest that the prefectures on the Pacific coast should consider measures to tackle the outflow of population and immediate restoration, and the inland prefectures should prepare for the acceptance of refugees.

10.6 Conclusion

By constructing a multi-regional NEG model of Japan, this chapter investigated the impact of megathrust earthquakes, such as the Great East Japan Earthquake that occurred in 2011, and a Nankai megathrust earthquake, which is predicted to occur in the near future. First, we derived the parameters of elasticity of substitution and the distance resistance by calibration with Japanese prefectural data. Second, a simulation analysis showed that the predicted labor distribution approached the actual distribution, where the correlation coefficient was about 0.95. Third, we confirmed that, as transportation costs decrease, labor distribution changes from dispersion to agglomeration in the metropolitan areas and then to re-dispersion in rural areas. Finally, adapting the damage data of megathrust earthquakes to this model, we predicted the impact of such an earthquake on the regional potential (i.e., indirect utility levels) and labor distribution among the prefectures. The results showed that the indirect utility levels in the affected prefectures fell by 1%, and the long-run labor population declined by 3% in the worst case. A further objective of this study is to explain the relationship between the multiple equilibria and recovery of affected regions. As Davis and Weinstein (2008) studied, large impacts of disasters can change the regional multiple equilibria structure and render the affected regions impossible to restore to the original state. Thus, if disasters hit core regions and if the resulting labor migration changes core regions to peripheries, then restoration of the environment may be ineffective for the recovery of core regions. When we consider the effectiveness of a restoration policy, it is desirable to develop the argument based on the multiple equilibria in the NEG framework.

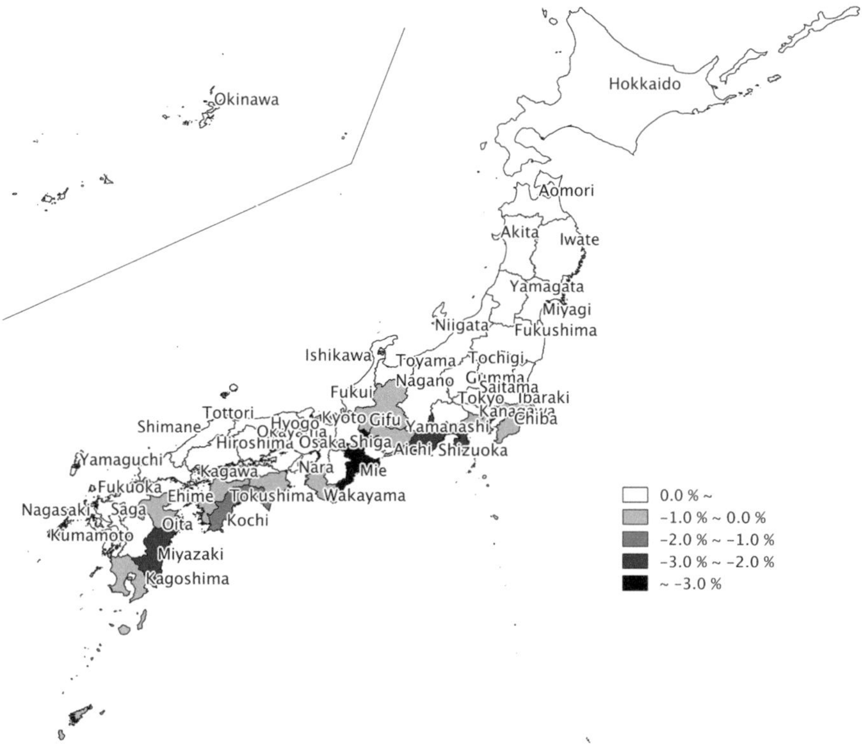

Fig. 10.3 Change of population following a Nankai Megathrust Earthquake. (**a**) Case 1, (**b**) Case 2, (**c**) Case 3, (**d**) Case 4

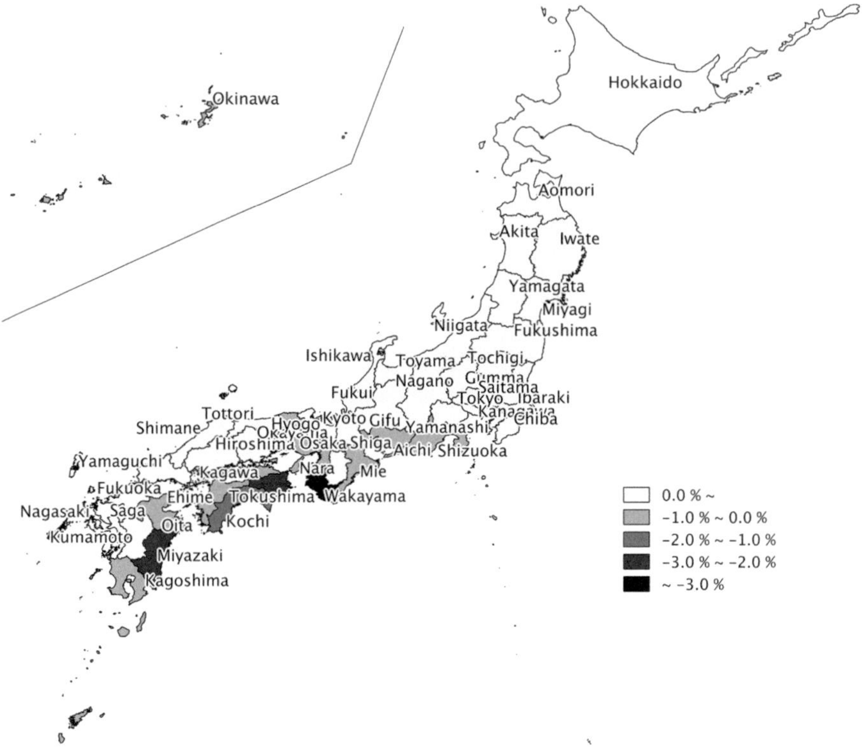

Fig. 10.3 (continued)

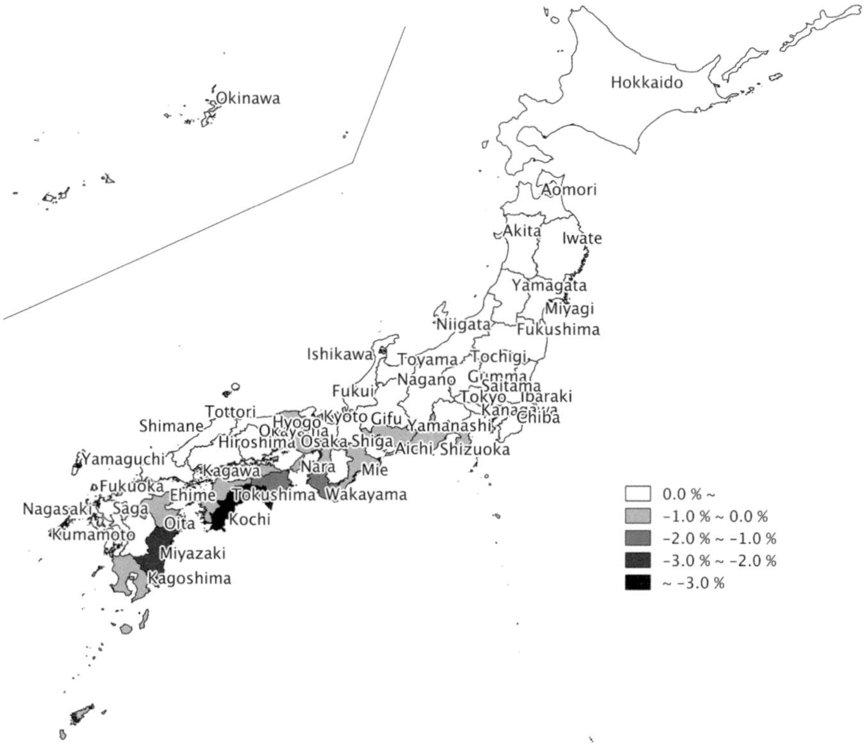

Fig. 10.3 (continued)

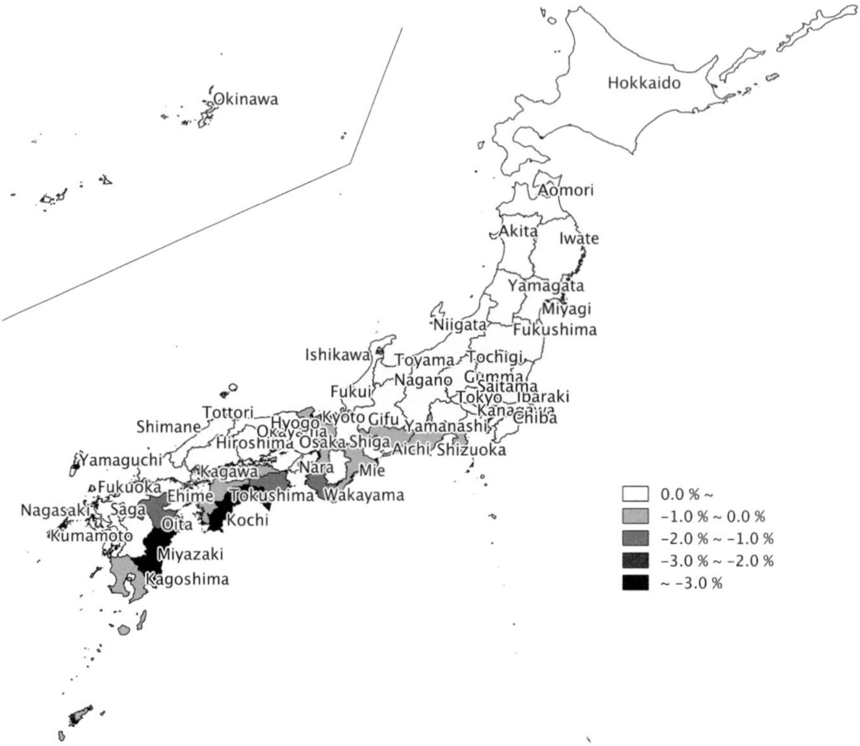

Fig. 10.3 (continued)

Appendix. Estimated Damage from a Nankai Megathrust Earthquake

Table 10.8 Percentage of damage in case 1

		Non-restoration		Partial restoration		
		Agricultural fields	Buildings	Agricultural fields	Buildings	
Flood depth		Over 1 cm	Over 30 cm	50% of area over 1 cm	Over 1 m	Percentage of victims
08	Ibaraki	0.16	0.12	0.08	0.07	0.00
12	Chiba	1.37	1.05	0.68	0.63	0.00
13	Tokyo	0.11	0.08	0.06	0.04	0.00
14	Kanagawa	1.55	1.11	0.77	0.61	0.00
22	Shizuoka	4.03	3.60	2.01	2.67	0.72
23	Aichi	4.08	3.08	2.04	1.32	0.01
24	Mie	6.11	5.62	3.05	4.34	0.11
27	Osaka	1.33	0.79	0.66	0.25	0.00
28	Hyogo	0.25	0.18	0.13	0.08	0.00
30	Wakayama	2.46	2.22	1.23	1.74	0.48
33	Okayama	0.17	0.11	0.08	0.04	0.00
34	Hiroshima	0.27	0.19	0.14	0.11	0.00
35	Yamaguchi	0.47	0.34	0.23	0.18	0.00
36	Tokushima	3.60	2.97	1.80	1.47	0.10
37	Kagawa	1.91	1.28	0.96	0.48	0.00
38	Ehime	1.01	1.51	0.51	0.62	0.01
39	Kochi	2.38	2.25	1.19	1.92	0.18
40	Fukuoka	0.10	0.08	0.05	0.05	0.00
41	Saga	0.81	0.68	0.41	0.40	0.00
42	Nagasaki	0.08	0.07	0.04	0.04	0.00
43	Kumamoto	1.81	1.63	0.91	1.19	0.00
44	Oita	4.68	4.35	2.34	3.53	0.04
45	Miyazaki	1.07	0.96	0.53	0.67	0.00
46	Kagoshima	2.88	2.34	1.44	1.10	0.00

Table 10.9 Percentage of damage in case 3

Flood depth		Non-restoration: Agricultural fields	Non-restoration: Buildings	Partial restoration: Agricultural fields	Partial restoration: Buildings	Percentage of victims
		Over 1 cm	Over 30 cm	50% of area over 1 cm	Over 1 m	
08	Ibaraki	0.11	0.09	0.06	0.01	0.00
12	Chiba	0.51	0.52	0.26	0.08	0.00
13	Tokyo	0.09	0.08	0.05	0.00	0.00
14	Kanagawa	0.65	0.63	0.33	0.05	0.01
22	Shizuoka	0.79	0.82	0.39	0.29	0.00
23	Aichi	2.02	2.11	1.01	0.20	0.04
24	Mie	2.45	5.17	1.22	0.80	0.00
27	Osaka	3.71	1.57	1.86	0.25	0.00
28	Hyogo	0.51	0.27	0.25	0.09	1.30
30	Wakayama	4.87	3.69	2.44	3.04	0.00
33	Okayama	0.30	0.12	0.15	0.07	0.00
34	Hiroshima	0.24	0.19	0.12	0.04	0.00
35	Yamaguchi	0.44	0.33	0.22	0.07	0.25
36	Tokushima	6.87	4.70	3.44	2.50	0.00
37	Kagawa	2.38	1.43	1.19	0.16	0.01
38	Ehime	0.94	0.87	0.47	0.39	0.33
39	Kochi	3.66	2.33	1.83	2.29	0.00
40	Fukuoka	0.09	0.08	0.05	0.03	0.00
41	Saga	0.84	0.68	0.42	0.05	0.00
42	Nagasaki	0.08	0.07	0.04	0.01	0.00
43	Kumamoto	1.83	1.64	0.92	0.58	0.03
44	Oita	4.29	4.30	2.15	2.28	0.00
45	Miyazaki	1.09	0.92	0.55	0.34	0.00
46	Kagoshima	2.74	2.13	1.37	0.06	1.97

Table 10.10 Percentage of damage in case 4

		Non-restoration		Partial restoration		
		Agricultural fields	Buildings	Agricultural fields	Buildings	
Flood depth		Over 1 cm	Over 30 cm	50% of area over 1 cm	Over 1 m	Percentage of victims
08	Ibaraki	0.11	0.08	0.05	0.00	0.00
12	Chiba	0.44	0.20	0.22	0.05	0.00
13	Tokyo	0.09	0.03	0.05	0.00	0.00
14	Kanagawa	0.46	0.15	0.23	0.03	0.01
22	Shizuoka	0.66	0.38	0.33	0.24	0.00
23	Aichi	1.70	0.39	0.85	0.15	0.03
24	Mie	2.27	1.26	1.13	0.63	0.00
27	Osaka	3.75	0.88	1.87	0.25	0.00
28	Hyogo	0.46	0.17	0.23	0.07	0.19
30	Wakayama	3.40	2.41	1.70	1.59	0.00
33	Okayama	0.33	0.17	0.16	0.09	0.00
34	Hiroshima	0.24	0.10	0.12	0.04	0.00
35	Yamaguchi	0.51	0.19	0.25	0.08	0.20
36	Tokushima	6.10	3.38	3.05	1.82	0.00
37	Kagawa	2.55	0.72	1.28	0.17	0.01
38	Ehime	0.97	0.64	0.49	0.45	0.47
39	Kochi	5.08	4.51	2.54	3.82	0.00
40	Fukuoka	0.10	0.05	0.05	0.03	0.00
41	Saga	0.93	0.46	0.47	0.08	0.00
42	Nagasaki	0.09	0.05	0.05	0.01	0.00
43	Kumamoto	2.04	1.39	1.02	0.71	0.02
44	Oita	4.00	2.83	2.00	2.02	0.00
45	Miyazaki	1.22	0.81	0.61	0.43	0.00
46	Kagoshima	2.96	1.15	1.48	0.11	0.93

Table 10.11 Percentage of damages in case 5

		Non-restoration		Partial restoration		
		Agricultural fields	Buildings	Agricultural fields	Buildings	
Flood depth		Over 1 cm	Over 30 cm	50% of area over 1 cm	Over 1 m	Percentage of Victims
08	Ibaraki	0.14	0.10	0.07	0.01	0.00
12	Chiba	0.47	0.38	0.24	0.07	0.00
13	Tokyo	0.08	0.99	0.04	0.00	0.00
14	Kanagawa	0.40	0.30	0.20	0.03	0.01
22	Shizuoka	0.65	0.55	0.32	0.23	0.00
23	Aichi	1.64	1.09	0.82	0.17	0.03
24	Mie	2.55	2.19	1.28	0.75	0.00
27	Osaka	3.34	2.24	1.67	0.23	0.00
28	Hyogo	0.42	0.30	0.21	0.06	0.16
30	Wakayama	3.36	3.04	1.68	1.66	0.00
33	Okayama	0.33	0.27	0.16	0.10	0.00
34	Hiroshima	0.24	0.17	0.12	0.04	0.00
35	Yamaguchi	0.54	0.37	0.27	0.09	0.18
36	Tokushima	5.17	4.46	2.59	1.36	0.00
37	Kagawa	2.52	1.76	1.26	0.18	0.01
38	Ehime	1.15	1.02	0.57	0.57	0.41
39	Kochi	4.94	4.75	2.47	3.66	0.00
40	Fukuoka	0.09	0.08	0.05	0.03	0.00
41	Saga	1.01	0.84	0.51	0.10	0.00
42	Nagasaki	0.10	0.08	0.05	0.02	0.00
43	Kumamoto	2.38	2.16	1.19	1.10	0.15
44	Oita	5.11	4.78	2.55	2.86	0.00
45	Miyazaki	1.44	1.31	0.72	0.61	0.00
46	Kagoshima	3.18	2.56	1.59	0.15	0.95

References

Bosker M, Brakman S, Garretsen H, Schramm M (2007) Looking for multiple equilibria when geography matters: German city growth and the WWII shock. J Urban Econ 61(1):152–169

Brakman S, Garretsen H, Schramm M (2006) Putting new economic geography to the test: freeness of trade and agglomeration in the EU regions. Reg Sci Urban Econ 36(5):613–635

Breinlich H (2006) The spatial income structure in the European Union – what role for Economic Geography? J Econ Geogr 6:593–617

Davis DR, Weinstein DE (2008) A search for multiple equilibria in urbanindustrial structure. J Reg Sci 48(1):29–65

Fujita M, Krugman P, Venables AJ (1999) The spatial economy: cities, regions and international trade. The MIT Press, Cambridge, MA

Hanson G (2005) Market potential, increasing returns and geographic concentration. J Int Econ 67 (1):1–24

Head K, Mayer T (2006) Regional wage and employment responses to market potential in the EU. Reg Sci Urban Econ 36(5):573–594

Helpman E (1998) The size of regions. In: Pines D, Sadka E, Zilcha I (eds) Topics in public economics: theoretical and applied analysis. Cambridge University Press, Cambridge

Ihara R (2011) Weber problem in the NEG: a case study of Asia. Ann Reg Sci 47(1):37–50

Ihara R (2014) Impact of the Great East Japan earthquake and the nankai megathrust earthquakes on labor distribution in Japan: an NEG approach. In: Tokunaga S, Okiyama M (eds) Model analyses on the recovery and regeneration from earthquake disasters. Bunsei-do (in Japanese)

Ihara R, Nakamura R and Morita M (2013) Impacts of the Great East Japan Earthquake on regional potentials and labor distribution in Miyagi Prefecture. RIETI Discussion paper series 13-J-053 (in Japanese)

Knaap T (2006) Trade, location, and wages in the United States. Reg Sci Urban Econ 36:595–612

Nakamura R, Ihara R and Morita M (2010) Demand and supply potentials with the interregional input-output structure: the NEG approach in Japan. RIETI Discussion paper series 10-J-031 (in Japanese)

Ottaviano GJP, Pinelli D (2006) Market potential and productivity: evidence from Finish regions. Reg Sci Urban Econ 36:636–657

Redding SJ, Venables AJ (2004) Economic geography and international inequality. J Int Econ 62 (1):53–82

Stelder D (2005) Where do cities form?: a geographical agglomeration model for Europe. J Reg Sci 45(4):657–679

Chapter 11
Evaluating Dynamic, Regional, and Economic Impacts of the Tokai Earthquake

Hiroyuki Shibusawa

Abstract Natural disasters result in economic losses. In this study, a dynamic spatial computable general equilibrium model is constructed to investigate the negative economic impacts of an earthquake. In our model, Japan is subdivided into 47 regions, which are all connected by transportation networks. The model is of a decentralized economy with utility-maximizing consumers and value-maximizing firms in a dynamic context. The model embodies both spatial and dynamic aspects, i.e., commodity flows among regions and the dynamics of regional investments. The model is calibrated for the 47 regions' economies using a multi-regional input–output table for Japan. We estimate the impacts of a hypothetical earthquake, which is expected to occur in the near future, on the regional economy (taking the Tokai region of Japan as a case study). The simulation results show both dynamic and spatial economic impacts before and after an earthquake. This study suggests that the economic impacts of such a disaster should be evaluated based on both ex-ante and ex-post criteria.

Keywords Disaster protection • Indirect economic impacts • Tokai earthquake • Dynamic spatial CGE modeling

11.1 Introduction

In this study, we set up a dynamic spatial computable general equilibrium (DSCGE) model to evaluate the economic impacts of an earthquake on the Tokai region of Japan. Our model is of a decentralized economy with utility-maximizing consumers and value-maximizing firms in a dynamic context. The model embodies both the spatial interactions among regions and the dynamics of regional investments.

A numerical simulation model is developed of an inter-regional, inter-sectoral economy that subdivides Japan into 47 regions, which are all connected by transportation networks such as roads, railways, seaways, and airways. The model is

H. Shibusawa (✉)
Toyohashi University of Technology, Toyohashi 441-8580, Japan
e-mail: hiro-shibu@tut.jp

S. Tokunaga, B.P. Resosudarmo (eds.), *Spatial Economic Modelling of Megathrust Earthquake in Japan*, New Frontiers in Regional Science: Asian Perspectives 11,
https://doi.org/10.1007/978-981-10-6493-7_11

calibrated for the Japanese regional economy. As a case study, the dynamic impacts of an earthquake in the Tokai region are analyzed through numerical simulation. In our hypothetical scenario, the primary physical damage caused by an earthquake is simply the reduction in industrial capital stock. The dynamic, optimizing production sector would invest before and after an earthquake to protect against economic losses or to repair crucial damage. Regional economic growth would be more sensitive to a disaster.

The Tokai region is located at the center of Japan and faces the northern end of the Philippine Sea plate. It is a potential location for a large earthquake. According to past studies, an earthquake is expected to occur in the Tokai region (Mogi 1970; Sato 1970; Ando 1975). In this region, earthquakes with a magnitude of 8 recur every 100–150 years. The most recent earthquake occurred in the Tonkai and Nankai regions in 1944. The occurrence of great earthquakes is periodic (Mogi 1985).

The computable general equilibrium (CGE) analysis is a major tool in economics, regional science, and engineering. It is also widely recognized as a policy evaluation method (Shoven and Whalley 1992; Kehoe et al. 2005; Borglin 2004, Dixit and Jorgenson 2013). There is a vast literature reporting applications of static CGE models, but few studies have employed dynamic and spatial frameworks (Oosterhaven 2003; Donaghy 2009). There are two streams of dynamic CGE modes: dynamic-recursive and forward-looking (Dixon and Jorgenson 2013). Recent trials of dynamic spatial or multi-regional CGE modeling have been undertaken by Ciesecke (2002, 2003), McGregor et al. (1995), and McKibbin and Wilcoxen (1992). Most dynamic spatial models are based on recursive dynamics. These models assume ad-hoc saving and investment functions. Due to computational difficulties, spatial CGE modeling with forward-looking dynamics remains a research frontier (Shibusawa et al. 2009; Shibusawa and Miyata 2011; Bröckera and Korzhenevych 2013).

The economic impacts of disasters have been analyzed using computational modeling, such as the input–output model, mathematical programming, CGE, and econometric models (Ellson et al. 1984; Rose et al. 1997; Sohn et al. 2003; Lee and Jang 2003; Okuyama and Chang 2004; Rose and Liao 2005). These studies mostly focused on assessing the economic impacts of damage to the public infrastructure such as transportation links, electric utility lifelines, water facilities, and telecommunications networks.

The CGE model gives us an excellent framework for analyzing disaster impacts and policy responses both across and between industries, households, and government. To assess the distributional impacts of a disaster in multi-regional settings, the spatial CGE model approach, which disaggregates the world or a country into a number of regions or counties, has also been developed. The models were characterized by the optimizing behavior of individual consumers and firms, subject to market balances and resource constraints, in a static framework. Spatial interactions between regions are internalized by transportation networks and trade costs. The spatial CGE (or CGE) model is a powerful tool, but it has some disadvantages regarding disaster analysis. The major reason for these disadvantages is that the economy is always in equilibrium.

In Japan, several disasters have been assessed, including the Tokai, Tonankai, and Niigata-Chuetsu earthquakes. Empirical studies have adequately estimated direct and indirect economic losses based on actual data and input–output models (Toyoda and Kochi 1997; Taniguchi 2007). On the other hand, spatial CGE models have been used to capture spatial and distribution impacts (Koike and Ueda 2005; Tsuchiya et al. 2003; Tatano and Tsuchiya 2008; Xie et al. 2014). In such practical studies, to incorporate a disequilibrium phenomenon after a disaster in the SCGE models, short-run and long-run equilibriums were defined in non-perfect competitive regional market conditions, and the model was solved in a static environment. Those solutions were compared based on a hypothetical scenario. Distributional impacts across economic institutions and between regions, caused by a direct economic loss in specific regions after an earthquake, were adequately estimated.

In a disaster analysis, another important issue has largely been neglected in the SCGE literature, namely, the indirect effects of disaster protection before an earthquake. In general, a natural disaster inevitably involves both indirect and distributional effects before and after its occurrence. Natural disasters have been recorded throughout history, and the periodic characteristics of disasters have been investigated and widely recognized. Economic assessments before and after a possible disaster are expected to be resolved simultaneously in an analytical framework.

In this study, a dynamic spatial CGE model is developed based on dynamic macroeconomic theory with a multi-regional and multi-sectoral specification (Abel and Blanchard 1983). Regional investment is endogenously determined by the behavior of value-maximizing firms, which involves capital adjustment functions. The dynamic impact of an earthquake is evaluated using our dynamic spatial CGE model, which is based on our previous work (Shibusawa et al. 2009; Shibusawa and Miyata 2011). Our model is calibrated using a multi-regional input–output table for Japan. A steady-state solution is derived as a base case. By numerical simulation, we assess the economic impacts of an earthquake in the Tokai region using our hypothetical scenarios. Two occurrences, one unpredicted and one predicted, are assumed, and two solutions, which are in a non-steady-state, are compared with the base case. In addition, we consider the impacts of technical progress after an earthquake.

Our contributions are as follows. First, because industrial investment is determined by a firm's dynamic optimizing behavior, we can assess dynamic changes in investment before and after an earthquake. We derive the optimal investment before an earthquake from our model. Although we do not fully resolve the disequilibrium phenomena related to a disaster, we derive industrial investment responses both before and after an earthquake as a non-steady-state solution. A dynamic analytical framework highlights the estimation of indirect and distributional economic impacts before and after an earthquake. Second, as our model involves transportation networks, we can also evaluate dynamic distributional impacts through intra- and inter-regional trade before and after an earthquake. Lastly, we describe the methodological advantage of DSCGE modeling for application to periodically predicted disasters. This may contribute to an understanding of artificial and non-natural disasters caused by human activity, such as global warming and other environmental issues.

This study is organized as follows. We first describe basic assumptions in Sect. 11.2. In Sect. 11.3, we outline the dynamic spatial CGE model and describe the optimization behaviors of firms and households. To obtain market prices, we also define equilibrium conditions. Section 11.4 provides simulation results. Two cases are compared with a base case, which is a steady-state solution for a 51-year period. Section 11.5 summarizes the study and offers some concluding remarks.

11.2 Basic Assumptions

We have subdivided the world into economic regions and industries, including general and transportation industries and households. The economy is endowed with labor and capital. Labor is mobile across industries but not regions, and capital is immobile across industries and regions. Goods and factor prices are determined in perfectly competitive regional markets. The commodity trade among regions in a country generates demand for transportation services, and unit transportation costs are endogenous. Commodities are perfect substitutes, i.e., trade among regions is calculated by trade coefficients. The movement of commodities among regions is enabled by road, rail, sea, and air transportation networks. The modal share is also given. The model is solved for a rational expectations equilibrium, assuming perfect competition and foresight. However, we assume that firms place priority on the investment–savings balance. Then levels of investment are determined by the firm's optimizing behavior.

The model is finitely set up in discrete time. $\mathrm{T} = \{1, 2, \cdots, \overline{T}\}$ denotes a planning period index, and $\overline{T}$ is the final planning period. The world is divided into a home country and a foreign country, with the home country subdivided by region. $\mathrm{R} = \{1, 2, \cdots, \overline{R}\}$ denotes a regional index in the home country. There are three kinds of industries, i.e., general, transportation, and distribution. The general industry involves domestic and foreign trade between regions. $\mathrm{I} = \{1, 2, \cdots, \overline{I}\}$ denotes a sector index for the general industry. $\mathrm{M} = \{I+1, I+2, \cdots, \overline{M}\}$ is a sector index for the transportation industry. All regions interact with each other via transportation networks, which are defined by nodes and links. For each pair of two regions, only one transport path connects them. The transport link distance is exogenously given.

11.3 Method

The model is based on dynamic macroeconomic theory with multi-region and multi-sector specifications. Each region has production and household sectors. Commodity trade flows are determined by trade and modal coefficients. We consider the problems related to optimization behavior in the production and household sectors in this economy.

11.3.1 Production Sectors

Each sector of the general and transportation industries maximizes its present cash flow value in each period, in addition to maximizing the asset value of its industrial capital in the final period. These sectors that operate with constant returns to scale technology, choose optimal investment and labor employment strategies. The behavior of the production sector $j \in \mathrm{I} \cup \mathrm{M}$ in region $r \in \mathrm{R}$ is given as

$$\max_{\{K_j^r(t), L_j^r(t), X_j^r(t), Z_j^r(t)\}} \sum_{t \in \mathrm{T}} \rho_j(t) NC_j^r(t) + \rho_j(\overline{T}+1) \Phi_j^r \left(K_j^r(\overline{T}+1) \right),$$

subject to

$$K_j^r(t+1) = (1-\delta_j) K_j^r(t) + \Delta K_j^r \left(Z_j^r(t) \right),$$

where $NC_j^r(t)$ is the cash flow of the production sector j in region r at period t and is defined as

$$NC_j^r(t) = p_j^{Or}(t) Y_j^r \left(K_j^r(t), L_j^r(t), X_j^r(t) \right) - w^r(t) L_j^r(t) - \sum_{i \in \mathrm{I} \cup \mathrm{M}} p_i^{Dr}(t) X_{ij}^r(t)$$
$$- \sum_{i \in \mathrm{I} \cup \mathrm{M}} p_i^{Dr}(t) G_i^r \left(Z_{ij}^r(t) \right).$$

$\rho_i(t) = \frac{1}{(1+\rho_i)^{t-1}}$ represents the discount factor for period $\boldsymbol{t}$, and $\boldsymbol{\rho_i}$ is the positive discount rate. $Y_j^r \left(K_j^r(t), L_j^r(t), \mathbf{X}_j^r(t) \right)$ is a production function of capital $K_j^r(t)$, labor $L_j^r(t)$, and a vector of intermediate input $\mathbf{X}_j^r(t) = \{X_{1j}^r(t), X_{2j}^r(t), \cdots, X_{Ij}^r(t)\}$. The value-added production function for labor and capital has a Cobb–Douglas form while the intensities of intermediate goods are fixed. The asset value for the final period $\Phi_j^r \left(K_j^r(\overline{T}+1) \right)$ is a linear function of capital stock in the final period. The capital stock $K_j^r(t)$, accumulated by an investment function $\Delta K_j^r\left(Z_j^r(t)\right)$, with constant returns to scale is a function of a vector of intermediate inputs for the investment $\mathbf{Z}_j^r(t) = \{Z_{1j}^r(t), Z_{2j}^r(t), \cdots, Z_{Ij}^r(t)\}$, assuming a Leontief-type technology. $\boldsymbol{\delta_j}$ is the depreciation rate. It is also assumed that the cost function of intermediate goods for investment $G_i^r\left(Z_{ij}^r(t)\right)$ has increasing returns to scale. It can be interpreted that this function reflects both the costs of intermediate goods and the costs of adjusting capital inputs.

In those sectors, there are two kinds of prices in each region, namely, the producer's prices $p_j^{Or}(t)$ and $p_i^{Dr}(t)$ in region $\boldsymbol{r}$. If a commodity $j \in \mathbf{I}$ is tradable between regions **o** and ***d***, then the producer's price in region ***o*** is represented by $p_j^{Oo}(t)$, and the purchaser's price is represented by $p_j^{Dd}(t)$. For transportation services, $j \in \mathbf{M}$, $p_j^{Or}(t)$, the unit price of the transportation services in region r. $\boldsymbol{w^r(t)}$ is the wage rate.

After paying wages to households, a sector has to decide how to distribute profit and financial investments. In this model, net investment is financed by new bonds. Let $B_j^r(t)$ be the number of bonds in period t and $r_j^r(t)$ be the interest rate. We simply assume that bonds are traded in each region. The initial number of bonds is normalized by $B_j^r(1) = K_j^r(1)$. In this case, the profit dividend is calculated as

$$\pi_j^r(t) = p_j^{Or}(t) Y_j^r\left(K_j^r(t), L_j^r(t), \mathbf{X}_j^r(t)\right) - r_j^r(t) B_j^r(t) - w^r(t) L_j^r(t) - \sum_{i \in \mathrm{I} \cup \mathrm{M}} p_i^{Dr}(t) X_{ij}^r(t)$$
$$- p_{Bj}^r(t) \delta_j K_j^r(t).$$

If the net investment is financed by issuing new bonds, it holds that

$$p_{Bj}^r(t) \Delta B_j^r(t) = \sum_{i \in \mathrm{I} \cup \mathrm{M}} p_i^{Dr}(t) G_i^r\left(Z_{ij}^r(t)\right) - p_{Bj}^r(t) \delta_j K_j^r(t),$$

where$\Delta B_j^r(t)$ is the number of new bonds issued by sector j in region r for period t. $p_{Bj}^r(t)$ is the price of the new bond. Therefore, the outstanding bond is given by

$$B_j^r(t+1) = B_j^r(t) + \Delta B_j^r(t),$$

with $B_j^r(1) = \overline{B}_j^r(1)$. $\overline{B}_j^r(1)$ is the initial bond at period 1. It is also assumed that the price of the new bond is set by $p_{Bj}^r(t) = q_j^r(t)$. $q_j^r(t)$ is the costate variable of the current-value Hamiltonian function $H_j^r(t)$, which is defined by

$$H_j^r(t) = NC_j^r(t) + q_j^r(t)\left(\Delta K_j^r\left(Z_j^r(t)\right) - \delta_j K_j^r(t)\right).$$

In this model, we assume that tradable goods are perfect substitutes. Profit from goods j from distribution activities in transportation services $\pi^r_{Disj}(t)$ is given by the following equation:

$$\pi^r_{Disj}(t) = p^{Dr}_j(\mathrm{t}) \sum_{o\in \mathrm{R}} \sum_{m\in \mathrm{M}} \mu^{or}_{jm}(t) F^{or}_j(t) - \sum_{o\in \mathrm{R}} \sum_{m\in \mathrm{M}} \left(p^{Oo}_j(t) + p^{Tor}_{jm}(t)\right) \mu^{or}_{jm}(t) F^{or}_j(t),$$

where

$$p^{Tor}_{jm}(t) = \kappa_{jm}(t) p^{Or}_m(t) D^{or}_m(t).$$

$\mu^{or}_{jm}(t)$ is the given modal share of transport mode m from region o to region r for goods j, which holds for all transport modes,

$$\sum_{m\in \mathrm{M}} \mu^{or}_{jm}(t) = 1 \quad (o, r\in \mathrm{R}).$$

We define the fixed trade coefficients $\tau^{or}_j(t)$ from region o to region r for goods j, which also holds for all the origins,

$$\sum_{o\in \mathrm{R}} \tau^{or}_j(t) = 1 \quad (r\in \mathrm{R}).$$

Subsequently, the commodity flow $F^{or}_j(t)$ from region o to region r for goods j is calculated as

$$F^{or}_j(t) = \tau^{or}_j(t) \Big(\sum_{i\in \mathrm{I}\cup\mathrm{M}} X^r_{ji}(t) + C^r_j(t) N^r_j(t) + \sum_{i\in \mathrm{I}\cup\mathrm{M}} G_{ji}\left(Z^r_{ji}(t)\right)\Big)$$

Each term in the parentheses of the right-hand side of the above equation represents intermediate demand, household consumption, and investment goods, respectively. $p^{Tor}_{jm}(t)$ is the transportation cost of mode m from region o to region r for goods j. $D^{or}_m(t)$ is the distance between origin and destination, along with a given path for mode m. $\kappa_{jm}(t)$ is mode m's given unit transportation service for goods j. From the zero profit condition, since it holds that $\sum_{o\in \mathrm{R}} \sum_{m\in \mathrm{M}} \mu^{or}_{\mathrm{jm}}(t) \tau^{or}_j(t) = 1 (r\in \mathrm{R})$, the purchaser's price is given by

$$p^{Dr}_j(t) = \sum_{\mathrm{o}\in \mathrm{R}} \sum_{m\in \mathrm{M}} \left(p^{Oo}_j(t) + p^{Tor}_{jm}(t)\right) \mu^{or}_{jm}(t) \tau^{or}_j(t).$$

The purchaser's price is the average price of goods from all regions and modes, derived from the sum of producer prices and transportation costs, weighted by transport mode and inter-regional trade coefficients.

11.3.2 Household Sector

A representative household maximizes its utility subject to income constraints. The full income consists of wages and interest on bond holdings. The behavior of a household in region $r \in \mathbf{R}$ for period t is given as

$$\max_{\{\mathbf{C}^r(t)\}} \sum_{t \in \mathrm{T}} \rho(t) U^r\left(\mathbf{C}^r(t)\right)$$

subject to

$$w^r(t) + \sum_{i \in \mathrm{I} \cup \mathrm{M}} r_i^r(t) A_i^r(t) + d^r(t) + FA^r(t) - \sum_{i \in \mathrm{I} \cup \mathrm{M}} p_i^{Dr}(t) C_i^r(t) - \sum_{i \in \mathrm{I} \cup \mathrm{M}} p_{Bi}^r(t) \Delta A_i^r(t)$$
$$\geq 0.$$

$U^r(\mathbf{C}^r(t))$ is a Cobb–Douglas utility function for period t, and it is a function of consumption $\mathbf{C}^r(t) = \{C_1^r(t), C_2^r(t), \cdots, C_I^r(t)\}$. $A_i^r(t)$ is the number of bond holdings per household. $\Delta A_i^r(t)$ represents new bonds issued for industrial investments. The household can receive interest income but must pay to obtain a new bond. $FA^r(t)$ is the income transfer that provides a balance against a surplus or deficit in foreign and regional trade. $d^r(t)$ is the profit dividend, given by

$$d^r(t) = \sum_{i \in \mathrm{I} \cup \mathrm{M}} \frac{\pi_i^r(t)}{N^r(t)},$$

as the utility function is not identical among regions.

In this model, we assume that the level of investment is determined by a firm's optimization behavior. Firms place priority on the investment–savings balance. Therefore, the level of household savings is adjusted to coincide with the level of investment. In this case, new bonds per household is given by

$$\Delta A_i^r(t) = \frac{\Delta B_i^r(t)}{N^r(t)} \quad (i \in \mathrm{I} \cup \mathrm{M})$$

The dynamic equation of bond holdings is also given by

$$A_i^r(t+1) N^r(t+1) = \left(\Delta A_i^r(t) + A_i^r(t)\right) N^r(t) \quad (i \in \mathrm{I} \cup \mathrm{M}).$$

11.3.3 Equilibrium Condition

To obtain an equilibrium solution, the following market-clearing conditions should be satisfied in each region $r \in \mathbf{R}$.

1. Goods and Services Markets

   General Goods

$$Y_j^r\left(K_j^r(t), L_j^r(t), \mathbf{X}_j^r(t)\right) = \sum_{i \in \mathrm{I} \cup \mathrm{M}} X_{ji}^r(t) + C_j^r(t)N^r(t) + \sum_{i \in \mathrm{I} \cup \mathrm{M}} G_{ji}\left(Z_{ji}^r(t)\right) + \sum_{d \in \mathrm{R}} F_j^{rd}(t)$$
$$- \sum_{o \in \mathrm{R}} F_j^{or}(t) + E_j^r(t) - M_j^r(t) \quad (j \in \mathrm{I})$$

$E_j^r(t)$ is a given export from region r, and $M_j^r(t)$ is a given import to region r. Aggregate demand is the total of intermediate demand $X_{ji}^r(t)$, consumption $C_j^r(r)N(t)$, outflow to region d $F_j^{rd}(t)$, inflow from region o $-F_j^{or}(t)$, exports $E_j^r(t)$; and imports $-M_j^r(t)$.

Transportation Services

$$Y_m^r\left(K_m^r(t), L_m^r(t), \mathbf{X}_m^r(t)\right) = \sum_{i \in \mathrm{I}} \sum_{o \in \mathrm{R}} \kappa_{im}^r(t)\mu_{im}^{or}(t)D_m^{or}(t)F_{im}^{or}(t) \quad (m \in \mathrm{M})$$

$\kappa_{im}^r(t)\mu_{im}^{or}(t)D_m^{or}(t)F_{im}^{or}(t)$ represent transportation services, which are derived from the commodity flow of goods i $\mathrm{F}_{\mathrm{im}}^{\mathrm{or}}(\mathrm{t})$ from region o to region r, by mode m.

2. Labor

$$N^r(t) = \sum_{i \in \mathrm{I} \cup \mathrm{M}} L_i^r(t)$$

$N^r(t)$ is the total labor force (population) in each region, and it is exogenously given.

3. Capital

$$\Delta A_i^r(t)N^r(t) = \Delta B_i^r(t) \quad (i \in \mathrm{I} \cup \mathrm{M})$$
$$A_i^r(t)N^r(t) = B_i^r(t) = K_i^r(t)$$

with $\overline{A}_i^r(1)N^r(1) = \overline{B}_i^r(1) = \overline{K}_i^r(1) \quad (i \in \mathrm{I} \cup \mathrm{M})$

$\overline{A}_i^r(1)$ is the initial number of bond holdings for sector i of a household. $\overline{B}_i^r(1)$ is the initial bond, and $\overline{K}_i^r(1)$ is the initial capital stock for Period 1.

11.4 Numerical Applications

In the simulation model, the world is subdivided into 47 regions, which cover all of Japan's prefectures. The economy is divided into seven sectors. General industry is divided into three sectors ($I = \{1, 2, 3\}$), i.e., agriculture, manufacturing, and services. There are four types of transportation networks: road, railway, sea, and air. The transportation industry then is divided again into four sectors ($M = \{4, 5, 6, 7\}$). The network structure, which is defined by the distance between an origin and a destination, is given for each period. The simulation period is set at $\overline{T} = 51$. Population growth and technological progress are also fixed over time. Utility, production, and investment functions are specified for the simulation analysis, as shown in the Appendix. The simulation model is calibrated using the multi-regional input–output table for Japan (Hitomi and Bunditsakulchai 2008).

11.4.1 Scenarios

Three cases are examined to evaluate the dynamic impacts of an earthquake in the Tokai region. The primary physical damage is simply assumed in terms of a reduction in industrial capital stock in the Tokai region. The impact of technological progress is also evaluated. In this simulation, the Tokai region (our target area) covers 10 prefectures, Chiba, Tokyo, Kanagawa, Yamanashi, Nagano, Gifu, Shizuoka, Aichi, Mie, and Wakayama.

(a) Base Case

The base case represents business as usual, where there is no earthquake. In this case, a steady-state solution is derived, where it holds that $K_j^r(t+1) = K_j^r(t)$ and $q_j^r(t+1) = q_j^r(t)$. The rates of population growth and technical progress growth are both 0%.

(b) Case 1

We assume that an unpredicted earthquake occurs suddenly and hits the target area. In this simulation, the earthquake occurs in the 25th period. The level of physical damage is also assumed to be a reduction in capital stock. The estimated damage is shown in Table 11.1 (Central Disaster Prevention Council 2003; Taniguchi 2007). In this situation, no industry is able to make a new investment before the earthquake to protect itself.

(c) Case 2

In this case, it is assumed that an earthquake is accurately predicted. Here, the earthquake also occurs in the 25th period, and capital stocks are reduced in this period, as shown in Table 11.1. The amount of the reduction is the same as in Case

Table 11.1 Rate of loss in capital stocks in the Tokai Region

Prefecture number	Rate of loss
12 Chiba	0.016%
13 Tokyo	0.001%
14 Kanagawa	0.058%
19 Yamanashi	0.247%
20 Nagano	0.147%
21 Gifu	0.011%
22 Shizuoka	10.000%
23 Aichi	1.421%
24 Mie	0.237%
30 Wakayama	0.016%

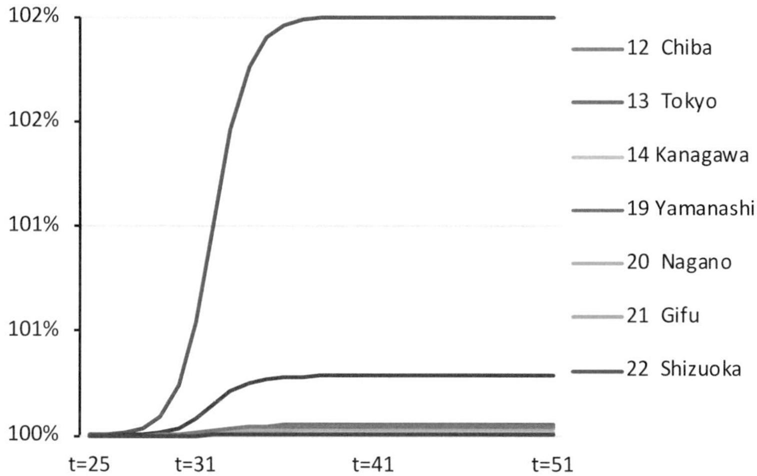

Fig. 11.1 Technological progress

1. In Case 2, each industry can make an additional investment to protect itself before the earthquake.

(d) Case1 and Case 2 with technological progress

After an earthquake, the industrial capital stock is rapidly accumulated by new investments. In general, the installation of new facilities contributes to productivity. In this case, technical progress is assumed. Figure 11.1 depicts a hypothetical path of technological progress after the earthquake, represented by a logistic curve. In Shizuoka prefecture, the productivity of all sectors gradually increases to 2%. The parameters of asset value for the final period also are adjusted. Two solutions, with either predicted or unpredicted technological progress, are compared.

11.4.2 Simulation Results

(a) Impacts on Capital Stock

The dynamic solutions for capital stocks and their value (the co-state variable) in the manufacturing sector are examined. We focus on two prefectures, Shizuoka and Saitama, as the most affected regions. Shizuoka is directly affected by the earthquake and suffers great losses. Saitama, which is adjacent to the Tokai region, in our scenario is indirectly affected by the earthquake through its transportation network. Figures 11.2 and 11.3 show the earthquake's dynamic impact on capital stock and its values. In our dynamic simulation, capital stock is accumulated by forward calculation while the co-state variable is solved by backward calculation. The change is defined as $\Delta x = \frac{x_{case} - x_{base}}{x_{base}} \times 100\%$.

In the Shizuoka prefecture, in Case 1, capital stock suddenly decreases during the 25th period but is gradually restored after the earthquake. Capital stock values are unchanged before the earthquake, but they increase unexpectedly during the 25th period due to the capital stock damage in Case 1. In Case 2, the capital damage is more rapidly repaired after the earthquake by an increase in investment that protects against the earthquake. In Case 2, capital stock values increase more before the earthquake compared with the base case and Case 1. This implies that the industrial sector would estimate the value of capital stock before the earthquake since it can accurately predict the occurrence of an earthquake.

The Saitama prefecture is severely affected by the earthquake and is indirectly affected by the Tokai region because of trade between prefectures. In Case 1, the Saitama prefecture experiences an increase in investment since output in the Tokai region decreases after the earthquake. In contrast, in Case 2, due to an increase in investment and output in the Tokai region before the earthquake, capital stock and output in the Saitama prefecture decrease, more than in the base case.

(b) Impacts on Gross Regional Product (GRP)

Figure 11.4 shows the earthquake's impact on GRP (Gross Regional Product) in the Tokai region. The change in the GRP is defined as $\Delta x = \frac{x_{case} - x_{base}}{x_{base}} \times 100\%$. In both cases, the earthquake occurs in the Tokai region during the 25th period. This figure shows the dynamic impacts of the earthquake in the Tokai region, which includes Chiba, Tokyo, Kanagawa, Yamanashi, Nagano, Gifu, Shizuoka, Aichi, Mie, and Wakayama prefectures (our target areas). In addition, the bold line depicts the change in total Japanese GRP, which estimates the economic impact of the earthquake. It may be useful to compare regional and national impacts. The industrial capital stocks in these prefectures are directly reduced by the earthquake. In both cases, the Tokai region suffers damage from the earthquake, and the amount of damage in the Shizuoka and Aichi prefectures appears to be greater than that of the whole of Japan. In particular, the Shizuoka prefecture sustains crucial damage.

In Case 2, production sectors can make additional investments before the earthquake. It is assumed that the disaster hits in the 11th period and that the

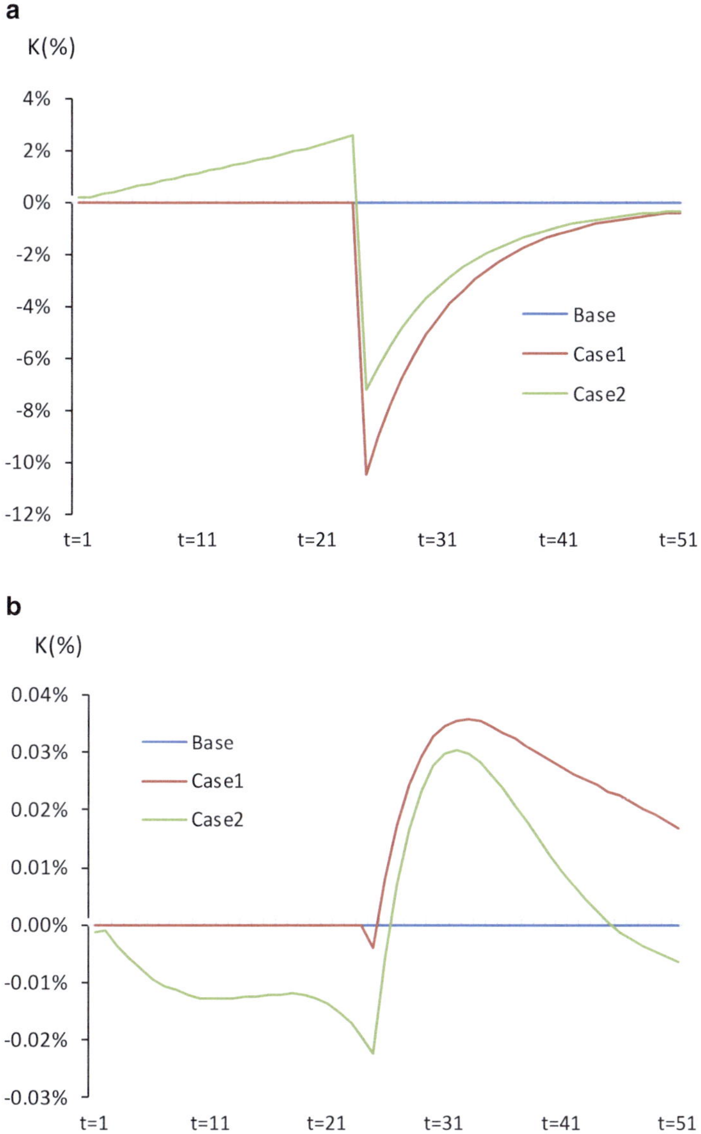

Fig. 11.2 Capital stock

earthquake is accurately predicted. Before the earthquake, the Tokai region experiences a positive impact, due to increased investment designed to protect the region from earthquake damage. In particular, GRP in the Shizuoka prefecture is largely influenced by additional investment before the disaster, and the change in GRP

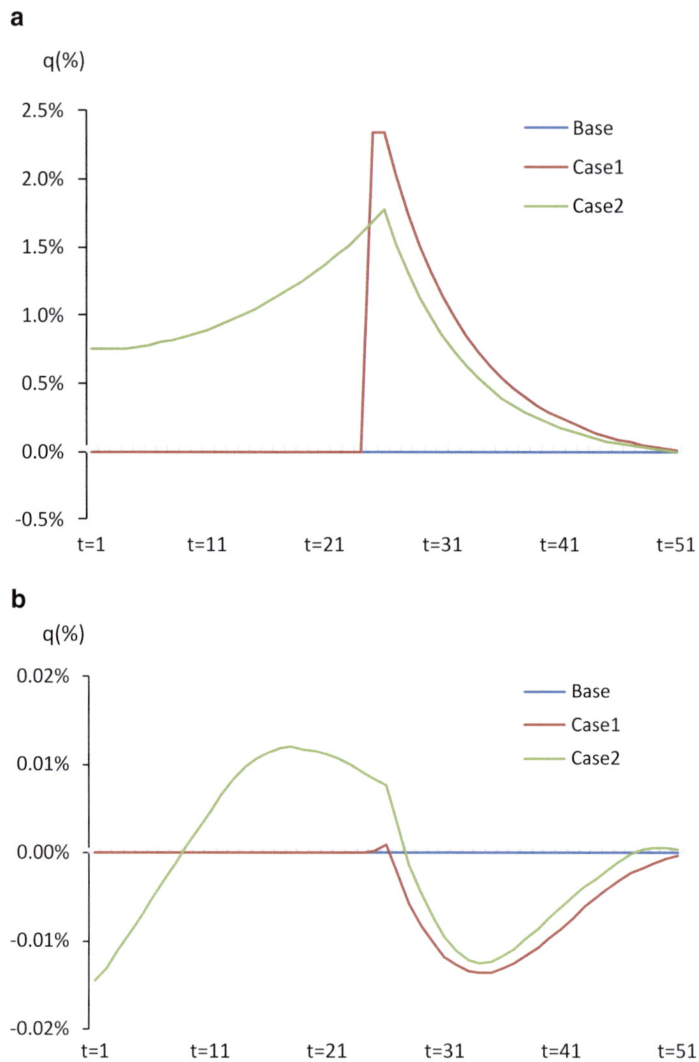

Fig. 11.3 Value of capital stock

would be greater than that of Japan. After the disaster, the Tokai region in Case 2 is more rapidly restored by investments from other prefectures, compared with Case 1.

Figure 11.5 shows the results of these changes in GRP in every region except Tokai. In Case 1, most prefectures far from the Tokai region experience positive changes in their GRPs after the earthquake, although Japan's total GRP is negative (depicted by the bold line). In Case 2, most prefectures, except Tokai, are negatively affected before the earthquake because of increased investments designed to

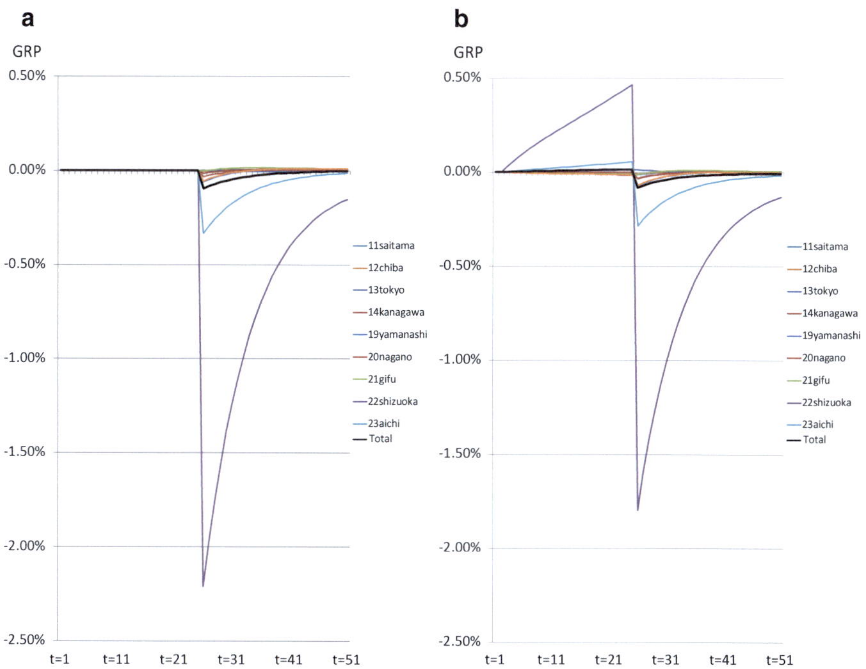

Fig. 11.4 GRP in the Tokai region

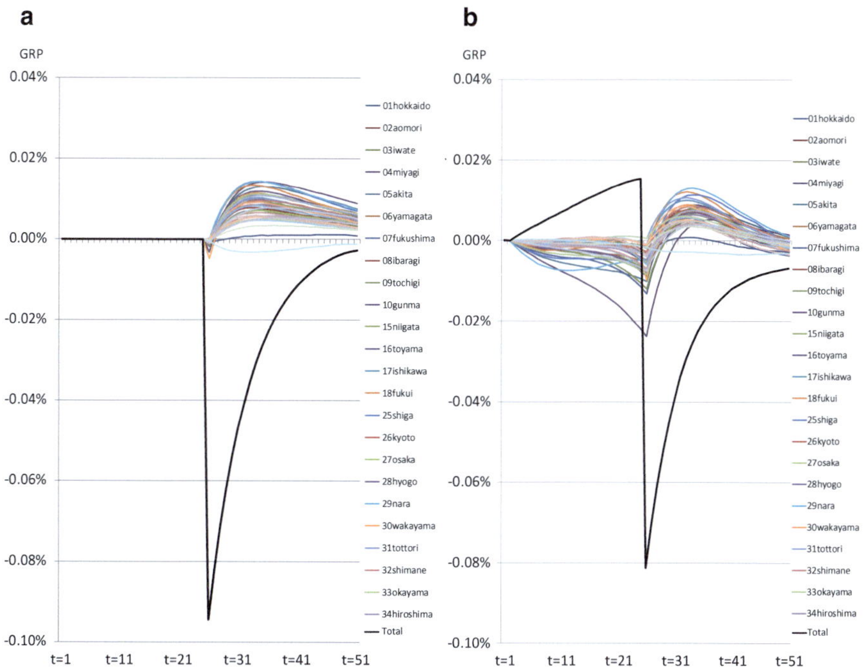

Fig. 11.5 GRPs in non-Tokai regions

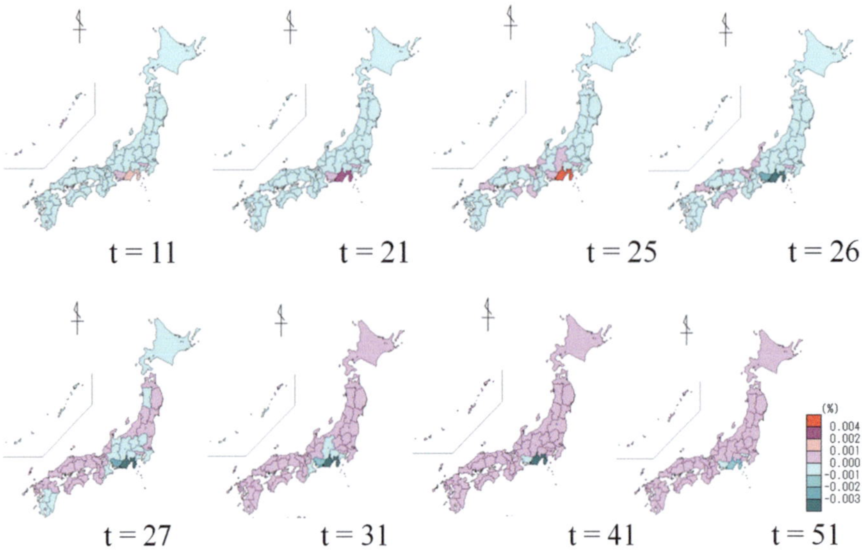

Fig. 11.6 Geographical and dynamic effects (%)

protect the Tokai region from earthquake damage, even though Japan's GRP change is positive. On the other hand, after the earthquake, most prefectures experience a positive impact, owing to an increase in investment to repair the damage.

Geographic and dynamic impacts of an earthquake in Case 2 are shown in Fig. 11.6. In this figure, blue and red indicate negative and positive impacts, respectively, on GRP compared with the base case. The Tokai region experiences positive changes in its GRP before the earthquake and sustains great damage after the earthquake. In other regions, conversely, most prefectures experience a decline in GRP before the disaster, but the situation becomes positive after the earthquake.

(c) Impacts on Commodity Flows

Here, we examine the earthquake's dynamic impacts on commodity flows between regions. Changes in investment before and after the earthquake inevitably involve changes in intra- and inter-trade commodity flows through transportation networks. The intra- and inter-trade of manufactured goods during the 23th and 27th periods are shown in those figures.

Figure 11.7 shows Case 2 changes in intra-trade commodity flows of manufactured goods for all transport modes, both before and after the earthquake. The change is defined as $\Delta x = (x_{case} - x_{base})$. Intra-trade commodity flows in Chiba, Tokyo, Kanagawa, Yamanashi, Nagano, Gifu, Shizuoka, Aichi, Mie, and Wakayama prefectures, shown as red lines, mostly increase both before and after the earthquake. The change in Shizuoka before the earthquake is particularly noticeable. In other prefectures, intra-trade commodity flows rise in most prefectures after

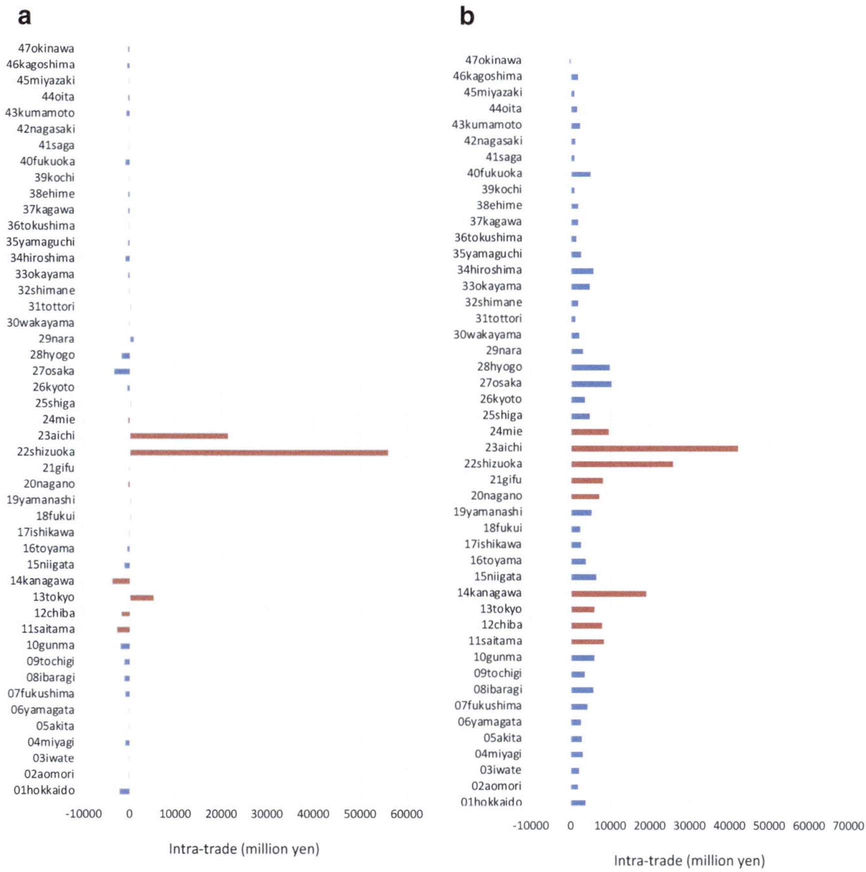

Fig. 11.7 Changes in Intra-trade commodity flows

the earthquake, while they decrease in coms prefectures before the earthquake. In Case 1, after the earthquake, similar changes can be seen, i.e., increases in intra-trade commodity flows in the Tokai region and other prefectures.

Commodity flows between prefectures is shown in Fig. 11.8, which presents changes in inter-trade commodity flows of manufactured goods for all transport modes between prefectures. We observe that in Case 2, changes in commodity flows before the earthquake are smaller than after the earthquake. Three kinds of major changes can be seen in the figures. The first is a noticeable increase in commodity inflows to the Tokai region from other prefectures. The second is a noticeable increase in inter-trade commodity flows between prefectures in the Chubu region, which is adjacent to the Tokai region. The last major change is an increase in inter-trade commodity flows between prefectures in the Kanto region, where the Tokyo metropolitan area is located and economic activity is

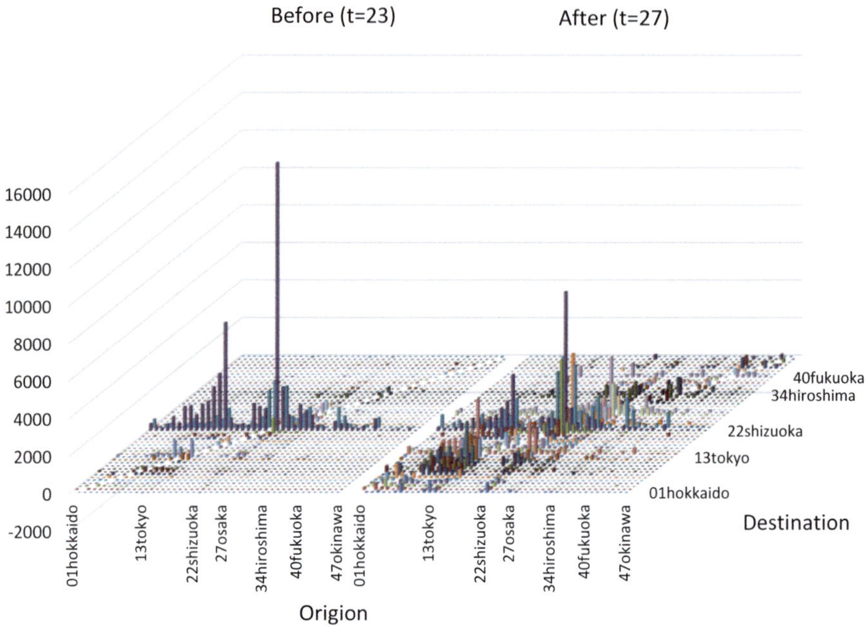

Fig. 11.8 Changes in Inter-trade commodity flows

concentrated. It shows that commodity flows in the Chubu and the Kanto regions seem to be influenced by an increase in investment in the Tokai region.

(d) Impacts of technological progress

Figure 11.9 shows the impact of technological progress on GRP in the Shizuoka prefecture, for Case 1 and Case 2. Two solutions are compared, one with and one without technological progress. In Case 1, with technological progress, GRP in the Shizuoka prefecture is restored more rapidly than in Case 2, which lacks technological progress. In the 41st period, the change in GRP turns from negative to positive. This productivity enhancement contributes to the regional economy in the Shizuoka prefecture.

In Case 2, although technological progress occurs after the earthquake, the disaster prevents an increase in investment. GRP is more rapidly restored in Case 2, which does not have technological progress. In the 40th period, GRP turns from negative to positive. Technological progress contributes to the restoration of the regional economy in a disaster region.

Figure 11.10 shows the impacts of technological progress on Japan's GRP (GDP) in Case 1 and Case 2. Total GRP (GDP) turns from negative to positive in the 45th period in Case 1 and in the 48th period in Case 2. Japan's entire economy is restored more slowly than in the regional economies experiencing a disaster.

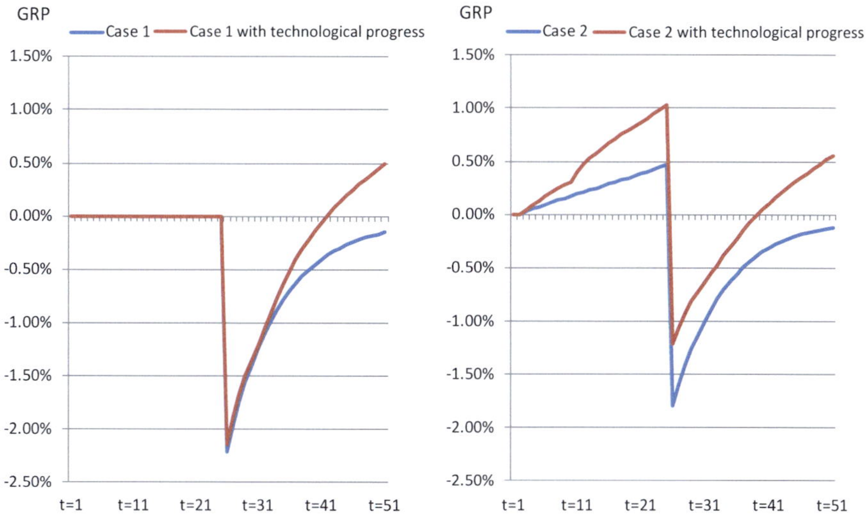

Fig. 11.9 GRP in the Shizuoka prefecture

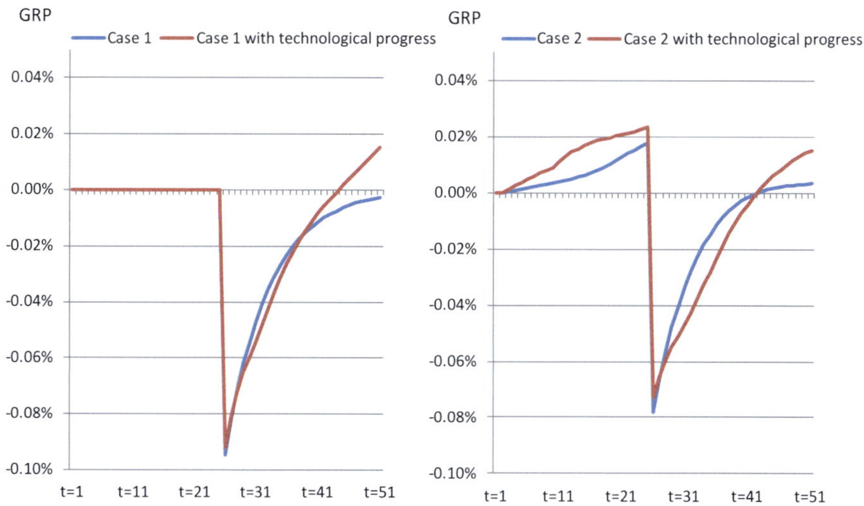

Fig. 11.10 Total GRP of Japan

11.5 Concluding Remarks

In this study, we described a dynamic spatial general equilibrium model. A decentralized economic system linked with transportation networks was constructed in a dynamic framework. The main purpose of the study was to assess the impacts of a disaster in the Tokai region on regional economies in Japan. We presented the results of three simulations: no earthquake, an unpredicted earthquake, and a predicted earthquake. We estimated dynamic and spatial impacts, i.e., industrial investments and commodity flows between regions for before and after the earthquake. Indirect effects, both before and after a disaster, were solved simultaneously. The two cases were compared with a base case. The results showed the importance of investment in terms of protecting the regional economy in the event of a disaster, i.e., an ex-ante evaluation. Our results suggest that any disaster analysis should evaluate the economic impacts of a disaster based on both ex-ante and ex-post criteria. We also found that to rapidly restore regional economies in disaster regions, productivity enhancements using capital inflows and new technology is important.

Many aspects of this study require further investigation. The policy impacts of public investment to repair damage should be considered. We should also consider the impact of the damage on transportation links. As another ex-ante criterion, an insurance system should be employed to relieve the effects of the damage. The investment–saving balance should be endogenously determined by both firms and consumers. This approach may provide fruitful results for comparison with deterministic and stochastic models. Basic assumptions in the decentralized model should be relaxed to internalize regional policies such as the tax-subsidy system and regulation.

Acknowledgement I would like to thank Y. Miyata for collaboration to the early stage of this work. This work was supported by JSPS KAKENHI Grant Number 16 K13141. This is also supported by Joint Usage and Research Center, Research Center for San-En-Nanshin Regional Collaboration, Aichi University.

Appendix

Production function ($j \in \mathrm{I} \cup \mathrm{M}, r \in \mathrm{R}$)

$$Y_j^r\left(K_j^r(t), L_j^r(t), \mathbf{X}_j^r(t)\right) = \min\left\{\frac{V_j^r(t)}{a_{Vj}^r}, \frac{X_{1j}^r(t)}{a_{1j}^r}, \frac{X_{2j}^r(t)}{a_{2j}^r}, \ldots, \frac{X_{Ij}^r(t)}{a_{Ij}^r}\right\}$$

$$V_j^r(t) = A_j^r K_j^r(t)^{\alpha_r^j} L_j^r(t)^{1-\alpha_r^j}$$

a_{ij}^r is the input coefficient and a_{Vj}^r is the value added ratio. $V_j^r(t)$ denotes the value added function. A_j^r is the production-efficiency parameter. α_r^j is the expenditure allocation ratio for capital.

Investment function $(j \in \mathrm{I} \cup \mathrm{M}, r \in \mathrm{R})$

$$\Delta K_j^r\left(\mathbf{Z}_j^r(t)\right) = \min\left\{\frac{Z_{1j}^r(t)}{\phi_{1j}^r}, \frac{Z_{2j}^r(t)}{\phi_{2j}^r}, \cdots, \frac{Z_{Ij}^r(t)}{\phi_{Ij}^r}\right\}$$

Utility function $(r \in \mathrm{R})$

$$U^r\left(\mathbf{C}^r(t)\right) = \prod_{i \in \mathrm{I}} C_i^r(\mathrm{t})^{\beta_i^r}, \quad \sum_{i \in \mathrm{I}} \beta_i^r = 1$$

β_i^r is the household expenditure allocation ratio for goods i.

Investment Cost function $(i \in \mathrm{I}, j \in \mathrm{I} \cup \mathrm{M}, r \in \mathrm{R})$

$$G_{ij}^r\left(Z_{ij}^r(t)\right) = G_{ij}^r\left(\phi_{ij}^r \Delta K_j^r(t)\right) = \phi_{ij}^r \cdot c_{Kj} \Delta K_j^r(t)^{\omega_j} \quad (\omega_j > 1)$$

cK_j and ω_j are the parameters for the cost function.

Asset function $(j \in \mathrm{I} \cup \mathrm{M}, r \in \mathrm{R})$

$$\Phi_j^r\left(K_j^r(\overline{T} + 1)\right) = \eta_j^r K_j^r(\overline{T} + 1).$$

η_j^r is the parameter for the assent function.

References

Abel AB, Blanchard OJ (1983) An inter temporal model of saving and investment. Econometrica 51:675–692

Ando M (1975) Possibility of a major earthquake in the Tokai District, Japan and its pre-estimated Seismotectonic effects. Tectonophysics 25:9–85

Borglin A (2004) Economic dynamics and general equilibrium. Springer, Berlin

Bröckera J, Korzhenevych A (2013) Forward looking dynamics in spatial CGE modeling. Econ Model 31:389–400

Central Disaster Prevention Council (2003) Estimated damages from the Tokai Earthquake. Cabinet office in Japan 1–18 (in Japanese)

Ciesecke J (2002) Explaining regional economic performance: an historical application of a dynamic multi-regional CGE model. Pap Reg Sci 81:247

Ciesecke J (2003) Targeting regional output with state government fiscal instruments: a dynamic multi-regional CGE analysis. Aust Econ Pap 42:214–223

Dixit PB, Jorgenson DW (2013) Handbook of computable general equilibrium modelling, vol 1A and 1B. Elsevier, Amsterdam

Donaghy KP (2009) CGE modeling in space. In: Capello R, Nijkamp P (eds) Regional dynamics and growth, advances in regional economics. Edward Elgar Publishing, Cheltenham, pp 731–799
Ellson RW, Milliman RJ, Roberts RB (1984) Measuring the regional economic effects of earthquakes and earthquake perditions. J Reg Sci 24:559–579
Hitomi K and Bunditsakulchai P (2008) Development of multi-regional input output table for 47 Prefectures in Japan. Socio-Economic Research Center, Rep. No. Y07035, CRIEPI (in Japanese).
Kehoe TJ, Srinivasan TN, Whalley J (2005) Frontiers in Applied General Equilibrium Modeling. Cambridge University Press, Cambridge
Koike A and Ueda T (2005) Economic damage assessment of catastrophe by using spatial general computable general equilibrium analysis. Proceedings of the 19th Pacific Regional Science Conference
McGregor PG, Swales K, Yin YP (1995) Migration equilibria in regional economies: a multi-period CGE analysis of an improvement in local amenities. In: CJM J, den Bergh V, Nijkamp P, Rietveld P (eds) Recent advances in spatial equilibrium modeling: methodology and applications. Springer, Berlin
McKibbin W and Wilcoxen P GCUBED (1992) A dynamic multi-sector general equilibrium growth model of the global economy. Brookings Discussion Papers in International Economics, No. 97
Mogi K (1970) Recent horizontal deformation of the earth's crust and tectonic activity in Japan (1). Bull Earth Res Inst, Tokyo Univ 48:413–430
Mogi K (1985) Temporal variation of crustal deformation during the days preceding a thrust-type great earthquake – the 1944 Tonankai earthquake of magnitude 8.1, Japan. PAGEOPH 122:765–780
Okuyama Y, Chang SE (2004) Modeling spatial and economic impacts of disasters. Springer, Berlin
Oosterhaven J, Knaap T (2003) Spatial economic impacts of transport infrastructure investments, transport projects, Programmes and policies. In: Pearman A, Mackie P, Nellthorp J (eds) Evaluation needs and capabilities. Ashgate, Aldershot, pp 87–110
Rose A, Liao S (2005) Modeling regional economic resilience to disasters: a computable general equilibrium analysis of water service distributions. J Reg Sci 45(1):75–112
Rose A, Benavides J, Chang S, Szczesniak P, Lim D (1997) The regional economic impact of an earthquake: direct and indirect effects of electricity lifeline disruptions. J Reg Sci 37 (3):437–458
Sato H (1970) Crustal movements associated with the 1994 Tonankai earthquake. J Geodetic Soc Japan 15:177–180. (in Japanese)
Shibusawa H, Miyata Y (2011) Evaluating the dynamic and spatial economic impacts of an earthquake: a CGE application to Japan. Reg Sci Inq 3(2):13–25
Shibusawa H, Yamaguchi M, Miyata Y (2009) Evaluating the impacts of a disaster in the Tokai region of Japan: a dynamic spatial CGE model approach. Stud Reg Sci 39:539–553
Shoven J, Whalley J (1992) Applying general equilibrium. Cambridge University Press, Cambridge
Sohn J, Kim TJ, Hewings JD, Lee JS, Jang SG (2003) Retrofit priority of transport network links under an earthquake. J Urban Plann Dev 129(4):195–210
Taniguchi H (2007) An risk evaluation of economic damage loss caused by earthquake in large cities. Annual meeting of Japanese Association for Applied Economics. Chuo University, Tokyo, pp 1–12. (in Japanese)
Tatano H, Tsuchiya S (2008) A framework for economic loss estimation due to seismic transportation network disruption: a spatial computable general equilibrium approach. Nat Hazards 44:253–265

Toyoda T, Kochi A (1997) Hanshin-Awaji Daishinsai ni yoru sangyo higai no suikei [estimation of economic damages in the industrial sector by the great Hanshin-Awaji earthquake]. J Econ Bus Adm 176(2):1–15. (in Japanese)

Tsuchiya S, Tatano H, Okada N (2003) Transport-related economic losses with the Tokai earthquake warning declaration by a spatial computable general equilibrium approach. J Soc Saf Sci 5:319–325. (in Japanese)

World Bank and United Nations (2010) Natural hazards, unnatural disasters: the economics of effective prevention, World Bank, Washington, DC

Xie W, Li N, Wu J-D, Hao X-L (2014) Modeling the economic costs of disasters and recovery: analysis using a dynamic computable general equilibrium model. Nat Hazards Earth Syst Sci 14:757–772

Chapter 12
Reconstruction and Rehabilitation After Large-Scale Natural Disasters: Lessons from the Indian Ocean Tsunami in Aceh and Nias, Indonesia

Budy P. Resosudarmo

Abstract The Indian Ocean Tsunami hit Aceh and Nias, Indonesia, on the morning of December 26, 2004. The giant wave destroyed all life and infrastructure within 10 km of the coastline in north and western Aceh, which then was a zone of armed conflict and all islands offshore. The death toll approached 167,000 people, with 128,000 missing and 811,000 internally displaced. This chapter analyses the impact of this huge natural disaster in a conflict zone. It documents the international response, the unexpectedly significant roles of NGOs on the Aceh and Nias economies and development in general, and the process of reconstructing and rehabilitating these areas. The reconstruction has been deemed successful. Among the important reasons for this success is that the Indonesian government wisely allowed local communities, leaders, and NGOs to participate in developing the master plan to reconstruct Aceh and Nias. Also, establishing the Badan Rehabilitasi dan Rekonstruksi Aceh-Nias to coordinate the reconstruction and rehabilitation was vital, as were funding commitments and willingness to deliver on them by government and other donors.

Keywords Natural disasters • Tsunami • Reconstruction • Rehabilitation • Development

12.1 Introduction

Survivors of the Indian Ocean tsunami of December 26, 2004, in northern and western Aceh and on the surrounding islands of Nias and Simeulu remember how an earthquake measuring 9.0 on the Richter Scale shook the earth at around 8 a.m.,

B.P. Resosudarmo (✉)
Arndt-Corden Department of Economics, Australian National University, Canberra ACT 0200, Canberra, Australia
e-mail: budy.resosudarmo@anu.edu.au

S. Tokunaga, B.P. Resosudarmo (eds.), *Spatial Economic Modelling of Megathrust Earthquake in Japan*, New Frontiers in Regional Science: Asian Perspectives 11,
https://doi.org/10.1007/978-981-10-6493-7_12

followed by another tsunami approximately 15 minutes later (Fig. 12.1). Waves up to 10 m hit the city of Banda Aceh in northern Aceh, and waves up to 12 m high hit the cities of Meulaboh, Calang, and Lamno in western Aceh, submerging areas up to 10 km from the coastline (Soehaimi et al. 2005). Those not personally affected had families, relatives, and friends swallowed by the tsunami. Many perished or remain missing.

Of the 10 countries hit by this tsunami, Indonesia suffered most. By mid-March 2005, the reported death toll was nearly 167,000 people, 128,000 were missing and 811,000 were internally displaced. Around 99.8% of the victims were in Aceh, particularly those living on the northern and western coasts. These numbers far exceed the three other countries most affected: Sri Lanka (30,000 dead, 6000 missing, and 500,000 displaced), India (11,000 dead, 6000 missing, and 648,000 displaced), and Thailand (5000 dead, 3000 missing, and 3000 displaced) (Athukorala and Resosudarmo 2005).

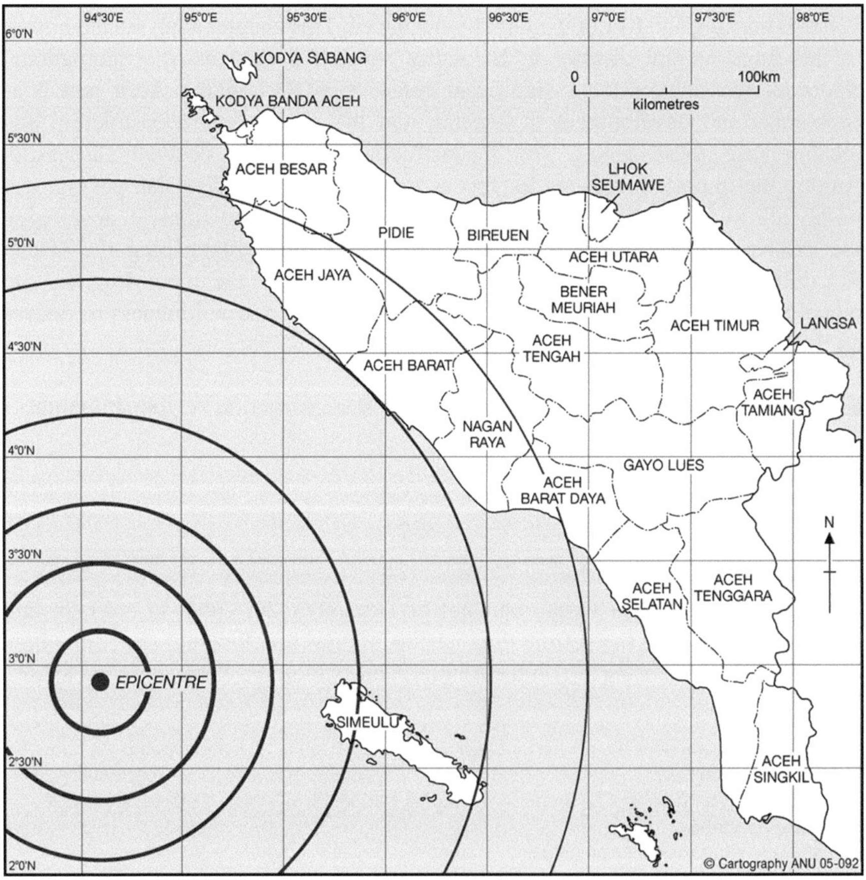

Fig. 12.1 Aceh's districts and the earthquake's epicenter

More than 99% of human loss and physical damage caused by the tsunami were in Aceh. In addition, Aceh had endured almost 30 years of conflict between the Free Aceh Movement (*Gerakan Aceh Merdeka* or GAM) and the Indonesian government, making it hard for many in Aceh to go about daily activities and to access education and health facilities and other public services. Aceh became one of the poorest regions in Indonesia. The tsunami compounded these difficulties and made rehabilitation and reconstruction more challenging after the tsunami.

Emergency relief from numerous domestic and international agencies and organizations was considerable and prompt. International donor response was remarkable and overwhelming. By mid-February, 34 countries and various organizations had made pledges of US$800 million to support relief and rehabilitation in Aceh and Nias. At the end of January, the Consultative Group for Indonesia (CGI) agreed to provide grants and loans of US$1.7 billion for reconstruction of Aceh. Most of these funds were distributed through non-governmental organizations (NGOs). This amount of international funding and the unexpectedly significant role played by NGOs had never before been experienced by Indonesia.

This chapter analyses the impact of this natural disaster on a conflict zone. It reports the international response, the unexpectedly significant effect of NGOs on the economies of Aceh and Nias and on development in general, and the process of reconstructing and rehabilitating these areas. This chapter reviews the impact of this disaster on the economy. It records the ongoing activities and challenges in reconstructing Aceh and Nias. Finally, it reports the final results of the reconstruction process and summarizes the current situation.

12.2 The Impact

The tsunami following this earthquake affected thousands of people in Aceh and Nias within minutes. Most Indonesians and the rest of the world knew only how badly the tsunami had hit Aceh one or 2 days afterward, whereas they heard about the impact of the tsunami in Sri Lanka and Thailand almost immediately. The conflict between GAM and the Indonesian government constrained the dissemination of information about Aceh to the world and other parts of Indonesia. When more complete information was provided, the world was astonished by the numbers of people killed, missing, and displaced in Aceh and by the amount of destruction.

12.2.1 Persons Lost and Internally Displaced

The reported number of victims rose daily. By the end of January 2005, the Indonesian government was able to publish a detailed report on the number of earthquake and tsunami victims (Table 12.1). Its report also mentions that around 5000 people were in hospitals and around 480,000 were in refugee camps.

Table 12.1 Regional distribution of victims

Region	Population	Number of people killed	Number of people lost	Total people affected	Percentage of people affected (%)
Aceh Province	4,104,187	120,663	116,126	714,517	17.4
1. City of Banda Aceh	269,091	78,417	64,552	182,478	67.8
2. District of Aceh Besar	306,718	58	43,902	141,907	46.3
3. City of Sabang	27,447	18	108	4529	16.5
4. District of Pidie	517,452	4646	2091	71,350	13.8
5. District of Bireun	350,964	1488	58	17,279	4.9
6. District of Aceh Utara	395,800	2217	233	22,816	5.8
7. City of Lhokseumawe	156,478	394	11	20,564	13.1
8. District of Aceh Timur	253,151	224	n.a.	13,934	5.5
9. City of Langsa	141,138	n.a.	n.a.	10,370	7.3
10. District of Aceh Tamiang	238,718	n.a.	n.a.	3100	1.3
11. District of Aceh Jaya	111,671	19,661	77	60,120	53.8
12. District of Aceh Barat	97,523	11,830	2911	93,558	95.9
13. District of Nagan Raya	152,748	493	865	16,127	10.6
14. District of Aceh Barat Daya	153,411	835	n.a.	14,799	9.6
15. District of Aceh Selatan	167,052	6	1086	17,290	10.4
16. District of Simeulu	76,629	22	1	18,167	23.7
17. District of Aceh Singkil	174,007	73	4	183	0.1
18. District of Aceh Tengah	158,641	192	227	4653	2.9
19. District of Aceh Tenggara	168,034	26	n.a.	26	0.0
20. District of Gayo Lues	67,514	27	n.a.	27	0.0
21. District of Bener Meuriah	120,000	36	n.a.	1240	1.0
North Sumatra Province	1,888,707	240	24	4266	0.2

(continued)

Table 12.1 (continued)

Region	Population	Number of people killed	Number of people lost	Total people affected	Percentage of people affected (%)
1. District of Nias	422,170	233	24	4257	1.0
2. District of Nias Selatan	275,422	1	n.a.	1	0.0
3. District of Tapanuli Tengah	272,333	1	n.a.	1	0.0
4. District of Serdang Bedagai	549,091	4	n.a.	6	0.0
5. District of Mandailing Natal	369,691	1	n.a.	1	0.0

Source: Bakornas PBP (2005a)

In numbers of people killed and missing, not all regions in Aceh and North Sumatra suffered equally. The city of Banda Aceh suffered most (Fig. 12.2), followed by the districts of Aceh Jaya and Aceh Barat. In percentage of population affected, Aceh Barat District suffered most, followed by Banda Aceh (Fig. 12.2), Aceh Jaya, and Aceh Besar, in that order. The impact of the earthquake and tsunami was concentrated in these four locations on the northern and western coasts of Aceh. Devastation across villages within these areas also varied, with 70% of the population killed or missing in some and others only losing a few.

After publishing its report, the Indonesian government continued to update the number of victims. A significant number of victims were women and children (Bakornas PBP 2005b). In August 2005, the government revised the number to approximately 125,000 people dead and 94,000 still missing (Indrawati 2005). In September 2005, BPS (Indonesia Central Statistics Agency) conducted the Aceh-Nias population census (BPS, Bappenas, UNFPA, CIDA, AusAID, and NZAid 2005a, b) and found 900,000 people displaced due to the tsunami. During the census, almost 260,000 were estimated as internally displaced persons (Nazara and Resosudarmo 2007). According to information gathered by the Ministry of Marine Affairs and Fisheries, by mid-January approximately 55,000 fishermen and aquaculture workers were among the dead—approximately half the total number of these workers in Aceh—and 14,000 were still missing (FAO 2005).

The human toll was the primary impact of the tsunami. As many of the victims worked in the fishery sector, many were women and children, and several villages lost up to 70% of their populations. The tsunami significantly changed the demographic structure in several places in Aceh. The question remains whether some local economies, villages, or even the general arc of development in Aceh will be ever recover. In addition, transfers of capital, particularly land, and property owned by the victims became an issue.

Source: Digital Globe (http://www.digitalglobe.com/tsunami_gallery.html)

Fig. 12.2 Northern Banda Aceh before and after the tsunami (Source: Digital Globe http://www.digitalglobe.com/tsunami_gallery.html)

12.2.2 Physical Damage

A quick survey through 14 districts most affected in Aceh found that approximately 27,000 houses were destroyed. The tsunami damaged infrastructure in 172 - sub-districts (*kecamatan*) encompassing 1550 villages and on island's north and west of Aceh (DepSos 2005). The Asian Development Bank (ADB) produced a different estimate, reporting the tsunami flattened 115,000 houses and severely damaged 150,000 (ADB 2005). Official reports also listed widespread destruction, including the loss of 100,000 houses, 3000 km of road, 14 seaports, 11 airports and air strips, 120 arterial bridges, 2000 school buildings, and 8 hospitals (BRR and International Partners 2006b; BRR 2006). The Food and Agriculture Organization (FAO 2005) reported that 40–60% of aquaculture ponds along the coast of Aceh and 36,000 to 48,000 ha of brackish water aquaculture ponds (mainly cultivating shrimp and milkfish) were seriously damaged. It is estimated that about 65–70% of the small-scale fishing fleet and associated gear in Aceh were destroyed, representing approximately 9500 units, of which 40% were canoes, 25% had outboard motors, and 35% had diesel inboard motors. FAO also reported that about 30,000 hectares of rice fields, about 10% of the area under rice cultivation in the province, were badly affected.

Using a standard assessment technique developed by the United Nations Economic Commission for Latin America and the Caribbean, the World Bank provided a comprehensive estimate of total damages and losses from the earthquake and tsunami. It assessed total damage and loss at US$4.45 billion, almost 100% of Aceh's GDP in 2003 (Table 12.2). Around 78% was borne by the private sector and the rest by the public sector. Devastation in the private sector was mostly houses and fishery-related production capital. The Bank estimated that 1.3 million homes and buildings were destroyed or heavily damaged. Public sector damage and losses were concentrated in the transportation sector. Approximately 120 km of road and

Table 12.2 Estimated damages

	World Bank's estimate			
Items	Private (US$ million)	Public (US$ million)	Total (US$ million)	LPEM-FEUI's estimate (Percentage of pre-disaster levels)
Housing	1408.4	28.7	1437.1	26.9–30.4%
Schools	9.0	119.4	128.4	30.0%
Hospital and health centers	23.2	68.6	91.8	24.0–33.3%
Roads and bridges	165.8	370.1	535.9	31.7%
Irrigation	132.1	89.1	221.2	24.9%
Agriculture and livestock	194.7	29.9	224.6	24.9–32.4%
Fisheries	508.5	2.5	511.0	40.0–70.0%[a]
Enterprises	428.9	17.7	446.6	27.4–27.9%
Environment	548.9	n.a.	548.9	28.3%
Governance and administration	n.a.	89.1	89.1	26.9%
Others	41.9	175.0	216.9	n.a.
Total	3461.4	990.1	4451.5	4.6 (US$ billion)

Note: [a]FAO's estimate
Sources: World Bank (2005a, b) and LPEM-FEUI (2005)

18 bridges, and approximately 85% of water sanitation facilities in urban areas along the north and west coasts of Aceh were ruined (World Bank 2005a).

Using a village level data system, the Institute for Economics and Social Research–Faculty of Economics University of Indonesia (LPEM-FEUI) estimated the tsunami destroyed one-third of the road network, schools, and hospitals (Table 12.2). It further estimated that damage from the tsunami might be US $4.6 billion (LPEM-FEUI 2005). Although LPEM-FEUI indicates the tsunami caused 25–30% of physical damages, it varies considerably across areas. Some areas lost up to 70% of their physical capital; others were not heavily damaged. Figure 12.2 illustrates one heavily damaged area in northern Banda Aceh. In areas suffering this degree of damage, identifying ownership of physical capital, particularly land, has become a major problem.

12.2.3 Economic Impact

Table 12.1 indicates that the impact of the Indian Ocean tsunami on North Sumatra province was concentrated in the district of Nias. There is little trade between Nias and other parts of North Sumatra, and its contribution to the area economy is small. The impact of its destruction on outside regions is expected to be trivial. Hence, this section focuses on the impact of the tsunami on Aceh's economy and on Indonesia's economy generally. First, however, it is important to understand the characteristics of Aceh's economy and its role in Indonesia before the tsunami.

Aceh is known for its rich natural resources, particularly oil and gas. Development of these sectors started in the mid-1970s. Table 12.3 shows that Aceh's

Table 12.3 Aceh's economy pre-tsunami

	1971	1983	1996	2002
GDP (including mining)				
GDP (in 1983 Rp billion)	452.1	3425.2	7055.6	5707.2
(percentage of national GDP)	1.5	4.8	4.1	3.1
GDP/Cap (in 1983 Rp,000)	224.9	1204.8	1788.4	1418.8
(ratio of national GDP/Cap)	0.9	2.7	2.1	1.7
Share of GDP				
Agriculture (percentage of GDP)	61.7	16.9	20.4	28.2
Manufacturing (percentage of GDP)	14.5	69.6	60.8	44.5
Service (percentage of GDP)	23.8	13.5	18.8	27.4
Poverty[a]				
Percentage of poor people	n.a.	11.6[b]	6.3	29.8
Percentage of national poor people	n.a.	15.1[b]	11.3	18.2

Source: BPS (various years)
[a]using BPS's poverty lines
[b]numbers are for 1990

economy in 1971 was dominated by agriculture. Aceh's gross domestic product (GDP) per capita nearly equaled average national GDP per capita.

During the mid-1970s, Aceh's oil and gas industry helped it to grow faster than most regions of Indonesia In 1983, Aceh produced nearly 5% of Indonesia's GDP, and its per capita GDP was three times the national average. Aceh became one of Indonesia's richest provinces along with Riau, East Kalimantan, and Jakarta. It is noteworthy, however, that the percentage of poor people in Aceh (11.6%) was scarcely less than the total percentage of Indonesia's poor.

In 1996, Aceh's GDP per capita remained among Indonesia's highest, although other provinces were catching up. Its contribution to national GDP declined to around 4%, and its GDP per capita was around twice the national average. Its manufacturing sector, particularly oil and gas, grew only at a moderate rate during 1983–1996. The encouraging news during this period was that the proportion of poor people declined significantly in Aceh from 11.6% in 1983 to 6.3% in 1996. Indonesia's poor then accounted for 11.3% of its population.

In 2002, Aceh's GDP per capita was below that in 1996, primarily because of the escalating conflict with GAM. More worrisome was that its proportion of poor rose to 29.8%—much above the national level and among the highest in Indonesia. The Ministry for the Development of Least Developed Regions had classified 11 districts in Aceh (50% of the province) as least-developed districts. The conflict also had destroyed and damaged about 900 schools, and school attendance had fallen dramatically. Healthcare had become less accessible because people feared to visit health facilities (Soesastro and Ace 2005; World Bank 2005a).

In 2003, Aceh's GDP was US$4.5 billion or 2.3% of Indonesia's GDP. Oil and gas and agriculture dominated Aceh's economy, contributing 43% and 32.2% of regional GDP, respectively. In agriculture, livestock (10%) and food crops (10%) contributed the highest share. The contribution of fisheries was relatively small at 6.5% (World Bank 2005a). Aceh's exports to other regions in Indonesia were only about 8% of its output, whereas exports abroad were 26% of output, and 66% was consumed within the province. Imports from other regions in Indonesia and from abroad were 6% and 4%, respectively, of total material inputs needed for Aceh's production sectors.

In sum, Aceh's contribution to Indonesia's economy was relatively small shortly before the tsunami. It typically has been much lower than that of Jakarta, West Java, Central Java, East Java, Riau, or East Kalimantan. Aceh's economy has been dominated by oil and gas since the 1970s. The contribution of the fishery sector to Aceh's economy has been relatively small. What stands out about Aceh's economy is that the percentage of poor has been high, nearly 30%.

As discussed, Aceh's agricultural sector was the most affected, particularly the fishery sector, in number of casualties and capital destroyed. Its oil and gas industry escaped the tsunami virtually unharmed (Soesastro and Ace 2005; World Bank 2005a). The characteristics of Aceh's economy would suggest that the impact of the tsunami on Aceh's economy would be moderate and the impact on Indonesia's economy small. However, the impact on percentages of the poor turned out to be severe, as significant numbers of people lost capital and jobs.

In the food crop sector, FAO (2005) observed that affected areas would lose two consecutive paddy seasons. The 2005 output of the fishery sector was expected to fall up to 50% for marine fishing and 41% for brackish water aquaculture. The livelihoods of 330,000 workers (600,000 including their families), mainly in the food crop and fishery sectors, were threatened.

On a macroeconomic level, the World Bank conducted a comprehensive analysis and predicted that Aceh's GDP would contract 7–28% from 2004. It further estimated that if Aceh's economy contracted moderately (i.e., around 14%,) per capita income would fall one-third, increasing unemployment from 9.4% (in 2004) to 27.5%, and the percentage of poor would increase 50%. The Bank estimated that Indonesia's 2005 GDP growth would be between 0.1% and 0.4% lower than the pre-tsunami forecast (World Bank 2005a). LPEM-FEUI estimated GDP would contract 22.3%. However, it predicted that the percentage of poor would increase up to 50%. The agency estimated that growth of Indonesia's economy would be slightly less, around 0.56%, compared to the situation without the tsunami (LPEM-FEUI 2005).

The average rate of inflation in Banda Aceh in January 2005 was 7.02%, whereas it was 1.43% for the whole country. The highest rates of inflation were for processed food and food products—19.26% and 11.24%, respectively. House rents were rising rapidly. The high rate of inflation did not persist. In February, it declined and stayed relatively close to the national rate thereafter (Fig. 12.3). Nonetheless price differentials between Banda Aceh and the national average have not returned to pre-tsunami levels.[1]

12.3 Disaster Management and Reconstruction

The first question upon hearing the extent of the devastation in Aceh and surrounding areas typically concerned how effective rescue efforts were in saving the victims' lives. After establishing that significant rescue efforts had been conducted, the next three questions would most likely be as follows. How did this disaster happen? How can the people of Aceh and Nias recover and their regions be reconstructed? What are the lessons learned so similar devastation can be avoided?

[1]The high rate of inflation in October 2005 was unrelated to the tsunami. It was induced by government's decision to raise the domestic price of fuel oil.

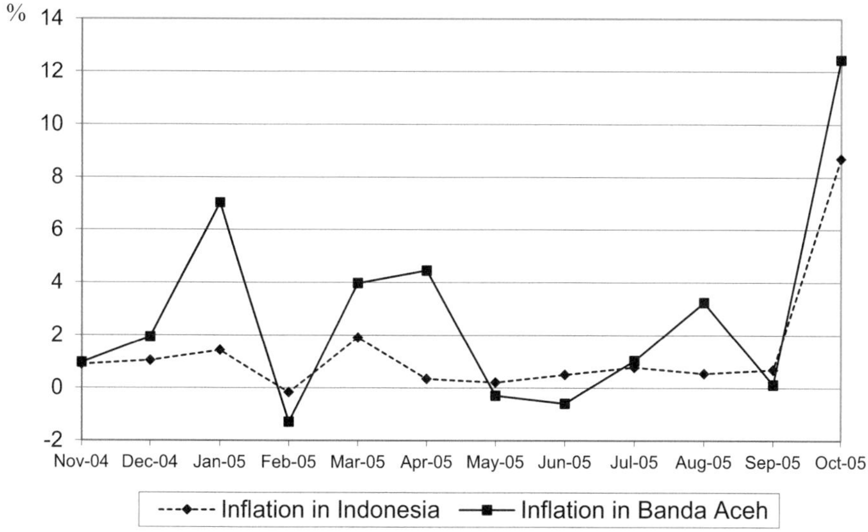

Source: CEIC Asia Database

Fig. 12.3 Monthly rates of inflation in Banda Aceh (Source: CEIC Asia Database)

12.3.1 Disaster Prevention, Early Warning System, and Mitigation

Seismic activity is so frequent in Indonesia that one would expect there are agencies responsible for monitoring it and developing mitigation procedures. The three main agencies responsible are the Meteorology and Geophysics Agency (*Badan Meteorologi dan Geofisika*), the Volcano and Geological Disaster Mitigation Directorate (*Direktorat Vulkanologi dan Mitigasi Bencana Geologi*) at the Ministry of Energy and Mineral Resources, and the Centre of Geological Research and Development (*Pusat Penelitian dan Pengembangan Geologi*). Other agencies, including the Agency for Assessment and Application of Technology (*Badan Pengkajian dan Penerapan Teknologi*), the Indonesian Science Institute (*Lembaga Ilmu Pengetahuan Indonesia*), and assorted Indonesia universities (Bandung Institute of Technology and the Surabaya Institute of Technology) research and monitor activity and design mitigation procedures. When natural disasters occur, the Indonesian Emergency Relief Coordination Agency (*Badan Koordinasi Nasional Penanggulangan Bencana dan Penanganan Pengungsi* or Bakornas PBP) and the Department of Social Services are responsible for relief and managing refugees.

Since the Flores tsunami in 1992, these agencies and universities, with the help of international institutions, have conducted further research and monitoring of tsunamis in Indonesia. The 1994, the Banyuwangi tsunami and the 1996 Biak tsunami encouraged such activities. A few years ago, these agencies jointly produced a map indicating where tsunamis are most likely. It was made available to the public upon request (Montgomery 2005).

Due to insufficient funding, however, a systematic and comprehensive tsunami monitoring system has never been developed. Critics have brought the absence of such an important system to the attention of national authorities and the public several times. Indonesia's leading national newspaper, *Kompas*, published a series of articles concerning earthquakes that could cause tsunamis and their likely impact on areas along the west coast of Sumatra, the south coast of Java, and Nusa Tenggara up to the Banda Sea (Kompas February 18, 2001, October 26, 2002, November 2, 2002, and December 21, 2002). These articles did not induce the central and regional governments to prioritize establishing a prevention strategy and monitoring and early warning systems for tsunami. Had such initiatives been undertaken, annihilation like that of the Indian Ocean tsunami likely could have been lessened, although perhaps only slightly in that instance because the epicenter of the earthquake was too close to Aceh and the tsunami was too big.

An interesting case is that of Simeulu Island 100 km northwest of Nias Island and close to the epicenter of the earthquake—closer, in fact, than anywhere on the coastline of Aceh. Simeulu Island was hit severely by the earthquake and tsunami. Approximately 5500 houses were destroyed and hundreds of people were injured, but only seven deaths were recorded (Kompas April 01, 2005). There are two main reasons for this low death toll. First, the coastal ecosystem—namely the coral reef, seagrass, and mangrove forests surrounding the island, particularly in the north—was relatively undisturbed and in good condition. It mitigated the force of the giant wave when it reached the coast (Montgomery 2005). Second, local island knowhow (*adat*) passed generation to generation for centuries, recognized the signs of a tsunami and specified how to escape it. Once someone recognizes indications a tsunami approaches, he should shout "Tsunami . . . tsunami . . . tsunami" and run to the nearest hills. Others hearing this warning should abandon what they are doing and also run to the hills while shouting the warning to others. This simple procedure proved effective (Wetlands International-Indonesia Programme 2005; Kompas April 1, 2005).

Evidence that coastal ecosystems mitigate the impact of tsunamis comes from other affected areas in Indonesia as well. For instance, the impact of the tsunami was less severe in areas along the west and east coasts of Aceh, where the coastal ecosystem remained in relatively good shape. Damage was much worse in coastal cities, where housing, tourism, and destructive fishing have disturbed it (Wetlands International-Indonesia Program 2005; Montgomery 2005).

Modern systems of tsunami detection, early warning, and mitigation were unavailable in Aceh and Nias during the 2004 tsunami and likely were unavailable in the rest of Indonesia. Simple traditional mitigation procedures imbedded in local culture, as on Simeulu Island, have been forgotten elsewhere or never existed. While previously there had been some debates whether or not Indonesia already needs to have as systematic method of disaster prevention, since this Indian Ocean tsunami disaster, it clear that Indonesia does need to build a systematic method of disaster prevention, monitoring, early warning, and mitigation. Its efforts should start by conserving coastal ecosystems. Efforts should expand to introducing simple natural disaster detection methods, early warning systems, and mitigation procedures—like those on Simeulu island—and embedding them in nationwide elementary education program and later installing modern monitoring technologies.

The Indonesian government did respond to the need for a natural disaster early warning system and mitigation procedures. Through Law No. 24/2007, the government developed a new agency, the National Agency for Disaster Management (*Badan Nasional Penanggulangan Bencana* or BNPB), to replace Bakornas PBP. Their mandate includes providing support to develop natural disaster courses in school curricula. In 2005, under the auspices of the Intergovernmental Oceanographic Commission of UNESCO and with collaboration of international partner institutes from Germany, the USA, China, and Japan, a large-scale project was undertaken to establish a modern Indonesian Tsunami Early Warning System. Its aim was to make a modern early warning system available by the early 2010s.

12.3.2 Rescue, Relief, and Outsiders' Early Responses

In the first days after the event, since little information was received by the outside world, rescue and relief operations were few and slow. When the outside world began to realize what had happen in Aceh and Nias, it responded quickly. In the first week after the event, despite some criticism, the Indonesian military conducted efficient rescue and relief operations and removed bodies. By the start of the second week, the numbers of domestic and international organizations reaching Aceh swell significantly. All did whatever they could wherever they saw need. There was probably little coordination among them, but simply being able to provide relief for the victims was more important. In the third week, the number of international organizations arriving in Aceh continued to increase. An estimated 250 domestic and international organizations sent representatives to the region, and countless others assisted in some form. Their contributions were appreciated and, in general, their performance was excellent (Indrawati 2005).

International response in collecting funding to support emergency relief was also prompt and gratifying. By mid-February 2005, just 2 months after the tsunami, the Office for the Coordination of Humanitarian Affairs (OCHA) recorded that pledges and commitments by 34 countries, and various organizations supporting relief and reconstruction in Aceh and Nias had approached US$800 million (OCHA 2005). Even then, pledged amounts were expected to increase. It is also important to note that the reported figure excluded donations in kind and soft loans for reconstruction of Aceh and Nias. For example, Australia had agreed to provide soft loans up to AU $500 million over 5 years. Totals reported by OCHA also excluded debt moratoriums or debt swaps offered by several countries, including Germany, France, and Italy. It may have been the most generous international response ever to a natural disaster in Indonesia (Athukorala and Resosudarmo 2005). Another source was the commitment by members of CGI on January 19–20, 2005. They agreed to contribute up to US$1.7 billion in 2005 for reconstruction of Aceh, approximately 35% of its estimated damages and losses. CGI's response was overwhelming. Of this amount, US$1.2 billion was to be grants and US$0.5 billion was project loans with little or no interest. Of the US$1.2 billion grants, only US$0.2 billion was to be distributed via the Indonesian government; the rest was to be distributed through NGOs. These stipulations revealed donors' concerns about the ability of the central government to employ funds transparently and without corruption. Considering the government's record, this was a reasonable concern. Nonetheless, there was also concern about the ability of NGOs to use these funds efficiently (Soesastro and Ace 2005). The Indonesian central government also responded appropriately to the disaster by announcing at the end of December 2004 that it would release approximately US$5 million to support rescue and relief. The government also announced it would conduct a three-phase operation in Aceh and Nias: (1) emergency rescue and relief, (2) rehabilitation and reconstruction of basic infrastructure, law, and order, and (3) rebuilding the economy and government. The first phase was completed by April 2005. The second was expected to take 2 years and the third three more years. Overall reconstruction and rehabilitation of Aceh and Nias was expected to be complete by yearend 2009. The Indonesian National Planning and Development Agency (Bappenas) was responsible for developing blueprints for the second and third phases. The 2005 government budget allocated approximately US $450 million for Aceh's rehabilitation and reconstruction, including funding received from international agencies. In 2006 the sum was approximately US $1.4 billion (Nazara and Resosudarmo 2007).

The exact amount of funding committed by international donors and the government for Aceh's rehabilitation and construction is difficult to calculate. By November 2005 the government recorded that 78 countries, 30 organizations, and many individual donors had pledged support (grants and soft loans) of US$6.1 billion (BRR and International Partners 2006b). By yearend 2006, the government revised the amount of funding to US$7.1 billion, consisting of about US$2.7 billion from the Indonesian government, US$2 billion from NGOs, and US$2.4 from international governments (BRR and International Partners 2006a, b). The World Bank predicted that rehabilitation and reconstruction up until 2009 would cost approximately US$6.1 billion. This estimate included damages and losses from the tsunami (US$4.5 billion), from the March 2005 earthquake in Nias (US $400 million), and US$1.2 billion for inflation. With total available funds of US $7.1 billion and the amount required being US$6.1 billion, the extra US$1 billion could be used to "build back better" and reduce poverty in Aceh and Nias (BRR and International Partners 2006a, b).

Hence, in general, the immediate domestic and international responses to the Indian Ocean tsunami—physical rescue efforts, relief activities, and pledges of funds for rehabilitation and reconstruction—were fast and extensive. The questions at that time concerned how successful actual rehabilitation and reconstruction would be and the quality of coordination within government agencies and NGOs and between government agencies and NGOs through which significant international funding was channeled. There was a more specific question whether coordination between the government and NGOs would be sufficient to reconstruct Aceh and Nias.

12.3.3 Initial Challenges

In early 2005, several concerns emerged about the process of rehabilitation and reconstruction in Aceh and Nias. The first challenge concerned coordination. The national government had appointed Bappenas the central agency for developing the recovery plan for Aceh and Nias. The main challenge for Bappenas was to develop a master plan that satisfied all agencies and organizations involved in reconstruction. At the time, dialogue between Bappenas and local governments was limited. In the absence of their direct involvement, many local governments felt excluded from the reconstruction process, which basically the central government dictated. Consequently, local governments wanted to design their own plans and programs that were incompatible with the Bappenas plan, leading to duplication of activities and inefficient use of funds (Nazara and Resosudarmo 2007).

There was also poor coordination between NGOs and Bappenas. Many NGOs resisted plans emanating exclusively from Bappenas. Several consortiums of NGOs developed their own reconstruction programs for Aceh and Nias. It was yet unclear how they would relate their plans to those of Bappenas or local governments. Coordination among (domestic and international) NGOs was also problematic. First, there were more than 100 international NGOs in Aceh alone. Second, many had high turnover of staff, with some staff working as briefly as 2 weeks. The new staff did not know what had been agreed during previous coordination meetings. This situation hardly contributed to coordinated and effective rehabilitation and reconstruction (Saldanha 2005).

The national government preferred at that time to establish a new special agency as coordinator of recovery activities. Local governments, communities, and businesses and NGOs feared that this new agency would constrain their involvement in rehabilitating and reconstructing Aceh. They were also suspicious that all construction bids would be tendered in Jakarta and won by a few large companies connected to officials in the central government and reconstruction activities would not meet local needs appropriately. Residents and NGOs demanded a more decentralized approach in rehabilitating and reconstructing Aceh that they believed would better accommodate local needs (Athukorala and Resosudarmo 2005).

The second challenge related to security conditions in Aceh. The ongoing 30 year-old secessionist rebellion of GAM had complicated the first phase of the operation in Aceh—rescue and relief—causing them to take much longer than in other tsunami-affected countries. There was concern that the ongoing conflict also would hamper rehabilitation and reconstruction.

Another challenge was to assure that commitments by international donors appeared in a timely manner. The fear was that commitments might not translate into fund flows for reasons beyond Indonesia's control. However, domestic aid absorption capacity was to play an important role. It was important for the Indonesian government and NGOs to communicate effectively with donors and to engage them in developing projects and programs that minimized mismatches between donor's interests and reconstruction priorities (Athukorala and Resosudarmo 2005).

The final challenge related to whether rehabilitation and reconstruction of Aceh and Nias would take into account the need for an integrated system of future disaster prevention, detection, early warning, and mitigation. International responses were not always generous. Would Indonesia develop more solid disaster management procedures and institutions?

12.3.4 Rehabilitation and Reconstruction in the First Two Years

Indonesia's central government seemed to understand the need for better coordination with local governments and NGOs and insistence on more decentralized system of rehabilitating and reconstructing Aceh. From March 2005 onwards, Bappenas consulted intensively over several weeks with community and political leaders in the affected areas, NGOs, and donors. Syiah Kuala University in Banda Aceh was given support to organize inputs from local communities and NGOs while central and local government agencies provided strategic and technical expertise.

In the meantime, donors were encouraged to contribute technical assistance and suggestions. A Master Plan resulted that was comprehensive and detailed, although the central government recognized no one plan could address every issue or cover every eventuality. It agrees that much would change and evolve as the huge rehabilitation and reconstruction program got under way. No strict blueprint was adopted for the rehabilitation and reconstruction of Aceh (Indrawati 2005; World Bank 2005b). For example, communities were provided opportunities to decide where, how, and by whom houses and other buildings were to be reconstructed. The central government only made sure that principle infrastructures such as main roads, electricity, and water sanitation were available.

Also, earlier plans called for tough zoning, mandatory setbacks from the sea, and relocating local markets etc. These plans were not implemented. People in Jakarta understood that locals were better equipped to make such decisions (Sen and Steer 2005). Strong conflicts among the central and local governments and among communities, business, and NGOs were minimized. This situation was encouraging, with the central government providing more opportunities for other regional stakeholders (local governments, communities, business, and NGOs) in developing plans and implementing them. Rehabilitation and reconstruction were successful, and this precedent of coordination and collaboration among stakeholders could be the blueprint for future regional development in Indonesia.

The central government's decisions to establish a special Aceh-Nias Rehabilitation and Reconstruction Agency (BRR, or *Badan Rehabilitasi dan Rekonstruksi Aceh–Nias*), as a single point for coordinating all agencies and donors in Aceh and Nias met no strong objection from local governments, communities, NGOs, and international donors. There were two main reasons for this. First, the central government had shown willingness to collaborate with these participants in developing the plan and to be flexible in implementing it. Second, the person elected to head this agency, Kuntoro Mangkusubroto, and his deputies had a reputation for honesty and ability. An Advisory Board of six ministers, heads of local governments, and local leaders was established to assure support from multiple stakeholders in Aceh. An independent Supervisory Board of prominent Indonesians was also set up to monitor and audit the entire reconstruction, making sure that BRR would work efficiently and without corruption (World Bank 2005b).

The central government also understood that to rehabilitate and reconstruct Aceh the 30 years of conflict with GAM had to end. GAM concurred. On August 15, the government of Indonesia signed a peace agreement with GAM. Under this agreement, by the end of 2005, the government significantly reduced its military and police presence in Aceh, and GAM handed over its weapons. GAM ceased attempts to separate from Indonesia in return for political, economic, and social rights. Political rights included the right to form local parties, stand for election, and participate in rehabilitation and reconstruction of Aceh. Since the signing of this agreement, armed conflicts between GAM and the Indonesian army have declined significantly (World Bank 2005b).

Nonetheless, the progress of reconstruction had been slow up until then (World Bank 2005a, b). Efforts to accommodate local needs in the plan and to establish BRR properly were taking time. The Ministry of Finance was criticized for tardiness in disbursing government budgets for rehabilitation and reconstruction in Aceh. Some personnel in line ministries (e.g., the Ministry of Public Works) who generally would be responsible for providing infrastructure felt marginalized by establishment of the BRR and were reluctant to work hard, particularly after discovering BRR staff received much higher salaries. In addition, many international donors obtained authorization from their parliaments for reconstruction funds at midyear and could only disburse funds thereafter. Some donors (e.g., Japan, some European countries, and the multi-lateral agencies such as the World Bank and ADB) preferred to distribute funds through Indonesian government budget systems. Also, the Indonesian budget year started January 1, and funds were not initially disbursed until the new budget year in 2006 (Sen and Steer 2005).

By end of September 2006, an estimated 7000 houses had been rebuilt, with 100,000 more needed. Reconstruction of necessary infrastructure such as roads, bridges, electricity, and drinking water facilities was also limited. Although collaboration and coordination among government agencies, communities, NGOs, and donors had been established, it was not yet at a level enabling fast housing reconstruction (World Bank 2005b; Kompas 17 October 2005). By yearend 2006, the big questions were whether the rehabilitation and reconstruction of Aceh and Nias would include an integrated system of future disaster prevention, detection, early warning, and mitigation and whether Indonesia would develop more solid disaster management procedures and institutions (Salim et al. 2005).

12.4 Results of the Reconstruction

In 2007, encouraging news about the situation in Aceh and Nias emerged. The peace agreement and the reintegration of ex-GAM personnel into the community started to show results. On December 11, 2006, for the first time in Indonesian history, the Acehese people voted directly to elect their own governor and district/municipality heads. The election proceeded successfully. There were no major conflicts, and electoral participation was high. A prominent ex-GAM member, Irwandi Yusuf, won election as Governor of Aceh from 2007 to 2012. The success of this election and that an ex-GAM member won were catalysts for a peaceful and democratic environment in Aceh after the peace agreement.

RALAS, a program of community-driven "mapping" showing who owned which land parcel (almost all land records were destroyed), proved successful. By yearend 2007, it had surveyed approximately 210,000 land parcels, and 105,000 land title certificates had been distributed by the National Land Administration Agency (*Badan Pertanahan National* or BPN). By the end of the program (June 30, 2009), RALAS had helped to restore 220,000 land titles and created more than 300,000 land maps (MDF 2012a, b). Almost all affected areas and beyond in Aceh have been mapped. The success of the RALAS program has been cited as enabling the faster building of houses since 2007.

A year after its establishment, BRR became much more effective and efficient. It established six regional offices and transitioned from sector-based to region-based management. This proved to be more practical in coordinating cross-sectoral projects. Moreover, decentralization brought BRR closer to all stakeholders. Its regional offices worked with local government to gain first-hand evaluations of the remaining needs of beneficiaries, while decentralized decision-making and the establishment of regional operational capacity improved implementation and developed local capabilities (BRR 2009). BRR leadership was able to improve the accountability of most projects under its authority and to act as coordinator with partners and the community. For example, BRR developed the Recovery of Aceh-Nias Database (RAN database), an open access data system to track off-budget reconstruction funding and to demonstrate who was doing what, when, and where. It also set up the BRR Knowledge Centre dedicated to the preservation of data and management of information related to the rehabilitation and reconstruction program in Aceh and Nias. As a result, donors were persuaded to disburse their pledges relatively on time. From the total US$7.1 billion pledged, US$6.7 billion was spent by the end of BRR program.

In implementing reconstruction and rehabilitation programs, BRR adopted an evolutionary approach taking into account various inputs and did not rigidly follow a single blueprint—namely, the central government's Master Plan. Under the new

Table 12.4 Progress of Aceh-Nias reconstruction and rehabilitation

Affected by the tsunami	Achievements
139,000 houses destroyed	140,304 permanent houses built
1089 places of worship destroyed	3781 places of worship built or repaired
104,500 SMEs destroyed	195,726 SMEs received assistance
	155,182 laborers trained
73,869 ha of agricultural land destroyed	69,979 ha of agricultural land reclaimed
13,828 fishing boats destroyed	7109 fishing boats built or donated
1927 teachers killed in Aceh	39,663 teachers trained
3415 school facilities destroyed	1759 school facilities built
517 health facilities destroyed	1115 health facilities constructed
2618 km of road destroyed	3696 km of road constructed
119 bridges destroyed	363 bridges constructed
22 ports destroyed	23 ports constructed
8 airports and runways destroyed	13 airports and runways constructed
669 government buildings destroyed	996 government buildings built

Source: BRR (2009)

approach, for example, communities were to be granted opportunities to participate in decisions about where, how, and by whom houses and other structures would be reconstructed. Although the central government would concentrate on providing principal infrastructure such as main roads, electricity, and water sanitation, it made a commitment to assure that local residents in Aceh and Nias were involved in deciding such matters. In this way, disagreements among the central government and local governments and communities were kept to a minimum. The combination of a peace in Aceh and the BRR's ability to gain support from national and local stakeholders reduced restrictions on the supply of labor and materials needed for the reconstruction to flow into the various regions in Aceh and Nias. There was inflation, but it was within the predicted range. Affordable labor and material were relatively available in Aceh and Nias for reconstruction and rehabilitation. The BRR developed its image as corruption free, accountable, effective, and favoring the local populace. It received support from local communities in reconstructing and rehabilitating the tsunami area. Local residents participated in restoring their communities.

By 2009, 140,000 new homes for tsunami victims, 1000 health clinics, 2000 schools, 3000 mosques, and 3000 km of roads were rebuilt (Table 12.4). BRR is acknowledged as the most successful natural disaster reconstruction and rehabilitation program in the developing world.

By the early 2010s, BNPB and several other institutions, including academic institutions, were training teachers, government staff, and others interested in natural disasters. Formal school curricula, starting from elementary school, have incorporated material on natural disasters (Antara News, 16 March 2011; Viva News, 21 April 2012). In 2011, construction of the Indonesian Tsunami Early

Warning System (InaTEWS) was completed. It is currently running and operated by the Indonesian Service for Meteorology, Climatology, and Geophysics (BMKG).

12.5 Conclusion

This Chapter has analyzed the impact of the Indian Ocean tsunami on Indonesia and the process of reconstruction and rehabilitation. The major problem in conducting this analysis was in obtaining accurate data and information. Accurate data on Aceh and Nias are hard to obtain and their quality is questionable. Several agencies provided estimates of the impact of this tsunami on Indonesia, but assuring their reliability is difficult. Various institutions produced reports on the situation in Aceh, but some are conflicting. Discerning the correct information is sometimes difficult. Notwithstanding difficulties in preparing this analysis, several conclusions are apparent.

First, the Indian Ocean tsunami killed thousands and affected almost a million people in Aceh and Nias. The majority of deaths were women and children. In several villages, almost everyone was killed. In many villages, everyone was affected. Houses, roads, electricity, and other infrastructure in the northern and western Aceh and surrounding islands were destroyed. Agriculture, particularly the fishery sector, was the most affected. The impact of the Indian Ocean tsunami on the local economy has been significant and serious, with the main concern being the increasing percentage of the poor people and unemployment in Aceh and Nias. The impact on Indonesia's economy, however, is probably small.

Second, emergency rescue and relief activities and outside responses were overwhelming. Although the rescue and relief were slow in the first days after the event due to limited information, they quickly accelerated. Around 250 domestic and international agencies were involved in rescue and relief. It is safe to say that they were able to complete their operations around April 2005. International commitments to provide funding to support relief, rehabilitation, and reconstruction were also prompt and gratifying. Never has such extensive recovery funding been allocated to a natural disaster in Indonesia. Another feature of this funding was that the majority of it was disbursed through NGOs, requiring close collaboration between the Indonesian government and NGOs to make reconstruction effective and avoid the dissipation of funds through corruption.

Third, the tsunami devastation forced the central government and GAM to reach a peace agreement. Its implementation was effective. Since then, a peaceful and democratic condition in Aceh has been maintained.

Fourth, the central government adopted a smart approach by allowing local communities, leaders, and NGOs to participate significantly in developing of the master plan to reconstruct Aceh and Nias. The central government also agreed that the master plan should be flexible, providing opportunities for communities to make decisions about where, how, and by whom houses and other structures were reconstructed. Thus, better collaboration among stakeholders was established. The rehabilitation and reconstruction effort was successful, and this precedent of strong coordination and collaboration among stakeholders could be the new paradigm for future regional development in Indonesia.

Fifth, establishment of BRR as the institution to coordinate reconstruction and rehabilitation was vital to the success of the effort in Aceh and Nias. A strong, solid, knowledgeable, and honest leadership team at BRR created an accountable, effective, and pro-local institution. Hence, the BRR received strong support from all stakeholders, including local communities.

Sixth, funding commitments and willingness to honor them by the government and other donors, including international donors, was another key feature in the reconstruction in Aceh and Nias.

Finally, natural disasters are frequent in Indonesia and seem likely to occur more frequently. Some could be even more damaging than those previously. Indonesia would be wise to be ready. Indonesia should develop solid systems and institutions for natural disaster prevention, detection, early warning, and mitigation. This could start by conserving coastal ecosystems, introducing simple natural disaster detection methods, early warning systems, and mitigation procedures—such as those on Simeulu island—embedding them in nationwide elementary education, and later providing modern monitoring technologies. Indonesia also should establish a significant emergency fund for natural disasters, as future international responses might be less generous than in the case of the Indian Ocean tsunami. So, far, there are signs Indonesia is willing to address these issues. Whether current efforts to deal with future large disasters will be sufficient remains to be seen.

References

ADB (Development Bank) (2005) ADB review, special issue: rebuilding lives after the tsunami, April. http://www.adb.org/review

Athukorala P, Resosudarmo BP (2005) The Indian ocean tsunami: economic impact, disaster management and lessons. Asian Econ Pap 4(1):1–39

Bakonas PBP (Indonesian Emergency Relief Coordination Agency) (2005a) Buletin Bakonas PBP. 31 Jan, Jakarta

Bakonas PBP (Indonesian Emergency Relief Coordination Agency) (2005b) Buletin Bakonas PBP. 14 Mar, Jakarta

BPS, Bappenas, UNFPA, CIDA, AusAID, and NZAid (2005a) Penduduk and Kependudukan Aceh Pasca Gempa dan Tsunami: Hasil Sensus Penduduk Nanggroe Aceh Darussalam 2005. BPS, Jakarta

BPS, Bappenas, UNFPA, CIDA, AusAID, and NZAid (2005b) Penduduk and Kependudukan Nias Pasca Gempa dan Tsunami: Hasil Sensus Penduduk Nias 2005. BPS, Jakarta

BRR (Badan Rehabilitasi dan Rekonstruksi Aceh-Nias) (2006) Membangun Tanah Harapan: Laporan Kegiatan Satu Tahun Badan Pelaksana Rehabilitasi dan Rekonstruksi Nanggroe Aceh Darussalam dan Nias April 2006. BRR, Banda Aceh

BRR (Badan Rehabilitasi dan Rekonstruksi Aceh-Nias) (2009) 10 management lessons for host governments coordinating post-disasters reconstruction by the executing agency for rehabilitation and reconstruction (BRR) of Aceh-Nias 2005–2009. BRR, Banda Aceh

BRR (Badan Rehabilitasi dan Rekonstruksi Aceh-Nias) and International Partners (2006a) Aceh and Nias two year after the tsunami. BRR, Jakarta

BRR (Badan Rehabilitasi dan Rekonstruksi Aceh-Nias) and International Partners (2006b) Aceh and Nias one year after the tsunami: the recovery effort and way forward. BRR, Jakarta

DepSos (Indonesian Department of Social Services) (2005) Perkembangan Korban Bencana NAD dan Sumut. 15 Jan, Jakarta

FAO (Food and Agricultural Organization) (2005) Special report: FAO/WFP food supply and demand assessment for Aceh province and Nias Island (Indonesia), 5 May. FTP.FAO.org/FI/Document/tsunami

Indrawati SM (2005) Toward recovery: lessons from Indonesian case in managing natural disaster crisis. Keynote speech in the 7th IRSA international conference on natural disaster's impact and challenges for recovery, Jakarta, 3–4 Aug

LPEM-FEUI (Institute for Economics and Social Research-Faculty of Economics University of Indonesia) (2005) Perhitungan Kebutuhan Dana Pembangunan Kembali Aceh. LPEM Internal Report, Jakarta

MDF (Multi-Donor Fund) (2012a) Sustainable futures: a legacy of reconstruction, Main report, vol 1. Multi Donor Fund for Aceh and Nias, Jakarta

MDF (Multi-Donor Fund) (2012b) Sustainable futures: a legacy of reconstruction, Project fact sheets, vol 2. Multi Donor Fund for Aceh and Nias, Jakarta

Montgomery RD (2005) Who will live and who will die in the next Sumatra tsunami? Do natural resource maps provide the answer? Paper presented in the 7th IRSA international conference on natural disaster's impact and challenges for recovery, Jakarta, 3–4 Aug

Nazara S, Resosudarmo BP (2007) Aceh-Nias reconstruction and rehabilitation: progress and challenges at the end of 2006. ADB Institute Dicussion Paper No. 70, ADBI

Office for the Coordination of Humanitarian Affairs (OCHA) (2005) Indian Ocean INDIAN OCEAN-Earthquake/Tsunami-December 2004: list of commitments/contributions and pledges as of 21-Feb 2005. http://www.reliefweb.int/

Saldanha JM (2005) Twin challenges of Post-TSUNAMI Aceh, Indonesia: can post-conflict reconstruction go together with post-disaster reconstruction? Paper presented in at the workshop on Post-Tsunami Asia, Oxford, 11 June

Salim W, Chan-Halbrendt C, Alexander B (2005) Effective disaster recovery planning: assessing the government of Indonesia plan using a synthesized disaster risk management framework. Paper presented in the 7th IRSA international conference on natural disaster's impact and challenges for recovery, Jakarta, 3–4 Aug

Sen K, Steer L (2005) Survey of recent development. Bull Indones Econ Stud 41(3):279–304

Soehaimi A, Kertapati EK, Surono S, Setiawan JH (2005) Gempa Bumi dan Tsunami NAD-SUMUT 26 Dec 2004. Paper presented in the GeoSeminar, 14 Jan, Pusat Penelitian and Pengembangan Geologi, Bandung

Soesastro H, Ace R (2005) Survey of recent development. Bull Indonesia Econ Stud 41(1):5–34

Wetlands International-Indonesia Programme (2005) Tsunami of Aceh and North Sumatra 26 Dec 2004. http://www.wetlands.or.id/

World Bank (2005a) Indonesia: preliminary damage and loss assessment: The Dec 26, 2004 Natural disaster. http://www.worldbank.org

World Bank (2005b) Rebuilding better Aceh and Nias: stocktaking of the reconstruction effort. http://www.worldbank.org

Appendix: The Two-Regional Computable General Equilibrium (2SCGE) Model

Basic Structure of the Static 2SCGE Model

The database used for the 2SCGE model is the two inter-regional social accounting matrix (SAM), which encompasses the inter-regional input–output table of competitive imports for the region comprising the four disaster-affected prefectures of Iwate, Miyagi, Fukushima, and Ibaraki and the non-disaster region comprising the prefectures plus Tokyo, Hokkaido, Osaka, and Kyoto except the above four prefectures.[1] The base data used in the SAM is a 2005 inter-regional input–output table of all 47 prefectures created jointly by Professor Yoshifumi Ishikawa and the Mitsubishi Research Institute.

Next, we constructed the two-regional computable general equilibrium (2SCGE) model with the addition of a recursive dynamic dimension.[2] The 2SCGE model comprises 15 agents (1 household, 11 industries, 1 company, 1 regional government, and 1 investment bank) in the two regions, 11 commodity markets, and the two production factor markets of labor and capital. To this, we add the two agents i.e., the central government and overseas sector. Then, labor and capital total endowments are exogenously fixed, and we assume no transfers outside the region, although both labor and capital can move between industries within the region. In addition, we make unemployment endogenous and social security benefits to the

[1]Please refer to Chapter 2 of the Tokunaga and Okiyama (eds) (2014) for more information on this SAM.

[2]For the regional CGE model, see Tokunaga et al. (2003), Hosoe et al. (2010), EcoMod Modeling School (2012), and Okiyama et al. (2014).

S. Tokunaga, B.P. Resosudarmo (eds.), *Spatial Economic Modelling of Megathrust Earthquake in Japan*, New Frontiers in Regional Science: Asian Perspectives 11,
https://doi.org/10.1007/978-981-10-6493-7

household sector endogenous in the 2SCGE model. Specifically, concerning the former we incorporated the following Eq. (A.3-33) of a Phillips curve-type formula into the 2SCGE model.[3]

$$\sum_{a\in A} PL^o \cdot L_a^o + \overline{LW^o} \cdot ER = PL^o \cdot \left(\overline{LS^o} - UNEMP^o\right) \qquad (A.3-33)$$

$$\left(\frac{PL^o \cdot PCINDEX^o}{PLZ^o \cdot PCINDEXZ^o} - 1\right) = phillips^o\left(\frac{UNEMP^o/\overline{LS^o}}{UNEMPZ^o/\overline{LS^o}} - 1\right) \qquad (A.3-58)$$

where $UNEMP^o(UNEMPZ^o)$ and $PCINDEX^o(PCINDEXZ^o)$ represent the unemployment (initial unemployment) and consumer commodity price index of the Laspeyres formula (initial consumer price index) of region o respectively, $phillips^o$ represents a Phillips parameter, L_a^o and $PL^o(PLZ^o)$ represent the labor and wage rate (initial wage rate) respectively, and $\overline{LS^o}, \overline{LW^o}$ and ER represent the labor endowment (exogenous), labor demand from the rest of world (exogenous), and exchange rate respectively.

Concerning the latter, we incorporated the following Eq. (A.3-30) in order to reflect an increasing aging population. That is, the allowance to the household sector from regional government ($TEGH^o$ in region o) is composed of income compensation to the unemployed person and social security expenditures (exogenous $\overline{TEPS^o}$) such as pension benefits that are tied to price changes.

$$TEGH^o = trep^o(PL^o \cdot UNEMP^o) + PCINDEX^o \cdot \overline{TEPS^o} \qquad (A.3-30)$$

Where $trep^o$ represents the replacement rate in region o.

Then, we explain the structure of each sector in the 2SCGE model: domestic production, households, savings and investment, trade, and regional and central government.

Domestic Production Sector

The domestic production sector has the nested structure of production shown in Fig. A.1-1. Each production sector $a(a \in A)$ in region $o(o \in S)$ is assumed to produce XD_a^o of 1 commodity $c(c \in C)$ and to maximize their profits and face a multi-level production function. In level 1 (A1), an industry in sector a constrained by the Leontief technology takes the production function using each intermediate input good XC_{ca}^o aggregated from 11 commodities and the added value KL_a^o. Since the producer price PD_a^o of sector a in region o holds true for the "zero profit condition," it follows that income = production costs. In level 2 (A2) on the right, we arrive at

[3]The 2SCGE model of Okiyama and Tokunaga (2014, Appendix 1) in which we assume that there is no relation between unemployment and inflation should be viewed as a simple 2SCGE model. In practice, however, this simple 2SCGE model (2014) seems realistic. This is because in most regions, a quasi-Phillips curve exists. Thus, we construct the new 2SCGE model with a Phillips curve-type formulation of Eq. (A.3-58) in a regional labor markets.

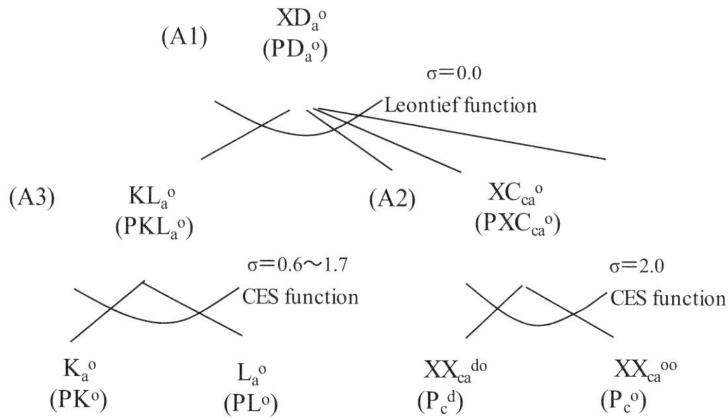

Fig. A.1-1 Structure of production sector

the intermediate goods aggregated for the 11 commodities from the composite commodity according to the Armington assumption XX_{ca}^{do} input from origin $d \in R$ in region d to destination $o \in S$ in region o, under the constraint of constant returns to scale CES technology, and the o intra-regional composite commodity according to the Armington assumption XX_{ca}^{oo}. In addition, we arrive at PXC_{ca}^{o}, the price of the intermediate goods aggregated from sector a of region o, by the income definition, considering the zero profit condition for intermediate goods. In addition, even the added value portion (A3) is derived from the labor L_a^o and capital K_a^o from sector a of the destination region o under the constraint of constant returns to scale CES technology in the same manner as the intermediate goods sector. Then, the producer price PD_a^o of sector a in region o holds true for the "zero profit condition." Thus, it can be derived from income = production costs. Since the return to capital PK^o and wage rate PL^o for the destination region o can move between industries within region o, they are identical for all industries in region o. In addition, we arrive at PXC_a^o, the price of the intermediate goods aggregated from sector a in region o, by the income definition, taking account of the zero profit condition for intermediate goods.

Household Sector

In the household sector, we have formulated the behavior maximizing the level of household utility UH^o as shown in Fig. A.1-2. At level 1 (B1), households in the destination region o maximize the linear-homogeneous Cobb–Douglas utility function for the goods HC_c^o aggregated under the budgetary constraints $CBUD^o$. In addition, at level 2 (B2), we derive the aggregated goods from the composite commodity according to the Armington assumption XH_c^{do} imported from region

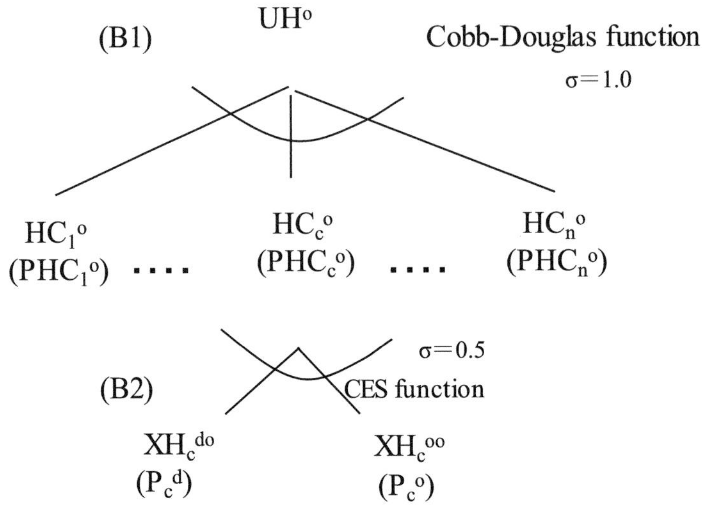

Fig. A.1-2 Structure of household sector

d in origin region $d \in R$ to destination $o \in S$ in region *o* under the constant returns to scale CES technology constraint and the intra-regional *o* composite commodity according to the Armington assumption XH_c^{oo}. In addition, we arrive at the price PHC_c^o of the aggregated goods *c*, by the income definition, taking account of the zero profit condition for such goods. Household income comprises employment income, capital income, social security benefits, property income, and receipts from other current transfers. In addition, the household budget $CBUD^o$ comprises payments of income tax from household income, household savings, social contributions, and payments from property income and other current transfers. Furthermore, household savings is calculated from household income, assuming a fixed propensity to save.

Savings and Investment Sector

Figure A.1-3 illustrates that the savings and investment sector has the same structure as the household sector. The 2SCGE model is closed to prior savings, and in terms of investment, an agent called a "bank" allots savings S^o to investment demand IC_c^o from 11 goods in accordance with the Linear-homogeneous Cobb–Douglas utility function. The savings as shown by Eq. (A.3-13) include the savings of household SH^o, company SN^o, regional government SLG^o, and central government SCG^o, in addition to income transfers SDB^o to offset inter-regional current account balances and overseas savings SF^o to offset a region's external current

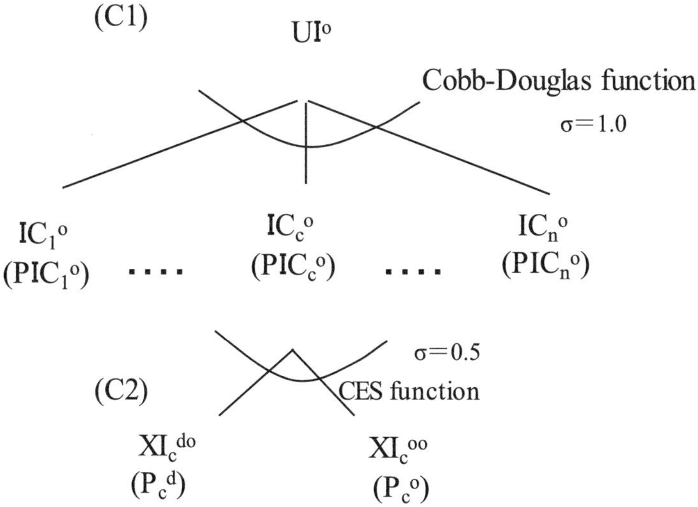

Fig. A.1-3 Structure of investment sector

account balances.[4] Hayashiyama et al. (2011), Okiyama et al. (2014), and others used a static inter-regional CGE model to indicate that income transfers are positive for savings in each region. In this study, however, we add a recursive dynamic dimension to the 2SCGE model by reflecting savings in the inter-regional current account balances as is. In other words, an inter-regional current account deficit negatively impacts a region's savings. This is because, as suggested by Hosoe et al. (2010), in the static model, a current account deficit limited to one period is considered "debt that need not be repaid," so there is no problem with deeming it to be an income transfer, whereas in a dynamic model "debt must be repaid at some point in time with interest," so an inter-regional current account deficit cannot be deemed an income transfer. In this study, however, we would deal with an income transfer as an inter-regional current account deficit for the following reason. This is because the D2SCGE model is a recursive dynamic model in the first reason, and a current account of a region comprising four disaster-affected prefectures will fall into the deficit from the surplus in the intensive reconstruction period because of the fiscal measures in the second reason. If an inter-regional current deficit is regarded as debt that needs to be repaid, the economic spinoff effect of the fiscal measures will be canceled out by the income transfer from the disaster-affected region to the non-disaster affected region. On the other hand, we need to take the inter-regional

[4]Since inter-regional transfers of income cannot be identified with the two-regional SAM, we created a dataset for the 2SCGE model that incorporates the difference between the amounts of imports and exports for each region when allocating the central government's savings to regional governments in each region.

current account between the disaster-affected region and the non-disaster affected region into consideration. This is because the inter-regional current account surplus of the disaster-affected region that had been kept as a supply power to the Kanto region from the Fukushima Nuclear Power Station cannot cancel out under the D2SCGE model, though production in the electricity, gas, and heat-water supply decreased deeply. Therefore, we must pay attention to the following point. The inter-regional current account surplus of the disaster-affected region that has kept more than one trillion yen could have a negative effect on the regional economy of the disaster area continuously. On the other hand, the 2SCGE model treats each region's external current account deficit (foreign savings) as an exogenous variable, and we do not regard this as a problem because the exchange rate is deemed to be an endogenous variable. In addition, at level 2 (C2), we arrive at the aggregated goods IC_c^o from the composite commodity according to the Armington assumption XI_c^{do} imported from region d in origin region $d \in R$ to region o in destination $o \in S$ under the constant returns to scale CES technology constraint and the intra-regional o composite commodity according to the Armington assumption XI_c^{oo}. In addition, we arrive at PIC_c^o the price of the aggregated goods c, by the income definition, taking account of the zero profit condition for such goods.

$$S^o = SH^o + SN^o + SLG^o + SCG^o + SDB^o + SF^o \cdot ER \qquad (A.3-13)$$

Structure of the Trade Sector

Although the trade sector includes exports and imports between each region and the foreign sector, as shown in Fig. A.1-4, trade also occurs through imports and exports between the regions. Specifically, the structure incorporated into the 2SCGE model allocates products produced in region d for the domestic market XDD_c^d and for export E_c^d, with (D1) derived by solving the problem of sales maximization under the constraint of the Constant Elasticity of Transformation function. In addition, for the composite commodity according to the Armington assumption, which is the composite commodity for domestic supply X_c^d comprising producer goods for the domestic market and imported goods M_c^d, (D2) is derived by solving the constrained optimization problem by minimizing its total costs subject to the CES function constraint. The prices for the domestic market PDD_c^d and the composite commodity according to the Armington assumption P_c^d are both derived from the "zero profit condition." The export price PE_c^d and the import price PM_c^d are calculated by multiplying the international price by exchange rate ER, but the import price includes customs duty tt_c and commodity tax on imported goods tm_c. The 2SCGE model fixes the foreign-currency denominated international price and, in the trade balance formula, it sets net overseas transfers of labor $\overline{NLW^o}$ and net overseas transfers of capital $\overline{NKW^o}$ as exogenous variables for foreign savings SF^o in region o, whereas the sum of property and transfer incomes BOP^o and the

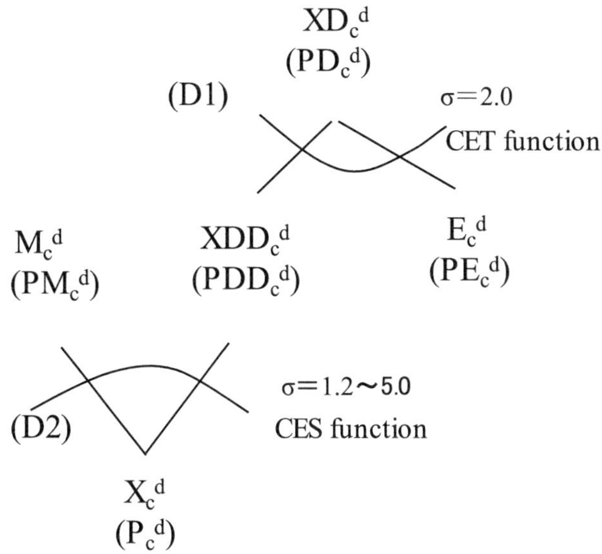

Fig. A.1-4 Structure of trade sector

common exchange rate for the regions are set as endogenous variables. So, we must take the following point into consideration. The exchange rate that is changed by the trade balances of the disaster-affected region has an impact on the price for the composite commodity in the non-disaster region.

Government Sector

Here, we explain the relationship between regional and central governments, which coexist. The central government itself does not engage in spending. Instead, its function is to reallocate the taxes it collects to the regional governments of the two regions, as well as to create savings by taking a fixed proportion of the tax revenue and allocating it to the savings sector of the two regions. Meanwhile, the regional governments in the two regions have budgets that contain the difference between the revenues generated from tax receipts and regional allocation tax grants, etc. and the subsidies disbursed to each production sector. These budgets are then multiplied by a fixed ratio to create savings, and expenditures comprise social security benefits to the household sector and transfers to other institutional sectors and the like.

Setting Elasticity of Substitution and Other Parameters

We estimated the function parameters for each sector with a calibration method that used the two-regional SAM data with 2005 as the benchmark year. However, estimating the function parameters requires that one of the parameters must depend on an external database. Thus, when describing the setting of these parameters, we referenced the values used in GTAP7.1 for the elasticity of substitution for labor and capital in the production sector and the elasticity of substitution for the CES-type (Armington) function of the trade sector. Then, referencing previous studies by Ban (2007), Hayashiyama et al. (2011), and others, we set the inter-regional elasticity of substitution for the production sector, the inter-regional elasticity of substitution for the household and investment sectors, and the elasticity of substitution of the CET-type function for the trade sector. These are summarized in Table A.1–1. In addition to setting the Phillips parameter for each region, we used time-series data from the past to arrive at an estimate of approximately −0.15 for the national figure. However, we used −0.45 for the disaster-affected region and −0.40 for the non-disaster region so that the labor markets in each region will not have full employment even in 2035.

How to Calculate Equivalent Variation

Suppose that we have two different policy regimes: the "benchmark equilibrium"(that will be denoted by the superscript "0" in the sequel) and the "proposed change"(i.e. the policy simulation, denoted by the superscript "1" in the sequel). We would like to measure the welfare effect of the "proposed change". We can use the monetary measure for this change in welfare: the equivalent variation (EV), which measures the household budget at current prices that would be equivalent to the "proposed change" in terms of its impact on utility.

In the "benchmark equilibrium" the commodity prices PD_c^0 of commodity c and household budget $CBUD^0$. Under the "proposed equilibrium" the prices are changed to PD_c^1 whereas the household budget turns out to be $CBUD^1$.

First we define the price index of the cost of living of the "proposed change" relative to the one in the "benchmark equilibrium": PCD_10.

$PCD10 = \prod_c \left(PD_c^1\right)^{\alpha H_c} / \prod_c \left(PD_c^0\right)^{\alpha H_c}$ where αH_c is cobb-Douglas preference parameter for private consumption goods.

Secondly we define the equivalent variation as follows: the equivalent variation is the difference between the household budget of the "proposed change", deflated by the price index of the cost of living, and the household budget of the "benchmark equilibrium".

Table A.1–1 List of elasticity of substitution setup in this model

Production sector	Elasticity of substitution between capital-labor in the CES function	Elasticity of transformation in CET function	Elasticity of substitution of ARMINGTON function	Elasticity of substitution between intermediate goods of different orgin in the CES function	Elasticity of substitution between final goods of different orgin in the CES function
Agriculture and forestry	0.6	2.0	2.7	2.0	0.5
Fisheries	0.6	2.0	1.2	2.0	0.5
Foods and Beverage	1.2	2.0	2.0	2.0	0.5
Electrical devices and parts	1.3	2.0	4.2	2.0	0.5
Motor vehicles and parts	1.3	2.0	2.8	2.0	0.5
Other manufacturing products	1.3	2.0	3.2	2.0	0.5
Construct	1.4	2.0	1.9	2.0	0.5
Electricity,gas and heat-water supply	1.3	2.0	2.8	2.0	0.5
Commerce	1.3	2.0	1.9	2.0	0.5
Transport	1.7	2.0	1.9	2.0	0.5
Financial and insurance,Real estate, Communcation, Public adminstration,Eduction and Servics etc.	1.3	2.0	1.9	2.0	0.5

$$EV = CBUD^1/PCD\_10 - CBUD^0$$

Then considering time period, the equivalent variation in year t EV_t is given by the following equation.

$EV_t = CBUD_t^1/PCD_t10 - CBUD^0 \times (1 + growth)^t$ where $CBUD^0 \times (1 + growth)^t$ is on a steady-state growth path that is a 0.6% potential growth rate [*growth*].

Sets, Variables, Parameters, and Equations of the Static 2SCGE Model

Sets

$a \in A$	Production sectors
$c \in C$	Commodities
$o, d \in S, R$	Regions (o is the region of origin and d is the region of destination)

Endogenous Variables

Price Variables

PL^o	Wage rate in region o
$(PLZ^o$	Initial wage rate in region o=1)
PK^o	Return to capital in region o
PD_a^o	Producer price of production sector a in region o
$(PDZ_a^o$	Initial producer price of production sector a in region o=1)
P_c^o	Price of composite commodity$c$ in region o from imports and domestic production
PXC_{ca}^o	Price of aggregate intermediate input of production sector a in region o
PKL_a^o	Price level of capital-labor bundle of production sector a in region o
PHC_c^o	Price of aggregate commodity c for household demand of region o
$(PHCZ_c^o$	initial price of aggregate commodity c for household demand of region o=1)
PIC_c^o	Price of aggregate commodity c for investment demand of region o
PDD_c^d	Price of domestic output $c$delivered to market being produced in region d
PE_c^d	Export price in domestic currency of domestic output c produced in region d

PM_c^d	Import price in domestic currency of commodities $c$produced in region d
$PCINDEX^o$	Consumer commodity price index of the Laspeyres formula of region o
($PCINDEXZ^o$	Initial consumer commodity price index of the Laspeyres formula of region o=1)
ER	Exchange rate

Quantity Variables

L_a^o	Labor demand by production sector a of region o
$UNEMP^o$	Unemployment in region o
($UNEMPZ^o$	Initial Unemployment in region o)
K_a^o	Capital demand by production sector a of region o
XD_a^o	Domestic production of production sector a in region o
XC_{ca}^o	Aggregate intermediate input $c$of production sector a in region o
XX_{ca}^{oo}	Composite commodity c produced in region o as intermediate input to production sector a in region o
XX_{ca}^{do}	Composite commodity c produced in region d as intermediate input to production sector a in region o
KL_a^o	Demand of capital-labor bundle of production sector a in region o
HC_c^o	Household demand in region o of aggregate commodity c
(HCZ_c^o	Initial household demand in region o of aggregate commodity c)
XH_c^{oo}	Household consumption demand in region o for composite commodity c being produced in region o
XH_c^{do}	Household consumption demand in region o for composite commodity c being produced in region d
IC_c^o	Demand for investment in region o by aggregate commodity c
XI_c^{oo}	Demand for investment in region o for composite commodity c being produced in region o
XI_c^{do}	Demand for investment in region o for composite commodity c being produced in region d
G_c^o	Regional government consumption demand in region o for composite commodity c
N_c^o	Company consumption demand in region o for composite commodity c
E_c^d	Exports of commodity c being produced in region d
M_c^d	Imports of commodity c being produced in region d
XDD_c^d	Domestic output c being produced in region d and supplied to markets
X_c^d	Domestic sales of composite commodity c from imports and domestic production in region d

DEX_c^{do}	Exports to region o composite commodity c being produced in region d
DIM_c^{do}	Imports to region o composite commodity c being produced in region d

Value Variables

Y^o	Household income of region o
HLS^o	Income of household's labor in region o
HKS^o	Income of household's capital in region o
$CBUD^o$	Consumption budget of the household in region o
($CBUDZ^o$	Initial consumption budget of the household in region o)
$LGRBU^o$	Budget of regional government in region o
$CGRBU$	Budget of central government
S^o	Total savings in region o
SH^o	Household savings in region o
SLG^o	Regional government savings in region o
SCG^o	Central government savings in region o
SN^o	Company savings in region o
SF^o	Foreign savings in region o (in foreign currency)
BOP^o	Balance of the capital account with property income sector and current transfer sector in region o
SDB^o	Balance of the regional capital account in region o
$TRDH^o$	Direct tax revenues from household in region o
$TRDN^o$	Direct tax revenues from company in region o
$DDTR^o$	Direct tax receipts to regional government in region o from direct tax sector
$GDDTR$	Direct tax receipts to central government from direct tax sector
$DIDR^o$	Indirect tax receipts to regional government in region o from indirect tax sector
$GDIDT$	Indirect tax receipts to central government from indirect tax sector
$TRPT^o$	Total indirect production tax revenues in region o
$TRTT^o$	Total tariff revenues in region o
$TRMT^o$	Total import tax revenues in region o
$TRCT^o$	Total sales tax revenues in region o
$TGGS^o$	Transfer from central government to regional government in region o
PTW	Expenditure from ROW to property income sector in foreign currency
CTW	Expenditure from ROW to current transfer sector in foreign currency
$TEGH^o$	Social benefits to the household sector in region o from regional government in region o

Exogenous Variables

Factor Variables (Quantity)

$\overline{LS^o}$	Labor endowment including the unemployment in region o (working population)
$\overline{KS^o}$	Capital endowment in region o
$\overline{NKS^o}$	Income of company's capital in region o
$\overline{LGKS^o}$	Income of regional government's capital in region o
$\overline{CGKS^o}$	Income of central government's capital from region o

Value Variables

$\overline{TEHG^o}$	Transfers from household in region o to regional government in region o
$\overline{TEHN^o}$	Transfers from household in region o to company in region o
$\overline{TEPC^o}$	Social benefits except unemployment benefits to the household sector in region o from regional government in region o
$\overline{TENH^o}$	Transfers from company in region o to regional government in region o
$\overline{TEGG^o}$	Transfers from regional government to regional government in region o
$\overline{TGG^o}$	Transfers from regional government in region o to central government
$\overline{NCG^o}$	Transfers from company in region o to central government
$\overline{LW^o}$	Labor demand in region o from ROW in foreign currency
$\overline{LWS^o}$	Labor endowment in region o to ROW in foreign currency
$\overline{KW^o}$	Capital demand in region o from ROW in foreign currency
$\overline{KWS^o}$	Capital endowment in region o from ROW in foreign currency
$\overline{PIWS}$	Receipts to ROW from property income sector in foreign currency
$\overline{CIWS}$	Receipts to ROW from current transfer sector in foreign currency
$\overline{HPI^o}$	Expenditure from household in region o to property income sector
$\overline{HPIS^o}$	Receipts from property income sector to household in region o
$\overline{NPI^o}$	Expenditure from company in region o to property income sector
$\overline{NPIS^o}$	Receipts from property income sector to company in region o
$\overline{GPI^o}$	Expenditure from regional government in region o to property income sector
$\overline{GPIS^o}$	Receipts from property income sector to regional government in region o

$\overline{HCT^o}$ Expenditure from household in region o to current transfer sector
$\overline{HCTS^o}$ Receipts from current transfer sector to household in region o
$\overline{NCT^o}$ Expenditure from company in region o to current transfer sector
$\overline{NCTS^o}$ Receipts from current transfer sector to company in region o
$\overline{GCT^o}$ Expenditure from regional government in region o to current transfer sector
$\overline{GCTS^o}$ Receipts from current transfer sector to regional government in region o

Price Variables

$\overline{PWE_c}$ World price of exports commodity c
$\overline{PWM_c}$ World price of imports commodity c

Parameters

Function Parameters

$aF1_a^o$ Efficiency parameter for capital-labor bundle in production sector a in region o (In level 1(A1) in Fig. A.1-1).
$aF2_a^o$ Efficiency parameter in the production function of sector a in region o (the added value portion (A3) in Fig. A.1-1).
$\sigma F2_a$ Elasticity of substitution between capital and labor in the CES function of sector a (the added value portion (A3) in Fig. A.1-1).
$\gamma F2_a^o$ CES distribution parameter in the production function of sector a in region o(the added value portion (A3) in Fig. A.1-1).
σT_c Elasticities of transformation in the CET function
σA_c Elasticity of substitution of the Armingtion function
σR_c Inter –regional elasticity of substitution for intermediate goods c in the CES function
σHI_c Inter –regional elasticity of substitution for final goods c in the CES function
γT_c CET distribution parameter regarding destination of domestic output c
γA_c CES distribution parameter of Armingtion function of commodity c
$\beta_{XX_{ca}^{do}}$ Shift parameter in the CES function of production sector a in region o of composite commodity c being produced in region d

$\beta_{XH_c^{do}}$ Shift parameter in the CES function of household demand in region o of composite commodity c being produced in region d
$\beta_{XI_c^{do}}$ Shift parameter in the CES function of investment demand in region o of composite commodity c being produced in region d
aT_c Shift parameter in the CET function of production sector of commodity c
aA_c Efficiency parameter of the Arimington function of commodity c
αH_c^o Cobb-Douglas preference parameter for the household consumption of commodity c in region o
αI_c^o Cobb-Douglas preference parameter for the investment of commodity c in region o

General Parameters

io_{ca}^o Technical coefficients of intermediate input
$hmps^o$ Household's marginal propensity to save in region o
$nmps^o$ Company's marginal propensity to save in region o
$lgmps^o$ Regional government's marginal propensity to save in region o
$cgmps^o$ Central government's marginal propensity to save in region o
shG_c^o Marginal share of consumption spending on commodity c for regional government in region o
shN_c^o Marginal share of consumption spending on commodity c for company in region o
tm_c^d Commodity tax rate on imports commodities c
tt_c^d Customs duty of commodities c
ts_c^d Sales tax of commodities c
tp_a^d Production tax rate of production sector a
sp_a^d Production subsidies rate of production sector a
htd^o Direct tax rate by household in region o
ntd^o Direct tax rate by company in region o
ddt^o Distribution rate of direct tax revenues to region o
idt^o Distribution rate of indirect tax revenues to region o
$phillips^o$ Phillips parameter in region o
$trep^o$ Replacement ratio in region o

Other Variables

UH^o Household utility in region o
UI^o Investment agent utility in region o
PCD^o Price index of the cost of livings in region o
EV^o Equivalent variation in region o
$RGRP^o$ Real gross regional products in region o

Equations of 2SCGE Model

Production Equations

(A1)

$$\begin{pmatrix} \text{Max.} \quad PD_a^o \cdot XD_a^o - PKL^o \cdot KL_a^o - \sum_{c \in C} PXC_{ca}^o \cdot XC_{ca}^o \\ \text{s.t.} \quad XD_a^o = \min\left[\frac{KL_a^o}{b_a^o}, \frac{XC_{1a}^o}{io_{1a}^o}, \frac{XC_{2a}^o}{io_{2a}^o}, \cdots\right] \end{pmatrix}$$

$$KL_a^o = b_a^o \cdot XD_a^o = \frac{XD_a^o}{aF1_a^o} \tag{A.3 – 1}$$

$$XC_{ca}^o = io_{ca}^o \cdot XD_a^o \tag{A.3 – 2}$$

$$PD_a^o \cdot XD_a^o = PKL_a^o \cdot KL_a^o + \sum_{c \in C} PXC_{ca}^o \cdot XC_{ca}^o \tag{A.3 – 3}$$

(A2)

$$\begin{pmatrix} \text{Max.} \quad PXC_{ca}^o \cdot XC_{ca}^o - \sum_{d \in R} P_c^d \cdot XX_{ca}^{do} \\ \text{s.t.} \quad XC_{ca}^o = \left[\sum_{d \in R} \beta_{XX_{ca}^{do}} \cdot XX_{ca}^{do\,\frac{\sigma R_c - 1}{\sigma R_c}}\right]^{\frac{\sigma R_c}{\sigma R_c - 1}} \end{pmatrix}$$

$$XX_{ca}^{do} = \left[\frac{1}{\beta_{XX_{ca}^{do}}} \cdot \frac{P_c^d}{PXC_{ca}^o}\right]^{-\sigma R_c} \cdot XC_{ca}^o \tag{A.3 – 4}$$

$$PXC_{ca}^o \cdot XC_{ca}^o = \sum_{d \in R} P_c^d \cdot XX_{ca}^{do} \tag{A.3 – 5}$$

(A3)

$$\begin{pmatrix} \text{Max.} \quad PKL^o \cdot KL_a^o - \left(PL^o \cdot L_a^o + PK^o \cdot K_a^o\right) \\ \text{s.t.} \quad KL_a^o = aF2_a^o \cdot \left(\gamma F2_a^o \cdot K_a^{o\,\frac{-(1-\sigma F2_a)}{\sigma F2_a}} + \left(1 - \gamma F2_a^o\right) L_a^{o\,\frac{-(1-\sigma F2_a)}{\sigma F2_a}}\right)^{\frac{-\sigma F2_a}{1-\sigma F2_a}} \end{pmatrix}$$

$$K_a^o = \gamma F2_a^{o^{\sigma F2_a}} PK^{o^{-\sigma F2_a}} \left(\gamma F2_a^{o^{\sigma F2_a}} PK^{o^{(1-\sigma F2_a)}} + (1-\gamma F2_a)^{\sigma F2_a} PL^{o^{(1-\sigma F2_a)}} \right)^{\frac{\sigma F2_a}{1-\sigma F2_a}} \cdot \left(\frac{KL_a^o}{aF2_a^o} \right) \quad \text{(A.3 – 6)}$$

$$L_a^o = \left(1-\gamma F2_a^o\right)^{\sigma F2_a} PL^{o^{-\sigma F2_a}} \left(\gamma F2_a^{o^{\sigma F2_a}} PK^{o^{(1-\sigma F2_a)}} + \left(1-\gamma F2_a^o\right)^{\sigma F2_a} PL^{o^{(1-\sigma F2_a)}} \right)^{\frac{\sigma F2_a}{1-\sigma F2_a}} \cdot \left(\frac{KL_a^o}{aF2_a^o} \right) \quad \text{(A.3 – 7)}$$

$$PKL_a^o \cdot KL_a^o = PL^o \cdot L_a^o + PK^o \cdot K_a^o \quad \text{(A.3 – 8)}$$

Household Equations

(B1)

$$\begin{pmatrix} \text{Max.} \quad UH^o = \prod_{c=1}^{n} HC_c^{o^{\alpha H_c^o}} \\ \text{s.t.} \quad CBUD^o = \sum_{c \in C} PHC_c^o \cdot HC_c^o \end{pmatrix}$$

$$PHC_c^o \cdot HC_c^o = \alpha H_c^o \cdot CBUD^o \quad \text{(A.3 – 9)}$$

(B2)

$$\begin{pmatrix} \text{Max.} \quad PHC_c^o \cdot HC_c^o = \sum_{d \in R} P_c^d \cdot XH_c^{do} \\ \text{s.t.} \quad HC_c^o = \left[\sum_{d \in R} \beta_{XH_c^{do}} \cdot XH_c^{do^{\frac{\sigma HI_c - 1}{\sigma HI_c}}} \right]^{\frac{\sigma HI_c}{\sigma HI_c - 1}} \end{pmatrix}$$

$$XH_c^{do} = \left[\frac{1}{\beta_{XH_c^{do}}} \cdot \frac{P_c^d}{PHC_c^o} \right]^{-\sigma HI_c} \cdot HC_c^o \quad \text{(A.3 – 10)}$$

$$PHC_c^o \cdot HC_c^o = \sum_{d \in R} P_c^d \cdot XH_c^{do} \quad \text{(A.3 – 11)}$$

Saving and Investment Equations

(C1)

$$\begin{pmatrix} \text{Max.} & UI^{o} = \prod_{c=1}^{n} IC_{c}^{o\,\alpha IC_{c}^{o}} \\ \text{s.t.} & S^{o} = \sum_{c\in C} PIC_{c}^{o} \cdot IC_{c}^{o} \end{pmatrix}$$

$$PIC_{c}^{o} \cdot IC_{c}^{o} = \alpha I_{c}^{o} \cdot S^{o} \tag{A.3 – 12}$$

$$S^{o} = SH^{o} + SN^{o} + SLG^{o} + SCG^{o} + SDB^{o} + SF^{o} \cdot ER \tag{A.3 – 13}$$

(C2)

$$\begin{pmatrix} \text{Max.} & PIC_{c}^{o} \cdot IC_{c}^{o} - \sum_{d\in R} P_{c}^{d} \cdot XI_{c}^{do} \\ \text{s.t.} & IC_{c}^{o} = \left[\sum_{d\in R} \beta_{XI_{c}^{do}} \cdot XI_{c}^{do\,\frac{\sigma HI_{c}-1}{\sigma HI_{c}}}\right]^{\frac{\sigma HI_{c}}{\sigma HI_{c}-1}} \end{pmatrix}$$

$$XI_{c}^{do} = \left[\frac{1}{\beta_{XI_{c}^{do}}} \cdot \frac{P_{c}^{d}}{PIC_{c}^{o}}\right]^{-\sigma HI_{c}} \cdot IC_{c}^{o} \tag{A.3 – 14}$$

$$PIC_{c}^{o} \cdot IC_{c}^{o} - \sum_{d\in R} P_{c}^{d} \cdot XI_{c}^{do} \tag{A.3 – 15}$$

Trade Equations

(D1)

$$\begin{pmatrix} \text{Max.} & PDD_{c}^{d} \cdot XDD_{c}^{d} + PE_{c}^{d} \cdot E_{c}^{d} \\ \text{s.t.} & XD_{c}^{d} = aT_{c}^{d}\left(\gamma T_{c}^{d} \cdot E_{c}^{d\,\frac{-(1-\sigma T_{c})}{\sigma T_{c}}} + (1-\gamma T_{c}^{d}) XDD_{c}^{d\,\frac{-(1-\sigma T_{c})}{\sigma T_{c}}}\right)^{\frac{-\sigma T_{c}}{1-\sigma T_{c}}} \end{pmatrix}$$

$$E_{c}^{d} = \gamma T_{c}^{d\,\sigma T_{c}} PE_{c}^{d\,-\sigma T_{c}} \left(\gamma T_{c}^{d\,\sigma T_{c}} PE_{c}^{d\,1-\sigma T_{c}} + (1-\gamma T_{c}^{d})^{\sigma T_{c}} \cdot PDD_{c}^{d\,1-\sigma T_{c}}\right)^{\frac{\sigma T_{c}}{1-\sigma T_{c}}} \left(\frac{XD_{c}^{d}}{aT_{c}^{d}}\right) \tag{A.3 – 16}$$

$$XDD_c^d = \left(1-\gamma T_c^d\right)^{\sigma T_c} PDD_c^{d^{-\sigma T_c}} \left(\gamma T_c^{d^{\sigma T_c}} PE_c^{d^{1-\sigma T_c}} + \left(1-\gamma T_c^d\right)^{\sigma T_c} \cdot PDD_c^{d^{1-\sigma T_c}}\right)^{\frac{\sigma T_c}{1-\sigma T_c}} \left(\frac{XD_c^d}{aT_c^d}\right) \quad (A.3-17)$$

$$\left(1+tp_c^d - sp_c^d\right) PD_c^d \cdot XD_c^d = PE_c^d \cdot E_c^d + PDD_c^d \cdot XDD_c^d \quad (A.3-18)$$

(D2)

$$\begin{pmatrix} \text{Max.} & PM_c^d \cdot M_c^d + PDD_c^d \cdot XDD_c^d \\ \text{s.t.} & X_c^d = aA_c^d \left(\gamma A_c^d \cdot M_c^{d^{\frac{-(1-\sigma A_c)}{AT_c}}} + \left(1-\gamma A_c^d\right) XDD_c^{d^{\frac{-(1-\sigma A_c)}{\sigma A_c}}}\right)^{\frac{-\sigma A_c}{1-\sigma A_c}} \end{pmatrix}$$

$$M_c^d = \gamma A_c^{d^{\sigma A_c}} PM_c^{d^{-\sigma A_c}} \left(\gamma A_c^{d^{\sigma A_c}} PM_c^{d^{1-\sigma A_c}} + \left(1-\gamma A_c^d\right)^{\sigma A_c} \cdot PDD_c^{d^{1-\sigma A_c}}\right)^{\frac{\sigma A_c}{1-\sigma A_c}} \left(\frac{X_c^d}{aA_c^d}\right) \quad (A.3-19)$$

$$XDD_c^d = \left(1-\gamma A_c^d\right)^{\sigma A_c} PDD_c^{d^{-\sigma A_c}} \left(\gamma A_c^{d^{\sigma A_c}} PM_c^{d^{1-\sigma A_c}} + \left(1-\gamma A_c^d\right)^{\sigma A_c} \cdot PDD_c^{d^{1-\sigma A_c}}\right)^{\frac{\sigma A_c}{1-\sigma A_c}} \left(\frac{X_c^d}{aA_c^d}\right) \quad (A.3-20)$$

$$P_c^d \cdot X_c^d = PM_c^d \cdot M_c^d + \left(1+ts_c^d\right) PDD_c^d \cdot XDD_c^d \quad (A.3-21)$$

$$PM_c^d = \left(1+tt_c^d + tm_c^d\right) \overline{PWM_c} \cdot ER \quad (A.3-22)$$

$$PE_c^d = \overline{PWE_c} \cdot ER \quad (A.3-23)$$

Inter-regional Trade Equations

$$DEM_c^{do} = \sum_{a \in A} XX_{ca}^{do} + XH_c^{do} + XI_c^{do} \qquad o \neq d \quad (A.3-24)$$

$$DIM_c^{do} \equiv DEX_c^{do} \quad (A.3-25)$$

$$SDB^o = \sum_{c,d \in C,R} P_c^d \cdot DIM_c^{do} - \sum_{c,d \in C,R} P_c^d \cdot DEX_c^{do} \quad (A.3-26)$$

Government Equations

$$P_c^o \cdot G_c^o = shG_c^o\left(LGRBU^o - SLG^o - TEGH^o - \overline{TEGG^o} - \overline{GPI^o} - \overline{GCT^o} - \overline{TGG^o}\right) \tag{A.3 – 27}$$

$$SLG^o = \text{lg}mps^o \cdot LGRBU^o \tag{A.3 – 28}$$

$$SCG^o = cgmps^o \cdot CGRBU^o \tag{A.3 – 29}$$

$$TEGH^o = trep^o(PL^o \cdot UNEMP^o) + PCINDEX^o \cdot \overline{TEPS^o} \tag{A.3 – 30}$$

Company Equations

$$SN^o = nmps^o\left(PK^o \cdot \overline{NKS^o} + \overline{TEHN^o} + \overline{NPIS^o} + \overline{NCTS^o}\right) \tag{A.3 – 31}$$

$$\begin{aligned} P_c^o \cdot N_c^o = shN_c^o\big(&PK^o \cdot \overline{NKS^o} + \overline{TEHN^o} + \overline{NPIS^o} - SN^o \\ &- \overline{TENH^o} - TRDN^o - \overline{NPI^o} - \overline{NCT^o} - \overline{NCG^o}\big) \end{aligned} \tag{A.3 – 32}$$

Market Clearing Equations

$$\sum_{a \in A} PL^o \cdot L_a^o + \overline{LW^o} \cdot ER = PL^o \cdot \left(\overline{LS^o} - UNEMP^o\right) \tag{A.3 – 33}$$

$$PL^o \cdot \left(\overline{LS^o} - UNEMP^o\right) = PL^o \cdot HLS^o + \overline{LWS^o} \cdot ER \tag{A.3 – 34}$$

$$\sum_{a \in A} PK^o \cdot K_a^o + \overline{KW^o} \cdot ER = PK^o \cdot \overline{KS^o} \tag{A.3 – 35}$$

$$PK^o \cdot \overline{KS^o} = PK^o\left(HKS^o + \overline{NKS^o} + \overline{LGKS^o} + \overline{CGKS^o}\right) + \overline{KWS^o} \cdot ER \tag{A.3 – 36}$$

$$X_c^d = \sum_{a \in A} \sum_{o \in S} XX_{ca}^{do} + \sum_{o \in S}\left(XH_c^{do} + XI_c^{do}\right) + N_c^d + G_c^d \tag{A.3 – 37}$$

$$\begin{aligned} \sum_{c \in C} M_c^o \cdot \overline{PWM_c} + \overline{LWS^o} + \overline{KWS^o} = &\sum_{c \in C} E_c^o \cdot \overline{PWE_c} + SF^o + \overline{LW^o} \\ &+ \overline{KW^o} + BOP^o \end{aligned} \tag{A.3 – 38}$$

$$\sum_{o \in S} BOP^o = (PIW + CTW) - \left(\overline{PIWS} + \overline{CTWS}\right) \tag{A.3 – 39}$$

$$\sum_{o\in S}\left(\overline{HPI^o}+\overline{NPI^o}+\overline{GPI^o}\right)/ER+PIW$$
$$=\sum_{o\in S}\left(\overline{HPIS^o}+\overline{NPIS^o}+\overline{GPIS^o}\right)/ER+\overline{PIWS} \quad (A.3-40)$$

$$\sum_{o\in S}\left(\overline{HCT^o}+\overline{NCT^o}+\overline{GCT^o}\right)/ER+CTW$$
$$=\sum_{o\in S}\left(\overline{HCTS^o}+\overline{NCTS^o}+\overline{GCTS^o}\right)/ER+\overline{CTWS} \quad (A.3-41)$$

$$CGRBU=\sum_{o\in S}\left(SCG^o+\overline{TGGS^o}\right) \quad (A.3-42)$$

$$\sum_{o\in S}(TRDH^o+TRDN^o)=\sum_{o\in S}DDTR^o+GDDTR \quad (A.3-43)$$

$$\sum_{o\in S}(TRTT^o+TRMT^o+TRPT^o)=\sum_{o\in S}DIDT^o+GDIDT \quad (A.3-44)$$

Definition Equations

$$DDTR^o=ddt^o(TRDH^o+TRDN^o) \quad (A.3-45)$$

$$DIDT^o=idt^o(TRTT^o+TRMT^o+TRPT^o) \quad (A.3-46)$$

$$TRTT^o=\sum_{c\in C}tt_c\cdot\overline{PWM_c}\cdot ER\cdot M_c^o \quad (A.3-47)$$

$$TRMT^o=\sum_{c\in C}tm_c\cdot\overline{PWM_c}\cdot ER\cdot M_c^o \quad (A.3-48)$$

$$TRCT^o=\sum_{c\in C}ts_c\cdot PDD_c^o\cdot XDD_c^o \quad (A.3-49)$$

$$TRPT^o=\sum_{a\in A}tp_a\cdot PD_c^o\cdot XD_c^o \quad (A.3-50)$$

$$TRDH^o=htd^o\cdot Y^o \quad (A.3-51)$$

$$TRDN^o=ntd^o\left(PK^o\cdot\overline{NKS^o}+\overline{TEHN^o}+\overline{NPIS^o}+\overline{NCTS^o}\right) \quad (A.3-52)$$

$$Y^o=PL^o\cdot HLS^o+PK^o\cdot HKS^o+\overline{TEHN^o}+TEGH^o+\overline{HPIS^o}$$
$$+\overline{HCTS^o} \quad (A.3-53)$$

$$CBUD^o = Y^o - TRDH^o - SH^o - \overline{TEHG^o} - \overline{TEHN^o} - \overline{HPI^o} - \overline{HCT^o} \quad (A.3-54)$$

$$LGRBU^o = PK^o \cdot \overline{LGKS^o} + \overline{TEHG^o} + \overline{TEGG^o} + DDTR^o + DIDT^o + \overline{GPIS^o} + \overline{GCTS^o} + TGGS^o - \sum_{a \in A} sp_a \cdot PD_a^o \cdot XD_a^o \quad (A.3-55)$$

$$CGRBU = \sum_{o \in S} \left(PK^o \cdot \overline{CGKS^o} + \overline{TGG^o} + \overline{NCG^o} + TRCT^o\right) + GDDTR + GDIDT \quad (A.3-56)$$

$$PCINDEX^o = \frac{\sum_{c \in C} PHC_c^o \cdot HCZ_c^o}{\sum_{c \in C} PHCZ_c^o \cdot HCZ_c^o} \quad (A.3-57)$$

$$\left(\frac{PL^o \cdot PCINDEX^o}{PLZ^o \cdot PCINDEXZ^o} - 1\right) = phillips^o\left(\frac{UNEMP^o/\overline{LS^o}}{UNEMPZ^o/\overline{LS^o}} - 1\right) \quad (A.3-58)$$

Equation Delivered to Other Variables

$$PCD^o = \prod_{a,c \in A,C} PD_a^{o^{\alpha H_c^o}} / \prod_{a,c \in A,C} PDZ_a^{o^{\alpha H_c^o}} \quad (A.3-59)$$

$$EV^o = \frac{CBUD^o}{PCD^o} - CBUDZ^o \quad (A.3-60)$$

$$RGRP^o = \sum_{c \in C} \left(HC_c^o + IC_c^o + G_c^o + N_c^o + E_c^o - M_c^o\right) + \sum_{c,d \in C,R} \left(DEX_c^{do} - DIM_c^{do}\right) \quad (A.3-61)$$

Variables, Parameters and Equations of the Dynamic 2SCGE (D2SCGE) Model

Variables and Parameters

Modified Variables and Parameters

PK_a^o	Return to capital of production sector a in region o
K_a^o	Capital stock of production sector a in region o
HKS_a^o	Household income from capital stock of production sector a in region o
$\overline{KS^o}$	Capital endowment in region o (deletion)
$\overline{NKS_a^o}$	Company income from capital stock of production sector a in region o
$\overline{LGKS_a^o}$	Income of regional government from capital stock of production sector a in region o
$\overline{CGKS_a^o}$	Income of central government from capital stock of production sector a in region o
$\overline{KW_a^o}$	Demand of Capital stock from ROW to production sector a in region o (in foreign currency)
$\overline{KWS_a^o}$	Endowment of Capital stock to ROW from production sector a in region o (in foreign currency)

Additional Variables and Parameters

IT_t^o	Total investment demand in region o in period t
$PKAVG_t^o$	Average return to capital in region o in period t
INV_{ta}^o	Actual investment of production sector a in region o in period t
$INVZ_a^o$	Initial actual investment of production sector a in region o in period t
aIT^o	A certain percentage of investment agent's utility in region o

Modified Equations and Additional Equations

Modified Equations

$$K_a^o = \gamma F2_a^{o^{\sigma F2_a}} PK_a^{o^{-\sigma F2_a}} \left(\gamma F2_a^{o^{\sigma F2_a}} PK_a^{o^{(1-\sigma F2_a)}} + (1-\gamma F2_a)^{\sigma F2_a} PL^{o^{(1-\sigma F2_a)}}\right)^{\frac{\sigma F2_a}{1-\sigma F2_a}} \cdot \left(\frac{KL_a^o}{aF2_a^o}\right) \tag{A.3 – 6'}$$

$$L_a^o = \left(1-\gamma F2_a^o\right)^{\sigma F2_a} PL^{o^{-\sigma F2a}} \left(\gamma F2_a^{o^{\sigma F2a}} PK_a^{o^{(1-\sigma F2a)}} + \left(1-\gamma F2_a^o\right)^{\sigma F2_a} PL^{o^{(1-\sigma F2a)}}\right)^{\frac{\sigma F2a}{1-\sigma F2a}} \cdot \left(\frac{KL_a^o}{aF2_a^o}\right) \quad (A.3-7')$$

$$PKL_a^o \cdot KL_a^o = PL^o \cdot L_a^o + PK_a^o \cdot K_a^o \quad (A.3-8')$$

$$SN^o = nmps^o\left(PK_a^o \cdot \overline{NKS_a^o} + \overline{TEHN^o} + \overline{NPIS^o} + \overline{NCTS^o}\right) \quad (A.3-31')$$

$$P_c^o \cdot N_c^o = shN_c^o\left(PK_a^o \cdot \overline{NKS^o} + \overline{TEHN^o} + \overline{NPIS^o} - SN^o - \overline{TENH^o} - TRDN^o - \overline{NPI^o} - \overline{NCT^o} - \overline{NCG^o}\right) \quad (A.3-32')$$

$$K_a^o + \overline{KW_a^o} \cdot ER = \left(HKS_a^o + \overline{NKS^o} + \overline{LGKS^o} + \overline{CGKS^o}\right) + \overline{KWS_a^o} \cdot ER \;\; (A.3-35' \text{ and } A.3-36') \quad (A3-37)$$

$$\sum_{c \in C} M_c^o \cdot \overline{PWM_c} + \overline{LWS^o} + \sum_{a \in A} \overline{KWS_a^o} = \sum_{c \in C} E_c^o \cdot \overline{PWE_c} + SF^o + \overline{LW^o} + \sum_{a \in A} \overline{KW_a^o} + BOP^o \quad (A.3-38')$$

$$TRDN^o = ntd^o\left(\sum_{a \in A} PK_a^o \cdot \overline{NKS_a^o} + \overline{TEHN^o} + \overline{NPIS^o} + \overline{NCTS^o}\right) \quad (A.3-52')$$

$$Y^o = PL^o \cdot HLS^o + \sum_{a \in A} PK_a^o \cdot HKS_a^o + \overline{TEHN^o} + TEGH^o + \overline{HPIS^o} + \overline{HCTS^o} \quad (A.3-53')$$

$$CBUD^o = Y^o - TRDH^o - SH^o - \overline{TEHG^o} - \overline{TEHN^o} - \overline{HPI^o} - \overline{HCT^o} \quad (A.3-54')$$

$$LGRBU^o = \sum_{a \in A} PK_a^o \cdot \overline{LGKS_a^o} + \overline{TEHG^o} + \overline{TEGG^o} + DDTR^o + DIDT^o + \overline{GPIS^o} + \overline{GCTS^o} + TGGS^o - \sum_{a \in A} sp_a \cdot PD_a^o \cdot XD_a^o \quad (A.3-55')$$

$$CGRBU = \sum_{a \in A, o \in S} PK_a^o \cdot CGKS_a^o + \sum_{o \in S} \left(\overline{TGG^o} + \overline{NCG^o} + TRCT^o\right) + GDDTR + GDIDT \quad (A.3-56')$$

Additional Equations: Capital Stock Equations

$$IT_t^o = aIT^o \cdot \prod_{c \in C} IC_{ct}^{o^{al_c^o}} \qquad (A.3-62)$$

$$PKAVG_t^o = \frac{\sum_{a \in A} PK_{at}^o \cdot K_{at}^o}{\sum_{a \in A} K_{at}^o} \qquad (A.3-63)$$

$$INV_{at}^o = INVZ_a^o \cdot \sqrt{PK_{at}^o / PKAVG_t^o} \qquad (A.3-64)$$

$$INV_{at}^o = IT_t^o \cdot \frac{INV_{at}^o}{\sum_{a \in A} INV_{at}^o} \qquad (A.3-65)$$

$$K_{at}^o = K_{at-1}^o + INV_{at}^o \qquad (A.3-66)$$

References

Ban K (2007) Development of a multiregional dynamic applied general equilibrium model for the Japanese economy - regional economic analysis based on a forward-looking perspective. RIETI discussion paper series, 07-J-043 (in Japanese)

EcoMod Modeling School (2012) Advanced techniques in CGE modeling with GAMS global economic modeling network, Singapore, January 9–13

Hayashiyama Y, Abe M, Muto S (2011) Evaluation of GHG discharge reduction policy by 47 Prefectures Multi-Regional CGE. J Appl Reg Sci 16:67–91. (in Japanese)

Hosoe N, Gasawa K, Hashimoto H (2010) Textbook of computable general equilibrium modeling. Palgrave Macmillan, New York

Okiyama M, Tokunaga S (eds) (2014) Structure of the two-regional computable general equilibrium (2SCGE) model. In: Tokunaga S, Okiyama M (eds) Reconstruction the disaster-affected Region of the Great East Japan Earthquake and recovery of the regional economy. Bunshindou, Tokyo, pp 296–307. (in Japanese)

Okiyama M, Tokunaga S, Akune Y (2014) Analysis of the effective Source of Revenue to Reconstruct the disaster- affected region of the Great East Japan Earthquake: utilizing the two–regional CGE model. J Appl Reg Sci 18:1–16. (in Japanese)

Tokunaga S, Okiyama M (eds) (2014) Reconstruction the disaster-affected region of the Great East Japan Earthquake and recovery of the regional economy. Bunshindou, Tokyo. (in Japanese)

Tokunaga S, Resosudarmo BP, Wuryanto LE, Dung NT (2003) An inter-regional CGE model to assess the impacts of tariff reduction and fiscal decentralization on regional economy. Stud Reg Sci 33(2):1–25

Index

A
Aceh, Indonesia, 4, 8, 313–333
Anselin, L., 31, 203
Armington assumption, 43, 136, 337, 338, 340
Armington composite commodities, 43
Asian Development Bank (ADB), 318, 319, 329
Automotive industry clusters, 45, 49, 118
Automotive parts and assembly, 116

B
Biogas electricity power plant, 6, 7, 41, 59, 222–226, 228–232
Board of Audit, 124, 128

C
Cabinet Office, 51, 128, 129, 133, 169, 260, 274, 275
Calibration, 130, 140, 174, 178, 265–267, 279, 342
CES function, 45, 51, 208, 261, 340, 343
CES function's inter-regional substitution elasticity, 106
Change in productivity for each industry, 141
Coagglomeration, 6, 47, 163–191
Cobb–Douglas utility function, 43, 45, 46, 296, 337, 338
Commodity-grade, 3, 50, 51, 53, 54, 58, 96, 101, 106, 109, 110, 113, 116, 118, 177, 179, 190
Computable general equilibrium (CGE) model, 45, 126, 193
Constant returns to scale CES technology, 293, 337, 338, 340
Consumer demand, 238, 245–250, 254, 255
Coordination of Humanitarian Affairs (OCHA), 325
Core parts, 54, 96, 106, 109, 110, 113
Core–periphery (CP) model, 46, 47
Corporate tax cuts, 178, 179, 185, 187, 190, 191

D
Depopulating society, 2, 7, 126–128
Development, 2–4, 6, 49, 164, 191, 194, 315, 317, 320, 328, 333
Diamond-shaped production pyramid structure, 96
Disaster-affected region, 2, 3, 6, 43, 50, 51, 54, 56, 58, 66, 71, 72, 77, 81, 86, 88–90, 93, 125, 130, 133, 135, 137, 139–141, 144, 147, 154, 155, 158–160, 169, 170, 180, 183, 185, 188, 190, 339–342
Disaster protection, 291
Disaster-unaffected region, 66, 71, 72, 77, 81, 83, 88
Dixit–Stiglitz model, 3, 46
Dynamic CGE Model, 6, 49, 135, 193–220
Dynamic spatial CGE modeling, 5, 291, 292
Dynamic two regional computable general equilibrium (D2SCGE) model, 126

E
Early warning system, 323–330
Economic and human damage, 14, 40, 59

S. Tokunaga, B.P. Resosudarmo (eds.), *Spatial Economic Modelling of Megathrust Earthquake in Japan*, New Frontiers in Regional Science: Asian Perspectives 11,
https://doi.org/10.1007/978-981-10-6493-7

Economic impact, vii, 2, 3, 6–8, 14, 41, 45, 49, 50, 59, 66, 96, 105, 125, 158, 220, 222, 223, 238, 239, 241, 243–246, 253, 255, 257, 289–295, 297, 298, 300, 304, 306, 308, 320–322
Economies of agglomeration, 6, 48, 177, 187, 191
Economic ripple effects, 6, 41, 169, 222–224, 226, 228–230, 232, 233
Economic welfare (equivalent variation), 109, 159
Efficiency parameters, 133, 140, 141, 170, 174, 178, 179, 309, 348, 349
Elasticity of substitution for labor and capital in the production sector, 52, 342
Elasticity of substitution for the CES-type (armington) function of the trade sector, 43, 91, 342
Elasticity of substitution of the CET-type function for the trade sector, 43, 342
Ellison, G., 48, 165, 203
Equivalent variation (EV), 54, 185, 215, 342

F
Fiscal measures for recovery, 29, 30, 45
Fisheries and seafood product manufacturers, 33, 195–199, 202–205
Fujita, M., vii, 46–48, 51, 174
Fukushima Prefecture, 7, 14, 15, 20, 22, 24, 26, 28–30, 38, 59, 66, 72, 81–83, 88, 89, 92, 93, 130, 160, 165, 194, 195, 238, 255, 270

G
Geographic concentration, 203, 205, 217–219
Geographically weighted regression (GWR) model, 3, 31, 33
Gerakan Aceh Merdeka (GAM), 315, 321, 327, 329, 330, 332
Glaeser, E.L., 48, 165, 203
Great East Japan Earthquake, vii, 1–4, 6–8, 14, 26, 31–40, 49, 59, 66, 67, 70–72, 77, 81–83, 88, 92, 93, 95–118, 123–160, 170, 195–205, 222, 237–257, 260, 261, 267–273, 278

H
Hosoe, N., 41, 174, 195, 335, 339

I
Indian Ocean Tsunami, 4, 5, 7, 8, 59, 313, 315, 317, 319–333
Indirect economic impacts, 291
Induced labor, 232
Induced production, 7, 225, 229–232
Industrial linkages, 66, 206, 222
Innovative TFP improvement, 3
Input–output (I–O) analysis, 49
Input–output (I–O) table, 3
Intensive reconstruction period, 51, 93, 125, 126, 128, 130, 140, 144–155, 158–160, 164, 170, 174, 180, 183, 339
Inter-regional input–output table, 7, 41, 67, 103, 160, 243, 244, 335
Inter-regional linkage effect, 71
Inter-regional social accounting matrix (SAM), 66, 335
Intra-regional linkage effect, 71
IPSS population projections, 128
Iwate Prefecture, 6, 15, 22, 26, 31, 35, 38, 41, 59, 66, 82, 96, 128, 130, 137, 160, 165, 194–196, 198, 199, 201, 203, 205, 217, 218, 221–233, 238, 239, 241–243, 255, 271, 335

J
Japan Institute for Labour Policy and Training, 128

K
Keynes, J.M., 8
Krugman P, 45, 47, 48

L
Labor distribution, 7, 260, 261, 263–267, 270–273, 275–279
Labor force population, 128, 137, 139, 140, 170
Labor participation rate, 136
Leontief technology, 42, 294, 336
Local allocation tax grants, 6, 51, 130, 158, 159, 169, 174, 177
Loss of capital stock, 4, 66, 81–93, 128–133, 140

M
Marshall, A., 48
Megathrust earthquake, 1–8, 59, 259–283

Miyagi Prefecture, 6, 15, 59, 169, 195, 197, 198, 203, 205, 217, 218
Moran's I, 31, 33, 38, 203–206
Multiplier effect, 66, 71–77, 82–93
Multiregional input–output model, 40

N
Nankai Trough Earthquake, 2, 4, 8
National Land Administration Agency (*Badan Pertanahan National*/BPN), 330
Natural disaster, 4, 8, 14, 40–49, 105, 106, 109, 113, 116, 118, 168, 169, 194, 195, 198, 217, 222, 233, 260, 291, 313, 315, 317, 320–322, 324–333
Negative supply shock, 4, 96, 97, 101, 105, 106, 109, 110, 113, 116, 118
Nested structure of production, 336
New automobile industry cluster, 51, 164, 166–169, 177, 178
New economic geography (NEG) model, 14, 46, 59
New food industry cluster, 51, 164–166, 174, 179, 183, 185, 187, 190, 191
New industrial agglomeration, 8, 14, 40, 48, 49, 51
New industry cluster, 3, 6, 8, 51, 54, 164, 166, 169, 170, 174, 178, 180, 185, 187, 190, 191
Nuclear accident in Fukushima Prefecture, 83, 93
Nuclear power plant disaster, 15, 26–30, 45, 59

P
Phillips curve-type formula, 336
Phillips parameter, 336, 342
Population aged 20–64, 2, 127
Population aged 65 and above, 2, 127
Population decline, 3, 7, 41, 50, 59, 144, 159, 183, 238, 239, 243–247, 251, 254, 255, 273, 279
Porter, M., 3, 6, 8, 49, 190
Production damage, 66, 72, 77, 83, 88, 89, 92
Production loss, 4–6, 50, 65–94, 218
Production recovery, 5, 6, 20, 24, 25, 45, 194–199, 201–203, 205, 207–213, 215, 216, 218, 219
Productivity, 6, 45, 51, 53, 54, 58, 59, 133, 139–142, 158, 160, 164, 169, 170, 174, 177–179, 183, 185, 191, 195, 201, 202, 209–211, 213, 219, 299, 306, 308
Public fiscal transfer amount, 154, 159
Pyatt, G., 70
Pyramid structure, 96, 104, 105, 118

R
Real gross regional product (GRP), 6, 133
Reconstruction, 124–126, 128, 130, 133, 135–137, 139, 140, 144, 147, 148, 154, 155, 158–160, 313, 315, 317, 319–333
Reconstruction Agency, 25, 28, 124, 125, 130, 239, 243, 270, 329
Reconstruction budget, 3, 4, 6, 59, 72, 93, 124, 125, 128–133, 140, 144, 147, 158, 159
Recovery support, 66, 77, 93, 219
Recursive dynamic regional CGE model, 6, 158
Regional Computable General Equilibrium (CGE) model, 67
Regional potential, 7, 260, 261, 267, 271–273, 275–279
Regional renewal, 4, 13–60
Rehabilitation, 59, 160, 313, 315, 317, 319–333
Renesas shock, 96
Round, J.I., 70

S
SAM multiplier analysis, 5, 66–71, 81, 82, 93
Schumpeter, J.A., 8
Steady-state growth rate, 135, 136
Supply chain disruptions, 3–5, 40, 49, 50, 53, 54, 96, 97, 101, 104–106, 109, 110, 113, 116, 118
Supply linkages, 194, 205–207, 213

T
Tangible fixed capital, 82, 83
Three-region SAM, 66, 67, 81, 82, 92
Tokai Earthquake, 2, 4, 5, 7, 45, 289–308
Tsunami, 1, 4, 5, 7, 14, 22, 23, 31, 33, 38, 71, 72, 77, 82, 83, 92, 93, 124, 165, 194, 195, 217, 222, 223, 270, 271, 274, 275, 313, 315, 317–322, 324–333
Two-regional computable general equilibrium (2SCGE) model, 3, 5, 14, 41, 43, 45, 49–59, 97, 105, 106, 109, 118, 126, 135, 136, 140, 335–359
Two-regional SAM, 40–45, 136, 335, 339

W
World Bank, 319, 321, 322, 326, 328, 329

MIX
Papier aus verantwortungsvollen Quellen
Paper from responsible sources
FSC® C105338

If you have any concerns about our products,
you can contact us on
ProductSafety@springernature.com

In case Publisher is established outside the EU,
the EU authorized representative is:
Springer Nature Customer Service Center GmbH
Europaplatz 3, 69115 Heidelberg, Germany

Printed by Libri Plureos GmbH
in Hamburg, Germany